Managing Centrarchid Fisheries in Rivers and Streams

Managing Centrarchid Fisheries in Rivers and Streams

Edited by

Michael J. Siepker
Iowa Department of Natural Resources
2325 Siewers Spring Road
Decorah, Iowa 52101, USA

and

Jeffrey W. Quinn
Arkansas Game and Fish Commission
213A Hwy 89 South
Mayflower, Arkansas 72106, USA

American Fisheries Society Symposium 87

Proceedings of the symposium

Managing Centrarchid Fisheries in Rivers and Streams

Held in Kansas City, Missouri
August 23–25, 2016

American Fisheries Society
Bethesda, Maryland
2019

A suggested citation format for this book follows.

Entire Book

Siepker, M. J., and J. W. Quinn, editors. 2019. Managing centrarchid fisheries in rivers and streams. American Fisheries Society, Symposium 87, Bethesda, Maryland.

Chapter within the Book

Fleming, B. P., and N. G. Smith. 2019. Spatial distribution and hybridization levels in Guadalupe Bass five years after remedial stocking. Pages 231–244 *in* M. J. Siepker and J. W. Quinn, editors. Managing centrarchid fisheries in rivers and streams. American Fisheries Society, Symposium 87, Bethesda, Maryland.

Front cover photos: (Top-left) Smallmouth Bass sampling on the Eleven Point River, MO; provided by Michael Siepker. (Top-right) An 18 inch Smallmouth Bass captured and released by a young angler in the Jacks Fork River of the Ozark Highlands ecoregion, MO; provided by Shannon Brewer. (Middle-left) Sampling for Shoal Bass in Little Uchee Creek, AL; provided by Steven Sammons. (Middle-right) Smallmouth Bass fingerlings stocked after being harvested from Andrew Hulsey Fish Hatchery, AR; provided by Mike Wintroath, Arkansas Game and Fish Commission. (Bottom-left) Smallmouth Bass from Crooked Creek, AR; provided by Mike Wintroath, Arkansas Game and Fish Commission. (Bottom-right) A large Guadalupe Bass caught in the lower Colorado River, the premiere destination for trophy-size Guadalupe Bass; provided by Timothy Birdsong.

Printed in the United States of America on acid-free paper.

Library of Congress Control Number 2019931782
ISBN 978-1-934874-52-3
ISSN 0892-2284

American Fisheries Society Web site address: *www.fisheries.org*

American Fisheries Society
425 Barlow Place, Suite 110
Bethesda, Maryland 20814
USA

Table of Contents

Preface

Although much of the angling effort targeting centrarchids occurs on lakes and reservoirs, an increasingly important and often overlooked component occurs on the scenic rivers and streams of North America. In 2016, about 13.1 million or 45% of all freshwater anglers fished in rivers and streams, up from 11.9 million in 2011 (U.S. Department of the Interior, U.S. Fish and Wildlife Service and U.S. Department of Commerce, U.S. Census Bureau 2012, 2018). Participation in kayak fishing in the United States has increased 15% annually, on average, from 2010 to 2017 (Outdoor Foundation 2018). These surveys clearly indicate river and stream fisheries are important to anglers. These stream fisheries also offer many other values to humanity that are priceless, including their aesthetic and recreational value and their ability to connect people to nature. Managing populations of stream centrarchids has increased in complexity as new species and genetic lineages are discovered. Management challenges vary from state to state, and this symposium and publication provided a forum for managers and researchers to share their experiences related to the management of these important fisheries.

Papers within this book were presented as part of a special symposium held in conjunction with the 146th annual meeting of the American Fisheries Society (AFS) during August 2016 in Kansas City, Missouri. A total of 34 papers were presented at the symposium concerning the biology and management of nine species of centrarchids. The symposium was sponsored by the Southeast Aquatic Resources Partnership, the Black Bass Committee of the Southern Division AFS, the North Central Division (NCD) of AFS, the Centrarchid Technical Committee of the NCD of AFS, the Arkansas Chapter of the American Fisheries Society, and the Mississippi Interstate Cooperative Resource Agreement (MICRA). Our thanks go out to all our sponsors and the AFS 2016 Program Committee for helping to make the symposium a success.

We offer our sincerest thanks to the Centrarchid Technical Committee of the NCD, MICRA, and the Arkansas Chapter for their donations that supported student travel to Kansas City. In order for students to be eligible for travel assistance, students were required to present at the symposium, moderate at least one session, and submit a resume and cover letter stating financial need and involvement in AFS activities. Amy Cottrell (Auburn) and Jonathan Watson (Maine) received travel awards.

At the request of Wes Porak, President-Elect of the Southern Division of AFS, the symposium was convened a second time at the 2017 meeting of the Southern Division of AFS. Twenty papers were presented during the day-long symposium in Oklahoma City, Oklahoma. We thank Wes and the entire Oklahoma City Program Committee for making sure everything was in place for a successful symposium.

Papers in this book cover a wide range of topics and a diversity of species, unique lineages, and hybrids. A total of 14 papers are published concerning the management of six species and two subspecies, with management of Smallmouth Bass fisheries a theme in seven papers. Papers are published that present the management results from 11 states ranging from Maine in the north to Florida in the south and as far west as Texas. Hybridization and hybrid fisheries were a topic for over 25% of the papers. Studies that evaluate life history aspects, habitat and distribution, food habits, and population demographics are well represented in the book. Several management techniques, including stocking, fish removals, dam removal, forage addition, and harvest regulations are evaluated by authors. Several chapters provide basic biological information for unique lineages that have been recently recognized. Therefore, this book provides a timely reference for personnel charged with managing these fisheries. During the sessions, three consistent themes emerged: (1) stream centrarchid populations are a product of existing habitat, (2) continuing to genetically distinguish centrarchids from one another will allow us to better manage and protect populations for both ecological integrity and economic value, and (3) human influences on the environment continue to affect stream-dwelling populations of all centrarchids; however, increasing human alterations of these ecosystems tends to increase our involvement in a management context as a way to counteract these influences.

Authors were given great latitude and allowed to use both common and scientific names for Florida Largemouth Bass *Micropterus salmoides floridanus* and Florida Bass *Micropterus floridanus*, since this provisional species is a topic of debate at this time. Following Tringali et al. (2015), we also allowed the use of the common name Choctaw Bass for this unique lineage of *Micropterus cf. punctulatus*. The naming conventions used in this book do not represent an endorsement of newly described provisional species by the editors or their employers. The American Fisheries Society Committee on the Names of Fishes has not yet formally recognized either as a species (Page et al. 2013).

We would like to thank AFS staff, especially Aaron Lerner and Kurt West, for their assistance during the development of the symposium and book. We also thank those who participated in and contributed to the symposium and book. Throughout the publication process, we contacted over 50 peer reviewers that helped make these papers what they are and we thank them for their assistance. Jeff Quinn would like to thank Bill Posey, Chris Racey, and the Arkansas Game and Fish Commission for encouraging and supporting this work. Michael Siepker thanks Mike Steuck, Joe Larscheid, and the Iowa Department of Natural Resources for support during the symposium planning and publication process.

References

Outdoor Foundation. 2018. 2018 Outdoor Recreation Participation Report. Washington, D.C. Available: https://outdoorindustry.org/resource/2018-outdoor-participation-report/. (January 2019).

Page, L. M., H. Espinosa-Pérez, L. T. Findley, C. R. Gilbert, R. N. Lea, N. E. Mandrak, R. L. Mayden, and J. S. Nelson. 2013. Common and scientific names of fishes from the United States, Canada, and Mexico, 7th Edition. American Fisheries Society, Special Publication 34, Bethesda, Maryland.

Tringali, M. D., J. M. Long, T. W. Birdsong, and M. S. Allen, editors. 2015. Black bass diversity: multidisciplinary science for conservation. American Fisheries Society, Symposium 82, Bethesda, Maryland.

U.S. Department of the Interior, U.S. Fish and Wildlife Service and U.S. Department of Commerce, U.S. Census Bureau. 2018. 2016 National Survey of Fishing, Hunting, and Wildlife-Associated Recreation. Available: https://www.census.gov/content/dam/Census/library/publications/2018/demo/fhw16-nat.pdf. (January 2019).

U.S. Department of the Interior, U.S. Fish and Wildlife Service and U.S. Department of Commerce, U.S. Census Bureau. 2012. 2011 National Survey of Fishing, Hunting, and Wildlife-Associated Recreation. Available: https://www.census.gov/prod/2012pubs/fhw11-nat.pdf. (January 2019).

Symbols and Abbreviations

The following symbols and abbreviations may be found in this book without definition. Also undefined are standard mathematical and statistical symbols given in most dictionaries.

A	ampere
AC	alternating current
Bq	becquerel
C	coulomb
°C	degrees Celsius
cal	calorie
cd	candela
cm	centimeter
Co.	Company
Corp.	Corporation
cov	covariance
DC	direct current; District of Columbia
D	dextro (as a prefix)
d	day
d	dextrorotatory
df	degrees of freedom
dL	deciliter
E	east
E	expected value
e	base of natural logarithm (2.71828…)
e.g.	(exempli gratia) for example
eq	equivalent
et al.	(et alii) and others
etc.	et cetera
eV	electron volt
F	filial generation; Farad
°F	degrees Fahrenheit
fc	footcandle (0.0929 lx)
ft	foot (30.5 cm)
ft^3/s	cubic feet per second ($0.0283\ m^3/s$)
g	gram
G	giga (10^9, as a prefix)
gal	gallon (3.79 L)
Gy	gray
h	hour
ha	hectare (2.47 acres)
hp	horsepower (746 W)
Hz	hertz
in	inch (2.54 cm)
Inc.	Incorporated
i.e.	(id est) that is
IU	international unit
J	joule
K	Kelvin (degrees above absolute zero)
k	kilo (10^3, as a prefix)
kg	kilogram
km	kilometer
l	levorotatory
L	levo (as a prefix)
L	liter (0.264 gal, 1.06 qt)
lb	pound (0.454 kg, 454g)
lm	lumen
log	logarithm
Ltd.	Limited
M	mega (10^6, as a prefix); molar (as a suffix or by itself)
m	meter (as a suffix or by itself); milli (10^{-3}, as a prefix)
mi	mile (1.61 km)
min	minute
mol	mole
N	normal (for chemistry); north (for geography); newton
N	sample size

NS	not significant
n	ploidy; nanno (10^{-9}, as a prefix)
o	ortho (as a chemical prefix)
oz	ounce (28.4 g)
P	probability
p	para (as a chemical prefix)
p	pico (10^{-12}, as a prefix)
Pa	pascal
pH	negative log of hydrogen ion activity
ppm	parts per million
qt	quart (0.946 L)
R	multiple correlation or regression coefficient
r	simple correlation or regression coefficient
rad	radian
S	siemens (for electrical conductance); south (for geography)
SD	standard deviation
SE	standard error
s	second
T	tesla
tris	tris(hydroxymethyl)-aminomethane (a buffer)
UK	United Kingdom
U.S.	United States (adjective)
USA	United States of America (noun)
V	volt
V, Var	variance (population)
var	variance (sample)
W	watt (for power); west (for geography)
Wb	weber
yd	yard (0.914 m, 91.4 cm)
α	probability of type I error (false rejection of null hypothesis)
β	probability of type II error (false acceptance of null hypothesis)
Ω	ohm
μ	micro (10^{-6}, as a prefix)
′	minute (angular)
″	second (angular)
°	degree (temperature as a prefix, angular as a suffix)
%	per cent (per hundred)
‰	per mille (per thousand)

American Fisheries Society Symposium 87:1–20, 2019

Importance of Temperature and Streamflow Variables for Explaining Variation in Relative Abundance of Age-0 Smallmouth Bass in Southwestern Wisconsin Streams

Justin M. Haglund*

Wisconsin Department of Natural Resources, Fisheries Management, 1500 N. Johns Street, Dodgeville, Wisconsin 53533, USA

John Lyons

UW Zoological Museum, 250 N. Mills Street, Madison, Wisconsin 53706, USA

Paul Kanehl

Wisconsin Department of Natural Resources, Office of Applied Science, 2801 Progress Road, Madison, Wisconsin 53716, USA

Abstract.—Identifying abiotic variables that influence fish recruitment patterns is crucial to understanding, assessing, and managing populations. Smallmouth Bass *Micropterus dolomieu* have been sampled from five streams in southwestern Wisconsin since 1989 with the goals of explaining variation and describing patterns of annual age-0 relative abundance. Summer water temperature and stream stage data have been collected annually since 2010 and United States Geological Survey modeled stream temperature and stage data were acquired from 1990–2009. Catch-per-unit-effort (CPUE) of age-0 fish was highly variable within and among streams and ranged from 0 to 48.54 fish/100 m across all streams. Random forest models with stepwise variable selection processes were used to determine the relative importance of stream temperature and stream stage variables in describing variation in CPUE from 2010–2016. July mean water temperature and maximum summer temperature explained 69.7% of the variation in CPUE of age-0 Smallmouth Bass. July mean temperature and maximum summer temperature

*Corresponding author: JustinM.Haglund@Wisconsin.gov

were positively related with CPUE of age-0 fish from 2010–2016; however, modeled July mean water temperatures and modeled maximum summer temperatures were not significantly correlated with CPUE from 1990–2008. We conclude that caution must be taken when using models to predict CPUE of age-0 Smallmouth Bass from temperature or flow variables, as variability in both recruitment patterns and climatic conditions may reduce model application over longer time frames.

Introduction

Recruitment refers to the addition of new individuals to populations or to successive life stages within populations (Caley et al. 1996) and can be considered the most important factor structuring fish populations (Casselman et al. 2002; Smith et al. 2005). Therefore, understanding fluctuations in recruitment is essential for successful management. Substantial year-to-year variation in recruitment is common for Smallmouth Bass *Micropterus dolomieu* and has been linked to abiotic factors such as temperature and precipitation (Pflieger 1975; Paragamian 1984). Changes in temperature and precipitation during critical times of the year can either favor or diminish year-class success, especially during late spring and early summer when spawning, hatching, and early life stage growth occurs (Larimore and Duever 1968). Thus, the ability to understand how abiotic factors affect recruitment can aid fishery managers in assessing and interpreting changes in populations from year to year, resulting in better management.

Water temperature has been known to influence recruitment of age-0 Smallmouth Bass. Smallmouth Bass tend to prefer waters that range from 21–26°C (Hile and Juday 1941; Scott and Crossman 1973) and are most commonly classified as a warmwater species (Lyons et al. 2009). Smallmouth Bass spawn when temperatures reach approximately 13–16°C (Christie and Regier 1973; Forbes 1981), and in mid-temperate latitudes this tends to occur during May or June. Growth of age-0 fish typically occurs from June through September when temperatures exceed 14°C (Lyons 1997). Serns (1982) found that water temperatures during June–August were positively correlated with numbers of age-0 Smallmouth Bass in the fall in Wisconsin lakes. Strong year classes were also most likely to be produced during warm summers in the Mississippi River, Minnesota (Swenson et al. 2002). Forney (1972) reported a positive relationship between June air temperature and year-class strength of Smallmouth Bass in Oneida Lake, New York. Casselman et al. (2002) also found consistent and strong positive correlations between year-class strength of Smallmouth Bass and mean temperature of individual summer months (July and August) in eastern Lake Ontario.

The effects of high stream flows on recruitment of Smallmouth Bass have been documented and tend to be mostly negative (Smith et al. 2005). This is because high flow events tend to scour nests and displace fry (Simonson and Swenson 1990; Sabo and Orth 1994; Lukas and Orth 1995; Orth and Newcomb 2002). High flows have also been shown to reduce survival (Paragamian 1984; Mason et al. 1991) and growth of age-0 Smallmouth Bass (Paragamian and Wiley 1987). Slower growth produces smaller age-0 bass entering winter, negatively impacting year-class production (Oliver et al. 1979) and recruitment to the fishery (Buynak and Mitchell 2002). Knowledge of how stream flows in the Driftless Area affect age-0 fish may help fishery managers assess population health during years with variable precipitation patterns.

In this study, we examined the relationship between relative abundance of age-0 Smallmouth Bass and 34 water temperatures and stream stage variables in five wadeable, southwestern Wisconsin streams. The first objective was to quantify relative abundance trends of age-0 Smallmouth Bass in each stream across multiple years. The second objective was to determine the water temperature and stream stage variables that explain the majority of the variation in relative abundance of age-0 fish from our 2010–2016 surveys. The third objective was to determine if stream temperature and flow variables important for explaining age-0 relative abundance during 2010–2016 were also important explanatory variables for predicting relative abundance during 1990–2008.

Methods

Study Sites

The Galena River, Little Platte River, Mineral Point Branch, Rattlesnake Creek, and Sinsinawa River comprise the five streams sampled in this study. All five streams are located near the southern edge of the Driftless Area in southwestern Wisconsin, which is dominated by karst topography characterized by large bluffs, sloping terrain, and river valleys with mixed agricultural and forested landscapes. Watersheds range from 66 to 204 km^2, have stream gradients between 1.7 and 4.0 m/km, sinuosities of 1.2–2.7, and are of stream order 4 or 5 (Table 1). The Galena and Little Platte Rivers contain some of the best Smallmouth Bass stream habitat in the area and support excellent fisheries (Lyons and Kanehl 2016). Rattlesnake Creek and Sinsinawa River are slightly smaller, have habitat typical of the area, and are capable of supporting fair to good fisheries. Mineral Point Branch is considered a "nursery" stream with good habitat for juvenile Smallmouth Bass but is too small to harbor large numbers of adult fish.

Sampling and Data Collection

Beginning in 1990, each stream was sampled annually at a single fixed site in August or September of each year, with the exception of 2007 when Little Platte and Mineral Point sampling was delayed due to flood events in late summer. All sites could be sampled effectively by wading and ranged in length from 1,200–1,920 m. Standardized single-pass, upstream electrofishing procedures were performed using a tow barge-mounted, generator-powered DC stream shocker with three anodes (Lyons and Kanehl 1993), with output to the water at or near 200 V and 6 A. All Smallmouth Bass and other game fish encountered were collected. Total length (TL) was measured in millimeters (mm) and weight was measured in grams (g). Age-0 TL was individually measured and aggregate weights of all age-0 fish were taken, while age-1 and older fish were individually measured and weighed. Scales were collected from each Smallmouth Bass greater than 100 mm for age determination. Smallmouth Bass aging was conducted to identify age-0 fish using scale-impression analysis (Smith 1954). Scale images were pressed on cellulose acetate slides and then projected on an Indus 456-D, model 800 Microfiche reader. One of the authors (Paul Kanehl) determined ages for all individuals throughout the entire study by counting annuli along the anterior scale radius. Scales are a common structure removed for aging analysis because of the nonlethal nature and accuracy at young ages, which allows for larger sample sizes and survival of fish (Maceina et al. 2007; Sotola et al. 2014).

Table 1. Major characteristics of the stream sites sampled in Wisconsin, including the Galena, Little Platte, Mineral Point, Rattlesnake, and Sinsinawa Rivers from 1989–2016.

Stream	Latitude and longitude	Mean width (m)	Mean thalweg	Watershed area (km^2)	Gradient (m/km)	Sinuosity	Stream order
Galena	N 42.57239 W 90.35664	17.3	0.70	182	1.7	2.71	5
Little Platte	N 42.72411 W 90.52849	12.7	0.58	204	1.7	1.46	5
Mineral Point	N 42.86512 W 90.20548	6.6	0.38	69	1.8	1.69	4
Rattlesnake	N 42.77423 W 90.92038	9.6	0.56	117	4.0	1.83	4
Sinsinawa	N 42.52578 W 90.47990	8.2	0.44	66	3.9	1.22	4

In 2010, crews began recording hourly water temperature and stream stage data using Onset HOBO water level data loggers. Loggers were placed at the same location at each site from year to year to document fluctuations in water temperature and stream stage throughout the year. Loggers were removed, downloaded, and returned to the stream each fall after stream surveys had concluded for the year. Beginning in 2012, crews also deployed loggers in the terrestrial area adjacent to the stream, above the floodplain, to record barometric pressure at each site. Logger depth in each stream was calculated based on differences between stream pressure and atmospheric barometric pressure of the immediate area surrounding the stream, using HOBOware Barometric Compensation Assistant Software. To normalize comparison of stream stage among streams, we converted calculated stream stage to a proportion of bankfull height and used this bankfull proportion as an index of stream stage. Bankfull height was calculated by physically measuring the distance from the streambed to the top of the streambank at each site. Modeled atmospheric pressure data provided by the U.S. Geological Survey (USGS; http://usgs-r.github.com/) was used during two years when atmospheric pressure data were not collected by our crew (2010–2011).

Data Analysis

To address objective 1, CPUE (fish/100 m of stream length) of age-0 Smallmouth Bass in each stream was quantified to determine temporal variation in recruitment during the study period. Time series analysis of the data were conducted per methods provided in Maceina and Pereira (2007). To assess changes over time we computed the Pearsons correlation coefficient between CPUE and year (α = 0.05). Kendall tau-b correlation coefficients were calculated for the ranked data, as one or a few high catch rates can be influential when using linear methods to assess time series data. Simple linear regression between CPUE and year along with the Durbin-Watson statistic was also computed to determine temporal autocorrelation, which is the correlation between two successive year's residuals. Finally, regression residuals were plotted against year and evaluated.

To address objective 2, random forest models were used to determine the relative importance of water temperature and stream stage variables in explaining relative abundance of age-0 Smallmouth Bass from 2010 to 2016. In total, 34 water temperature and stream stage predictor variables were selected and included in our model (Table 2). These included overall summer (May through August) and individual monthly mean, minimum, and maximum temperatures and stream stage variables. We also calculated the coefficient of variation (CV) of summer (May–August) water temperatures and stream stage, mean base flow, and seven day low flow (both as a percentage of bankfull height). Base flow and seven day low flow were calculated using the lfstat package in R (R Core Team 2014) from daily mean flows from May through August. All environmental variables were selected based on their ability to influence Smallmouth Bass in their first summer after hatching (Serns 1982; Smith et al. 2005).

We used the randomForest model (Breiman 2001; Cutler et al. 2007) to detect relationships between age-0 Smallmouth Bass and environmental predictor variables. The random forest package (Liaw and Wiener 2002) was used to analyze data in the R statistical program (R Core Team 2014). The random forest approach is a robust, nonparametric machine-learning method that is an extension of regression trees in which response data are partitioned into two nodes by the value of a continuous variable (Van Poorten et al. 2013). Optimization algorithms are used to maximize homogeneity within a

Table 2. Environmental predictor variables included in the random forest model. Mean values (range) and mean variable importance are reported and are in order of decreasing importance.

Variable	Mean	Range	Mean variable importance
Maximum summer temperature	28.5	23.8–32.7	24.00
July mean temperature	22.5	19.0–26.9	23.09
July maximum temperature	28.4	23.7–32.7	22.07
July minimum temperature	17.4	14.2–21.8	10.09
Mean summer temperature	20.5	19.0–23.3	7.53
August maximum temperature	26.6	22.2–30.3	3.21
June maximum temperature	26.6	22.7–30.0	2.70
Summer temperature CV	16.1	9.3–21.6	1.70
June maximum stream stage	0.9	0.4–2.0	1.56
May mean temperature	17.0	14.8–19.4	1.43
May minimum temperature	9.8	6.1–13.6	1.19
Minimum summer temperature	9.8	6.1–13.6	1.10
June minimum temperature	14.9	11.8–18.0	0.81
Summer stream stage CV	16.1	3.7–35.5	0.79
June mean temperature	20.7	18.5–23.2	0.76
Maximum summer stream stage	1.2	0.4–2.8	0.65
August minimum temperature	16.9	14.3–19.5	0.26
May maximum stream stage	0.6	0.3–1.3	0.03
July maximum stream stage	0.9	0.3–2.7	–0.02
Mean summer stream stage	0.4	0.3–0.7	–0.03
May mean stream stage	0.4	0.3–0.6	–0.03
May maximum temperature	24.2	22.0–26.9	–0.03
August maximum stream stage	0.7	0.3–2.1	–0.04
August minimum stream stage	0.3	0.2–0.5	–0.05
Mean summer baseflow	0.4	0.2–0.7	–0.05
Seven day low flow	0.4	0.2–0.6	–0.06
July mean stream stage	0.4	0.3–0.8	–0.06
August mean stream stage	0.4	0.2–0.7	–0.07
June mean stream stage	0.4	0.3–0.7	–0.07
June minimum stream stage	0.4	0.2–0.6	–0.09
Minimum summer stream stage	0.3	0.2–0.5	–0.11
July minimum stream stage	0.4	0.2–0.6	–0.11
May minimum stream stage	0.4	0.2–0.5	–0.12
August mean temperature	21.4	19.1–23.4	–0.12

node as well as heterogeneity between nodes (Cutler et al. 2007). First, *n* replicate bootstrap data sets are drawn and for each replicate a tree is created; then a random sample of *m* variables is drawn from all classification variables and is used to decide where to split the data (at each node of the tree; Van Poorten et al. 2013). Each node splits using the best predictor variable from a subset of variables, which are randomly chosen at that given node. Each tree is used to make predictions on the out-of-bag (OOB) data that was not included in the bootstrap data, and then a majority vote is used to calculate the predicted class of each observation based on OOB predictions from the entire forest (Breiman 2001; Cutler et al. 2007). Finally, accuracies and error rates are calculated for each observation based on OOB predictions (Van Poorten et al. 2013). This method is robust against over-fitting and has performed well when compared against other classifiers such as discriminant analysis, neural network, and support-vector machines due to its ability to detect and represent nonlinear and interactive relationships between variables (Breiman 2001). A full explanation on how random forests work is provided in Cutler et al. (2007). Once the random forest analysis was completed, we used a stepwise model selection process similar to Hansen et al. (2015) to select the best-fit model with the variables that explained the majority of the variation in CPUE of age-0 Smallmouth Bass during 2010–2016.

To address our third objective, regression analysis was used to compare our most important predictor variables to CPUE of age-0 Smallmouth Bass during 1990–2008 to determine if relationships were similar between the two time periods. Considering that stream temperatures were not monitored prior to 2010 in our study, modeled July mean water temperature and maximum summer temperature data for each year and stream reach were acquired from the USGS (Stewart et al. 2016). These modeled temperatures coincided well with actual stream temperature trends for the warmwater streams included in the model and prediction accuracy was high ($R^2 = 0.76$; $R^2 = 0.74$; Stewart et al. 2014).

Results

Age-0 CPUE

A wide range of CPUEs were observed throughout the course of this study, and all streams had both years of great reproductive success as well as complete reproductive failure. Annual CPUE of age-0 Smallmouth Bass in all streams was highly variable, ranging from 0 to 48.54 fish/100 m of stream length and CV's from 1.35 to 1.74 (Figure 1). The Sinsinawa River was the only stream to exhibit a significant increase in CPUE over the study period (Table 3). Kendall tau-b correlation coefficients were computed for the ranked data, although results did not differ between the analyses. Durbin-Watson statistics indicated that Mineral Point Branch was the only stream to exhibit a significant but moderate autocorrelation in CPUE ($P < 0.05$; Table 3).

Temperature

During 2010–2016 average summer (May–August) water temperature for all streams combined was 20.54°C, with a minimum average summer water temperature of 19.0°C in Mineral Point Branch during 2014, and a maximum average summer water temperature of 23.3°C in the Galena River during 2012. The coolest summer water temperatures were observed in 2010 and averaged 19.8°C among streams. We observed the warmest summer water temperatures in 2012, averaging 22.3°C, which was the hottest year on record since the beginning of the 20th century (Frankson et al. 2017).

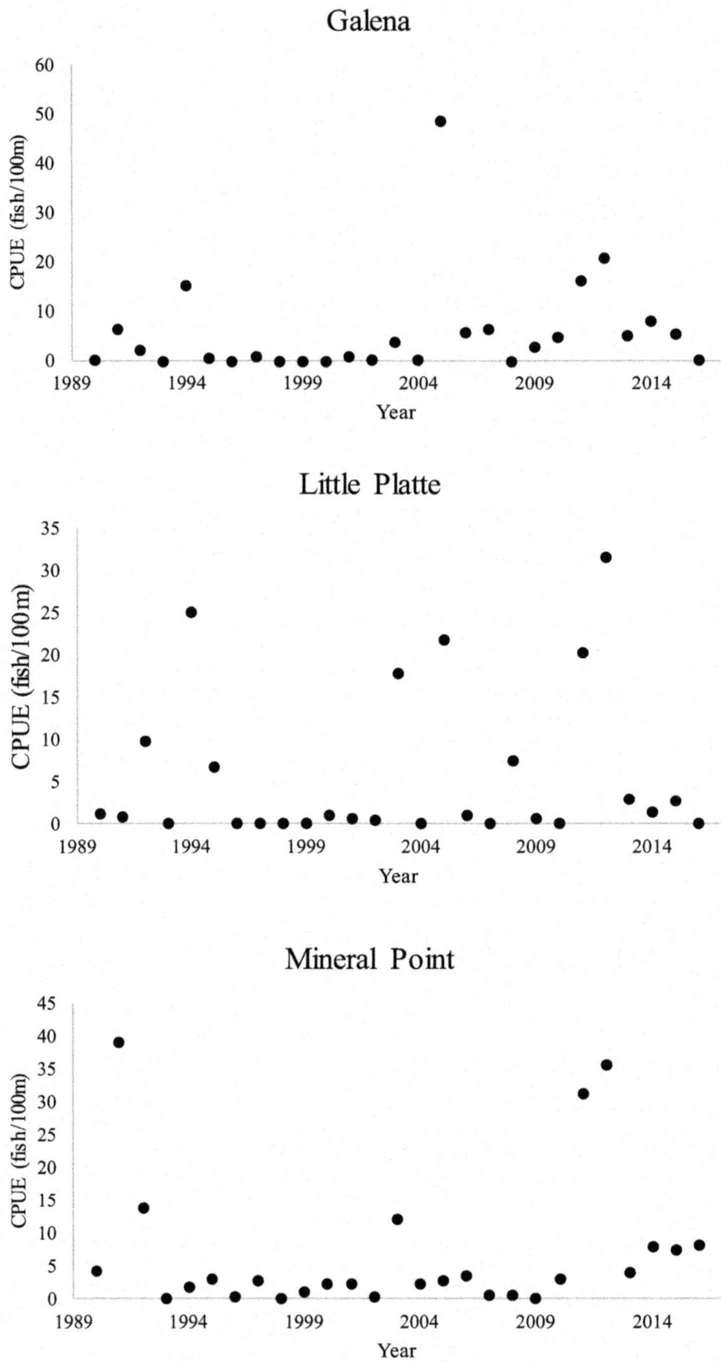

Figure 1. CPUE of age-0 Smallmouth Bass in the Galena, Little Platte, Mineral Point, Rattlesnake and Sinsinawa Rivers from 1990–2016.

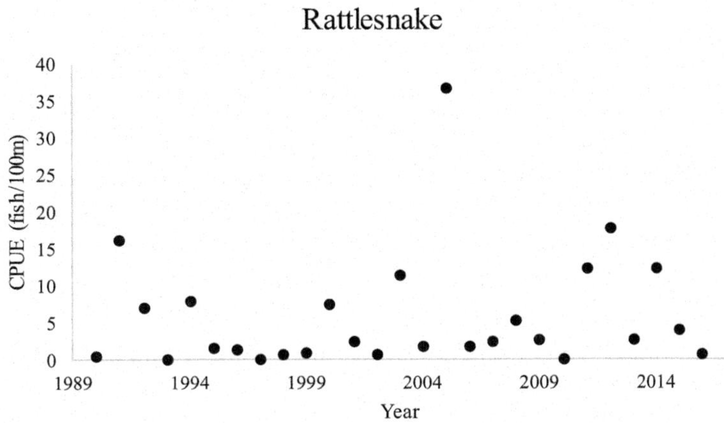

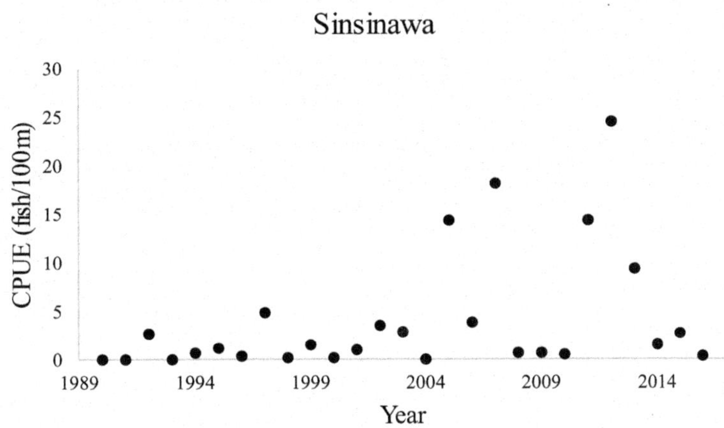

Figure 1. Continued.

Table 3. Results from Pearson correlation analysis between age-0 relative abundance and year along with Durbin-Watson statistics testing for the existence of first order autoregressive processes. The significance level was set at $P = 0.05$.

Stream	Pearson correlation coefficient	P-value	Durbin-Watson statistic	1st order autocorrelation
Galena	0.22	0.27	2.03	–0.04
Little Platte	0.1	0.61	1.95	0.01
Mineral Point	0.08	0.69	1.37	0.32
Rattlesnake	0.12	0.54	2.31	–0.17
Sinsinawa	0.39	0.04	1.66	0.14

Stream Stage

From 2010–2016 stream stage (proportion bankfull height) across all streams averaged 0.53 and ranged from a low of 0.42 in 2015 to a high of 0.71 in 2010. Little Platte River exhibited both the lowest and highest average summer stream stage ranging from 0.33 in 2014 (a year with near normal flows) to 0.91 during 2010. The most precipitation occurred in 2010, which was the second wettest year on record in Wisconsin since the beginning of the 20th century and the least precipitation occurred during 2012 (Figure 2; Frankson et al. 2017). Based on long term precipitation patterns, the period from 2010 to 2014 actually had the highest frequency of 2-in rain events since the year 1900, even though the period itself was relatively dry (Frankson et al. 2017).

Random Forest

Results indicated that July mean water temperatures and maximum summer temperatures were the most important variables explaining variation in relative abundance of age-0 Smallmouth Bass from 2010 to 2016, explaining 69.7% of the variation across all streams. Among 34 competing models, maximum variance of 73% was explained by a nine-parameter model; however, due to difficulty of interpretation, we decided to accept the two-parameter model because explained variance was similar (69 versus 73%; Table 4). Our final model results indicated that higher-than average numbers of age-0 Smallmouth Bass would most likely be produced when July mean water temperatures exceed 24°C and maximum summer water temperatures exceed 31°C, and lower than

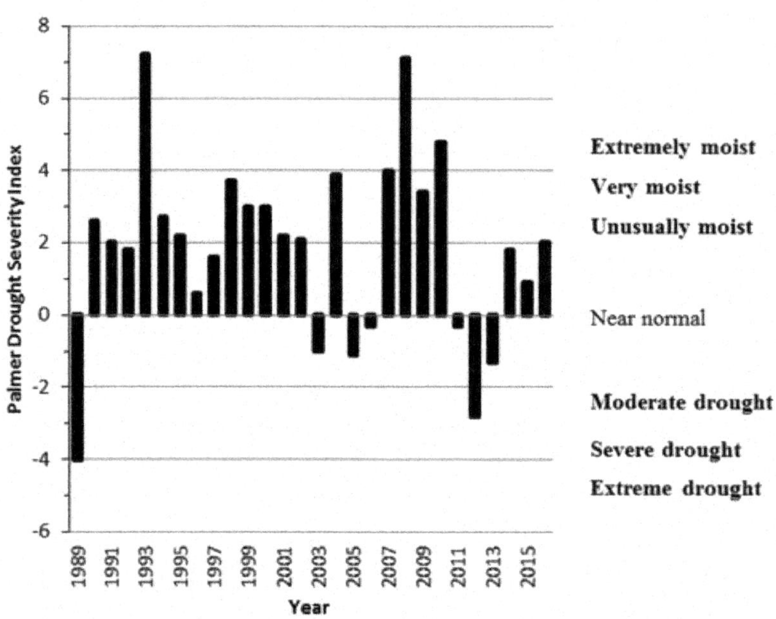

Figure 2. Southwestern Wisconsin Palmer Drought Severity Index during mid-summer from 1989–2016. Negative values refer to drought conditions, positive values refer to moist conditions. Historical data acquired from Wisconsin State Climatology Office (http://www.aos.wisc.edu/~sco/clim-watch/graphics/pdsi-ts-07-l.gif).

Table 4. Candidate models created by dropping the least important predictor variable from the overall random forest model at each step.

Model	Number of Predictors	Overall variance explained
1	34	0.70
2	33	0.70
3	32	0.70
4	31	0.70
5	30	0.70
6	29	0.70
7	28	0.70
8	27	0.70
9	26	0.71
10	25	0.71
11	24	0.70
12	23	0.71
13	22	0.71
14	21	0.71
15	20	0.71
16	19	0.71
17	18	0.71
18	17	0.71
19	16	0.72
20	15	0.72
21	14	0.72
22	13	0.71
23	12	0.72
24	11	0.72
25	10	0.72
26	9	0.73
27	8	0.71
28	7	0.71
29	6	0.72
30	5	0.70
31	4	0.69
32	3	0.65
33	2	0.69
34	1	0.52

average numbers would most likely be produced when July mean water temperatures were below 22.5°C and a maximum summer water temperature of 29°C was not exceeded (Figure 3).

Analyses of Water Temperature and Smallmouth Bass Relative Abundance

Maximum and minimum annual values for water temperature coincided with maximum and minimum values for CPUE of age-0 Smallmouth Bass during 2010–2016; CPUE in all streams combined was lowest in 2010, averaging 1.2 fish/100 m, and highest in 2012, averaging 20.9 fish/100 m. However, the relationships between July mean water temperatures and maximum summer water temperatures versus CPUE of age-0 Smallmouth Bass differed between 2010 and 2016 and 1990–2008. Simple regression analysis across all streams showed a significant, positive relationship between July mean water temperatures and age-0 CPUE from 2010 to 2016 ($F = 49.12$, $r^2 = 0.61$; $P < 0.01$; Figure 4). However, the relationship between modeled July mean water temperatures and age-0 CPUE during 1990–2008 was not significant ($F = 1.25$, $r^2 = 0.01$, $P = 0.27$; Figure 4). The relationship between maximum summer water temperatures and CPUE of age-0 Smallmouth Bass also differed between 2010–2016 and 1990–2008. Maximum summer water temperatures were positively related to age-0 CPUE from 2010 to 2016 ($F = 41.36$, $r^2 = 0.56$, $P < 0.01$; Figure 5), but no relationship was found for 1990–2008 data ($F = 0.98$, $r^2 = 0.01$, $P = 0.32$; Figure 5).

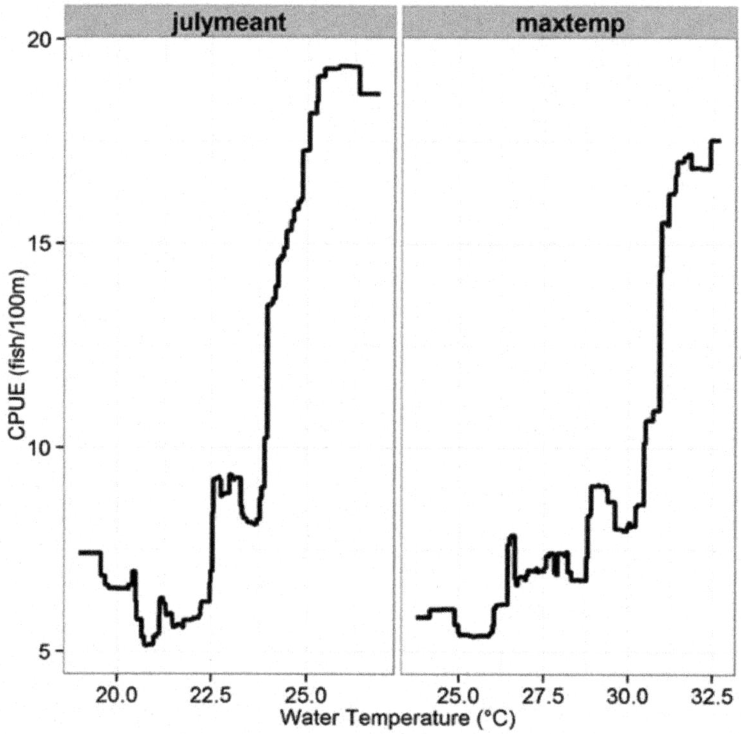

Figure 3. Partial dependence plots based on results from random forest modeling of age-0 Smallmouth Bass CPUE vs July mean water temperatures and age-0 Smallmouth Bass CPUE vs. summer maximum water temperatures in all streams combined for 2010–2016.

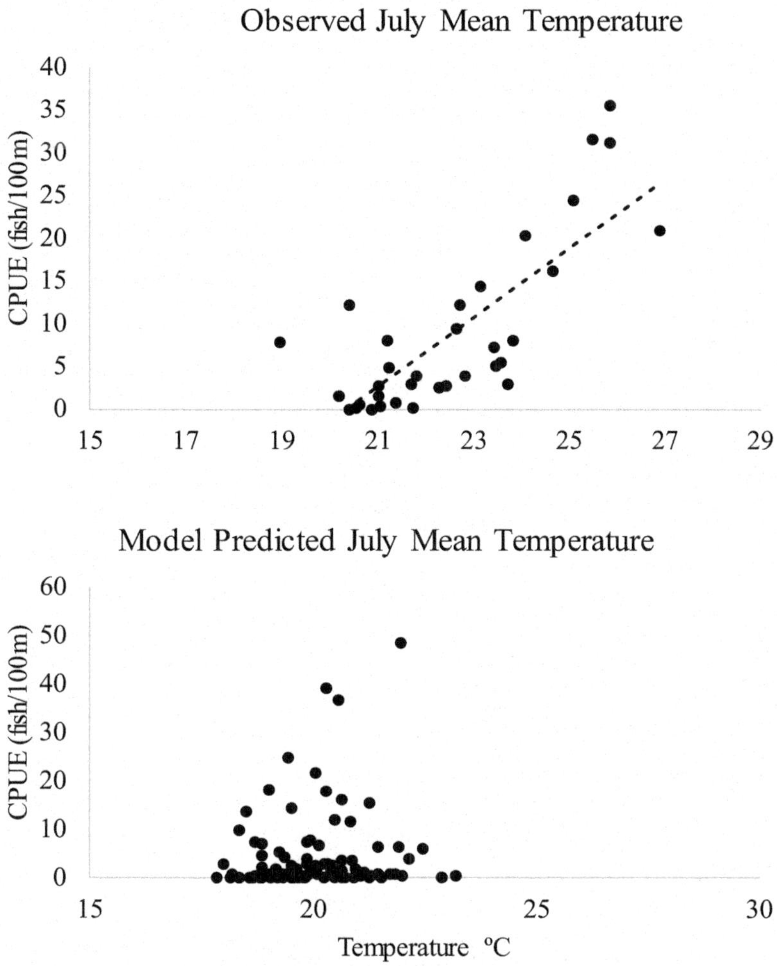

Figure 4. Scatter plots of age-0 Smallmouth Bass CPUE vs. observed July mean temperatures (2010–2016) and model predicted July mean temperatures (1990–2008). Dotted line refers to linear regression fitted to the data.

Discussion

Our analyses indicated that July mean water temperatures and maximum summer water temperatures were the most important variables predicting relative abundance of Smallmouth Bass in southwestern Wisconsin streams from 2010 to 2016. These warmer temperatures during summer may be essential to create warmwater conditions conducive for recruitment, growth, and ultimately survival of age-0 Smallmouth Bass, due to increased baseflows witnessed across the Midwest (Juckem et al. 2008; Lenhart et al. 2011). Warmer July temperatures were associated with higher age-0 CPUE, a finding consistent with results from other Smallmouth Bass populations (Forney 1972; Clady 1975; Serns 1982; Casselman et al. 2002; Swenson et al. 2002). Not only did July mean and maximum summer water temperatures outperform other temperature variables explaining age-0 CPUE, but they also had higher predictive potential than all the stream stage variables, which have been shown to influence Smallmouth Bass populations in

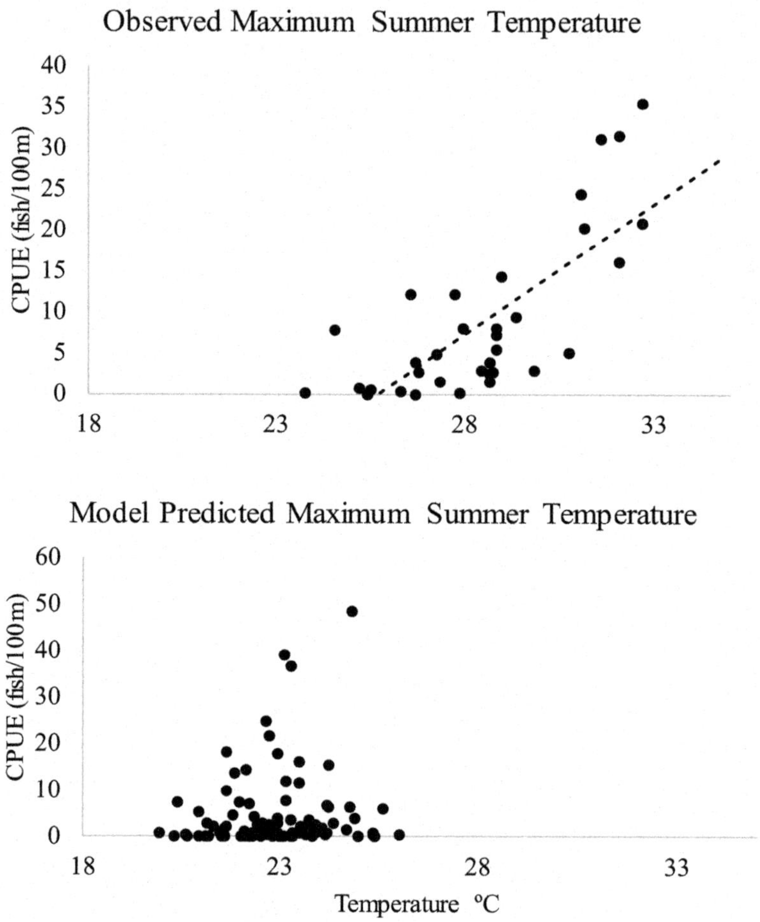

Figure 5. Scatter plots of age-0 Smallmouth Bass CPUE vs. observed summer maximum stream temperatures (2010–2016) and model predicted summer maximum stream temperatures (1990–2008). Dotted line refers to linear regression fitted to the data.

other studies (Lukas and Orth 1995; Smith et al. 2005).

Streamflows of both high and low magnitudes can influence recruitment of riverine Smallmouth Bass populations. Many studies have documented that high streamflows can negatively affect riverine Smallmouth Bass populations by scouring nests and displacing fry during the critical early life stages (Simonson and Swenson 1990; Sabo and Orth 1994; Lukas and Orth 1995; Orth and Newcomb 2002); however, our results did not associate age-0 CPUE with stream stage conditions. We hypothesized that low flow conditions or drought years would yield high numbers of age-0 Smallmouth Bass as opposed to low numbers of fish during flood years. Mason et al. (1991) documented that high flows influenced some of these streams in the mid-1980s. Although, findings from Smith et al. (2005) associated lower age-0 Smallmouth Bass abundance with low flows in Virginia streams. This could be due to a variety of factors including more competition, predation, reduction in invertebrates and lower abundance of riverine prey species (Paragamian and Wiley 1987; Brown and Coon 1994). Similar thresholds have been observed for other species such as stream-resident Brown Trout populations (Lobón-Cerviá 2007). In southwest Wisconsin, it seems that low flows coupled with warm summer temperatures provide Smallmouth Bass the best opportunity for survival and reproduction.

Overall, age-0 CPUE was highly variable from year to year, a finding well documented in other riverine bass populations (Paragamian 1984; Mason et al. 1991; Buynak and Mitchell 2002; Smith et al. 2005). Our CV's, 1.35–1.74, were much higher than a previous study by Smith et al. (2005), where CV's of Smallmouth Bass in Virginia streams ranged from 0.51 to 0.80. Our CV's are also much higher than inland lake and reservoir populations of crappies and Largemouth Bass, which are notorious for exhibiting boom and bust recruitment patterns (Allen and Pine 2000). However, the high variability we found coincides with findings previously discussed in Forbes (1989) and Mason et al. (1991) for southwestern Wisconsin streams prior to 1989 and it seems as though these episodic recruitment patterns are typical for streams in the region. Interestingly, CV's were much lower in 2010–2016 than they were in 1990–2008. CV's were lower in every stream and ranged from 0.78 to 1.38 as opposed to a CV range of 1.53–2.28 during 1990–2008. It seems there may be a linkage between varying CV's and inconsistencies in explanatory variables between time periods and the warmer, relatively drier conditions observed during the latter years may have contributed to less variable recruitment.

Patterns between CPUE of age-0 Smallmouth Bass and July mean water temperatures and summer maximum temperatures were not the same between the two time periods in our study and may be due to differences in climate patterns. The Palmer Drought Severity Index is a model of soil moisture based on supply and demand. Temperature and precipitation patterns contribute heavily to this model, although many other factors including evapotranspiration are important. We used this as a surrogate to demonstrate the relative climatic conditions over the course of our study period. As demonstrated during 1990–2008, drought periods were infrequent and long periods with above normal rainfall in Wisconsin were present, compared to the warmer and drier conditions that dominated in 2010–2016 (Figure 2). This trend may continue as temperatures have risen since 1900 and the 2000s have been the warmest historical period yet (Frankson et al. 2017). Patterns have shown that drought periods in Wisconsin are getting longer, although the precipitation events occurring are becoming more intense. This can have significant implications for riverine systems in the Driftless Area that are susceptible to flash flooding where displacement of young fish can occur. Many studies have examined riverine Smallmouth Bass populations and their relations to temperatures or streamflow; however, few if any have examined the effects and interactions of both multiple temperature and streamflow variables over longer time frames. Our observations and explanation of CPUE in relation to climatic conditions are predicated on the idea that when flows are low, water temperature is the key abiotic variable driving recruitment, but during wetter periods some combination of flow and temperature is important.

Both sampling efficiency and timing on an annual basis can introduce variability in catch data. In Wisconsin, the Smallmouth Bass spawning period typically begins in mid to late May and can last through late June and into early July during cool summers. Our sampling, which took place during late August or early September of each year, should accurately represent relative abundance of the age-0 Smallmouth Bass population in the given year (Lyons and Kanehl 1993). However, variation in sampling efficiency can add noise to the data and may exist due to environmental factors including variable streamflows, conductivity, and turbidity as well as technical factors that include electrofishing personnel and equipment used (Murphy and Willis 1996). Attempts were made to minimize these influences by sampling during optimal conditions, at the same times each year, using the same equipment.

Knowledge of the environmental effects on riverine Smallmouth Bass is important for the yearly assessment of populations, and these results can be used by management agencies as a benchmark for assessing age-0 catches on a year to year basis. This allows management staff to infer why population densities may be high, normal, or low in the given assessment year. Survival of these fish through their first summer is undoubtedly important to the ultimate number of Smallmouth Bass that recruit to the fishery. These fish that survive and recruit to the fishery can also be affected by environmental factors that contribute to the success or failure of year classes. These factors can include mortality caused by unpredictable episodic polluted runoff events, which are a long-standing problem in southwestern Wisconsin streams (Forbes 1989; Mason et al. 1991; Graczyk 1993), displacement and mortality of older fish caused by extreme weather events such as unusually cold winters or major floods (Shuter and Post 1990; Peterson and Kwak 1999; Swenson et al. 2002), or the limited carrying capacity of some streams for large year classes, especially smaller "nursery" streams such as Mineral Point Branch, and the consequent dispersal of large juveniles and adults to less crowded or more favorable habitats in larger streams (Humston et al. 2010). Overall, the knowledge of the environmental variables that have the greatest effect on these populations can contribute to a better understanding of both the biotic and abiotic condition of the fishery, leading to more successful management in the future.

Clearly, the response of Smallmouth Bass to climate factors is complex, and the abiotic factors most important in driving reproduction and recruitment success may vary over time and likely depend on other co-varying factors and antecedent conditions. Few, simple explanatory factors may explain Smallmouth Bass reproduction and recruitment patterns over short time frames, but are probably inadequate by themselves to account for the range of variation observed over longer time frames. This is especially true in the context of a changing climate where future climate conditions may have no analog in current or historical conditions. Despite nearly three decades of study, we still have much to learn about how habitat conditions such as flow and temperature drive Smallmouth Bass recruitment in southwestern Wisconsin streams.

Acknowledgments

We thank the many technicians, biologists, and scientists who have had the opportunity to work on this project since its origination in 1989. Without their continuing efforts, long-term-trend data like these would not be possible. We also thank the USGS for their stream model temperature data as well as their barometric pressure data. We thank Gretchen Hansen for her help with Random Forest coding in R and interpretation of statistical results. This study was funded in part by the Federal Aid in Sportfish Restoration program, Project F-95-P.

References

Allen, M. S., and W. E. Pine. 2000. Detecting fish population responses to a minimum length limit: Effects of variable recruitment and duration of evaluation. North American Journal of Fisheries Management 20:672–682.

Breiman, L. Random Forests. 2001. Machine Learning 45:5–32.

Brown, D. J., and T. G. Coon. 1994. Abundance and assemblage structure of fish larvae in the lower Missouri River and its tributaries.

Transactions of the American Fisheries Society 123:718–732.

Buynak, G. L., and B. Mitchell. 2002. Response of Smallmouth Bass to regulatory and environmental changes in Elkhorn Creek, Kentucky. North American Journal of Fisheries Management 22:500–508.

Caley, M. J., M. H. Carr, M. A. Hixon, T. P. Hughes, G. P. Jones, and B. A. Menge. 1996. Recruitment and the local dynamics of open marine populations. Annual Review of Ecology and Systematics 27:477–500.

Casselman, J. M., D. M. Brown, J. A. Hoyle, and T. H. Eckert. 2002. Effects of climate and global warming on year-class strength and relative abundance of Smallmouth Bass in eastern Lake Ontario. Pages 73–90 *in* D. P. Philip and M. S. Ridgeway, editors. Black bass: ecology, conservation, and management. American Fisheries Society, Symposium 31, Bethesda, Maryland.

Christie, W. J., and H. A. Regier. 1973. Temperature as a major factor influencing reproductive success of fish—two examples. Rappports et Procès-Verbaux Reunions Conseil International pour l'Exploration de la Mer 164:208–218.

Clady, M. D. 1975. Early survival and recruitment of Smallmouth Bass in northern Michigan. Journal of Wildlife Management 39:194–200.

Cutler, D. R., T. C. J. Edwards, K. H. Beard, A. Cutler, K. T. Hess, J. C Gibson, and J. J. Lawler. 2007. Random forests for classification in ecology. Ecology 88:2783–2792.

Forbes, A. M. 1981. Review of Smallmouth Bass (*Micropterus dolomieui*) spawning requirements and first year survival in lakes. Wisconsin Department of Natural Resources, Research Report 111, Madison.

Forbes, A. M. 1989. Population dynamics of Smallmouth Bass (*Micropterus dolomieu*) in the Galena (Fever) River and one of its tributaries. Wisconsin Department of Natural Resources, Technical Bulletin No. 165, Madison.

Forney, J. L. 1972. Biology and management of Smallmouth Bass in Oneida Lake, New York. New York Fish and Game Journal 19:132–154.

Frankson, R., K. Kunkel, and S. Champion. 2017. Wisconsin state summary. NOAA Technical Report NESDIS 149-WI.

Graczyk, D. J., editor. 1993. Surface-water hydrology and quality, and macroinvertebrate and Smallmouth Bass populations in four stream basins in southwestern Wisconsin, 1987–90. U.S. Geological Survey, Water-Resources Investigations Report 93–4024, Madison, Wisconsin.

Hansen, G. J. A., S. R. Carpenter, J. W. Gaeta, J. M. Hennessy, and M. J. Vander Zanden. 2015. Predicting Walleye recruitment as a tool for prioritizing management actions. Canadian Journal of Fisheries and Aquatic Sciences 72:661–672.

Hile, R., and C. Juday. 1941. Bathymetric distribution of fish in lakes of the northeastern highlands, Wisconsin. Transactions of the Wisconsin Academy of Science, Arts, and Letters 33:147–187.

Humston, R., B. M. Priest, W. C. Hamilton, and P. E. Bugas, Jr. 2010. Disperal between tributary and main-stem rivers by juvenile Smallmouth Bass evaluated using otolith microchemistry. Transactions of the American Fisheries Society 139:171–184.

Juckem, P. F., R. J. Hunt, M. P. Anderson, and D. M. Robertson. 2008. Effects of climate and land management change on streamflow in the Driftless Area of Wisconsin. Journal of Hydrology 355:123–130.

Larimore, R. W., and M. J. Duever. 1968. Effects of temperature acclimation on the swimming ability of Smallmouth Bass fry. Transactions of the American Fisheries Society 97:175–184.

Lenhart, C. F., H. Peterson, and J. Nieber. 2011. Increased streamflow in agricultural watersheds of the Midwest: implications for management. Watershed Science Bulletin, Spring 2011.

Liaw, A., and M. Wiener. 2002. Classification and regression by randomForest. R News 2:18–22.

Lobón-Cerviá, J. 2007. Numerical changes in stream-resident Brown Trout (Salmo trutta): uncovering the roles of density-dependent and density-independent factors across space and time. Canadian Journal of Fisheries and Aquatic Sciences Research 64:1429–1447.

Lukas, J. A., and D. J. Orth. 1995. Factors affecting nesting success of Smallmouth Bass in a regulated Virginia stream. Transactions of the American Fisheries Society 124:726–735.

Lyons, J. 1997. Influence of winter starvation on the distribution of Smallmouth Bass among Wisconsin streams: a bioenergetics modeling assessment. Transactions of the American Fisheries Society 126:157–162.

Lyons, J., and P. Kanehl. 1993. A comparison of four electroshocking procedures for assessing the abundance of Smallmouth Bass in Wisconsin streams. U.S. Forest Service, North Central Forest Experimental Station, St. Paul, MN, General Technical Report NC-159.

Lyons, J., and P. Kanehl. 2016. Status and trends in sportfish populations of southwestern Wisconsin warmwater streams. Wisconsin Department of Natural Resources performance report F-95-P, Madison.

Lyons, J., T. Zorn, J. Stewart, P. Seelbach, K. Wehrly, and L. Wang. 2009. Defining and characterizing coolwater streams and their fish assemblages in Michigan and Wisconsin, USA. North American Journal of Fisheries Management 29:1130–1151.

Maceina, M. J., J. Boxrucker, D. L. Buckmeier, R. S. Gangl, D. O. Lucchesi, D. A. Isermann, J. R. Jackson, and P. J. Martinez. 2007. Current status and review of freshwater fish-aging procedures used by state and provincial fisheries agencies with recommendations for future directions. Fisheries 32:329–340.

Maceina, M. J., and D. L. Pereira. 2007. Recruitment. Pages 121–185 *in* C. S. Guy and M. L. Brown, editors. Analysis and interpretation of freshwater fisheries data. American Fisheries Society, Bethesda, Maryland.

Mason, J. W., D. J. Graczyk, and R. A. Kerr. 1991. Effects of runoff on Smallmouth Bass populations in four southwestern Wisconsin streams. Pages 28–38 *in* D. C. Jackson, editor. Proceedings of the first international Smallmouth Bass symposium. Mississippi Agricultural and Forestry Experimental Station, Mississippi State University, Mississippi State.

Murphy, B. R., and D. W. Willis, editors. 1996. Fisheries Techniques, 2nd edition. American Fisheries Society, Bethesda, Maryland.

Oliver, J. D., G. F. Holeton, and K. E. Chua. 1979. Overwinter mortality of fingerling Smallmouth Bass in relation to size, relative energy stores, and environmental temperature. Transactions of the American Fisheries Society 108:130–136.

Orth, D. J., and T. J. Newcomb. 2002. Certainties and uncertainties in defining essential habitats for riverine Smallmouth Bass. Pages 251–264 *in* D. P. Philipp and M. S. Ridgway, editors. Black bass: ecology, conservation and management. American Fisheries Society, Symposium 31, Bethesda, Maryland.

Paragamian, V. L. 1984. Population characteristics of Smallmouth Bass in five Iowa streams and management recommendations. North American Journal of Fisheries Management 4:497–506.

Paragamian, V. L., and M. J. Wiley. 1987. Effects of variable streamflows on growth of Smallmouth Bass in the Maquoketa River, Iowa. North American Journal of Fisheries Management 7:357–362.

Peterson, J. T., and T. J. Kwak. 1999. Modeling the effects of land use and climate change on riverine Smallmouth Bass. Ecological Applications 9:1391–1404.

Pflieger, W. L. 1975. Reproduction and survival of the Smallmouth Bass in Courtois Creek. Pages 231–239 *in* R. H. Stroud and H. Clepper, editors. Black bass biology and management. Sport Fishing Institute, Washington, D.C.

R Core Team. 2014. R: A language and environment for statistical computing. R Foundation for Statistical Computing, Vienna, Austria. Available: http://www.R-project.org/. (July 2018).

Sabo, M. J., and D. J. Orth. 1994. Temporal variation in microhabitat use by age-0 Smallmouth Bass in the North Anna River, Virginia. Transactions of the American Fisheries Society 123:733–746.

Scott, W. B., and E. J. Crossman. 1973. Freshwater fishes of Canada. Fisheries Research Board of Canada, Ottawa, Bulletin 184.

Serns, S. L. 1982. Relation of temperature and population density to first-year recruitment and growth of Smallmouth Bass in a Wisconsin Lake. Transactions of the American Fisheries Society 111:570–574.

Shuter, B. J., and J. R. Post. 1990. Climate, population viability, and the zoogeography of temperate fishes. Transactions of the American Fisheries Society 119:314–336.

Simonson, T. D., and W. A. Swenson. 1990. Critical stream velocities for young-of-year Smallmouth Bass in relation to habitat use. Transactions of the American Fisheries Society 119:902–909.

Smith, S. H. 1954. Method of producing plastic impressions of fish scales without using heat. Progressive Fish-Culturist 16:75–78.

Smith, S. M., J. S. Odenkirk, and S. J. Reeser. 2005. Smallmouth Bass recruitment variability and its relation to stream discharge in three Virginia rivers. North American Journal of Fisheries Management 25:1112–1121.

Sotola, V. A., G. A. Maynard, E. M. Hayes-Pontius, T. B. Mihuc, M. H. Malchoff, and J. E. Marsden. 2014. Precision and bias of using opercles as compared to otoliths, dorsal spines, and scales to estimate ages of Largemouth and Smallmouth Bass. Northeastern Naturalist 21:565–573.

Stewart, J. S., S. A. Covert, N. J. Estes, S. M. Westenbroek, D. Krueger, D. J. Wieferich, M. T. Slattery, J. D. Lyons, J. E. McKenna Jr., D. M. Infante, and J. L. Bruce. 2016. FishVis, A regional decision support tool for identifying vulnerabilities of riverine habitat and fishes to climate change in the Great Lakes Region: U.S. Geological Survey Scientific Investigations Report 2016–5124. Available: http://dx.doi.org/10.3133/sir20165124. (July 2018).

Stewart, J. S., S. M. Westenbroek, M. G. Mitro, J. D. Lyons, L. E. Kammel, and C. A. Buchwald. 2014. A model for evaluating stream temperature response to climate change in Wisconsin. U.S. Geological Survey Scientific Investigations Report 2014–5186. Available: http://dx.doi.org/10.3133/sir20145186. (July 2018).

Swenson, W. A., B. J. Shuter, D. J. Orr, and G. D. Heberling. 2002. The effects of stream temperature and velocity on first-year growth and year-class abundance of Smallmouth Bass in the upper Mississippi River. Pages 101–113 *in* D. P. Philipp and M. S. Ridgway, editors. Black bass: ecology, conservation, and management. American Fisheries Society, Symposium 31, Bethesda, Maryland.

Van Poorten, B. T., S. P. Cox, and A. B. Cooper. 2013. Efficacy of harvest and minimum size limit regulations for controlling short-term harvest in recreational fisheries. Fisheries Management and Ecology 20:258–267.

Appendix A. Catch per unit effort (fish/100 m) of age-0 Smallmouth Bass in all streams sampled.

	CPUE (fish/100 m)				
Year	Galena	Little Platte	Mineral Point	Rattlesnake	Sinsinawa
1990	0.31	1.25	4.25	0.46	0.00
1991	6.35	0.75	39.08	13.80	0.00
1992	2.03	9.75	13.83	6.92	2.72
1993	0.00	0.00	0.00	0.00	0.00
1994	15.31	24.92	1.75	7.77	0.61
1995	0.57	6.67	3.00	1.46	1.22
1996	0.00	0.00	0.17	1.23	0.33
1997	2.20	0.08	2.67	0.00	4.78
1998	0.00	0.00	0.08	0.77	0.11
1999	0.00	0.08	1.00	0.85	1.56
2000	0.00	0.92	2.33	7.46	0.11
2001	0.89	0.67	2.17	2.38	1.06
2002	0.31	0.42	0.25	0.62	3.50
2003	3.91	17.75	12.00	11.46	2.83
2004	0.21	0.00	2.25	1.77	0.06
2005	48.54	21.75	2.75	36.69	14.33
2006	5.83	0.92	3.42	1.77	3.83
2007	6.41	0.10	0.60	2.38	18.11
2008	0.00	7.50	0.42	5.15	0.72
2009	2.86	0.50	1.30	2.69	0.61
2010	4.90	0.00	2.92	0.08	0.50
2011	16.20	20.17	31.17	12.23	14.39
2012	20.83	31.50	35.50	17.69	24.44
2013	5.00	2.83	3.92	2.62	9.33
2014	8.07	1.42	7.83	12.23	1.44
2015	5.47	2.58	7.25	3.92	2.61
2016	0.05	0.00	10.10	0.77	0.33
Mean	5.74	5.65	6.99	5.84	4.06
Max	48.54	31.50	39.08	36.69	24.44
Min	0.00	0.00	0.00	0.00	0.00
CV	1.74	1.58	1.52	1.35	1.55
SD	9.98	8.95	10.63	7.88	6.27

American Fisheries Society Symposium 87:21–37, 2019

First Summer Survival and Channel-Unit Habitat Use by the Neosho Subspecies of Smallmouth Bass

Shannon K. Brewer*
U.S. Geological Survey, Oklahoma Cooperative Fish and Wildlife Research Unit, Oklahoma State University, 007 Agriculture Hall, Stillwater, Oklahoma 74078, USA

Brandon Brown
Oklahoma Department of Wildlife
9097 North 34th Street West, Porter, Oklahoma 74454, USA

Thomas A. Worthington and Robert Mollenhauer
Oklahoma Cooperative Fish and Wildlife Research Unit, Oklahoma State University, 007 Agriculture Hall, Stillwater, Oklahoma 74078, USA

Anthony Rodger, Matt Skoog, and Jim Burroughs
Oklahoma Department of Wildlife
9097 North 34th Street West, Porter, Oklahoma 74454, USA

Abstract.—Unique genetic lineages of *Micropterus* species are increasingly recognized; however, little effort has been devoted to identifying their ecological relationships despite recognition of their conservation value by management agencies. Our study objectives were to determine young-of-year, first-summer survival, and examine overall channel-unit habitat use by the Neosho subspecies of Smallmouth Bass *Micropterus dolomieu velox* from two Ozark streams (Spring and Buffalo creeks). We completed snorkel surveys approximately every 2 weeks from June–September 2013. As anticipated, young-of-year mortality was high during the first 2 weeks of the sampling period (85% in Buffalo Creek and 99% in Spring Creek). Mortality stabilized by the end of July in both streams and was similar over subsequent 2-week periods (95% CI: 0.13%–2.38% and 0.72%–3.48%, in Spring Creek and Buffalo Creek,

*Corresponding author: shannon.brewer@okstate.edu

respectively). In Spring Creek, backwater habitats were unavailable, and young-of-year fish used both pool and run habitats throughout the study duration. However, we observed different habitat-use patterns in Buffalo Creek: young-of-year fish used pools and backwaters throughout the season, use of run habitats increased by late July, and increased use of backwater habitats followed an increase in late summer discharge. In general, there was substantial habitat use variability both within and between streams. Considering both stream reaches combined, young-of-year fish densities in riffle habitat were statistically lower than other channel units. We show that young-of-year Neosho Smallmouth Bass mortality is high during the first few weeks following swim up, and that backwater habitats may be important to early life stages under certain environmental conditions.

Introduction

Smallmouth Bass *Micropterus dolomieu* is endemic to central and eastern North America and is widely regarded as an important recreational and ecological fish species (Brewer and Orth 2015). Three clades, including two subspecies, are recognized for Smallmouth Bass. The northern subspecies *M. d. dolomieu* is the most geographically widespread, occurring east of the Missouri and Meramec rivers (Brewer and Orth 2015). The Neosho subspecies *M. d. velox* (hereafter referred to as Neosho Smallmouth Bass) is restricted to streams of the southwest Ozark Highland ecoregion of the Arkansas River catchment (Stark and Echelle 1998). Given its larger range, research efforts for Smallmouth Bass have largely favored the northern form, and very little is known about ecological differences between the subspecies (Brewer and Long 2015). Limited research has targeted streams occupied solely by Neosho Smallmouth Bass (i.e., outside the intergrade zone, Stark and Echelle 1998) and coarse-scale analyses have often considered the nominal and subspecies together (e.g., Brewer et al. 2007).

Although research is limited, Neosho Smallmouth Bass are thought to share both adult spawning tendencies and young-of-year characteristics with Northern Smallmouth Bass (Brewer and Long 2015). A single study found that stream-dwelling Neosho Smallmouth Bass spawned in calm, near-shore areas close to cover, resembling the lotic spawning habitat of the northern form (Dauwalter and Fisher 2007). Young-of-year Smallmouth Bass exhibit rapid growth (George and Hadley 1979), high levels of cannibalism (Dong and Deangelis 1998), and high mortality (up to 100%; Shuter et al. 1980; Lyons 1997; Curry et al. 2005), but subspecies differences are unknown.

Habitat use and dispersal patterns are poorly understood for stream-dwelling young-of-year Smallmouth Bass, and information is lacking almost entirely for the Neosho subspecies (but see Fore et al. 2007). Studies indicate that juvenile habitat use among Smallmouth Bass populations is not consistent (Pert et al. 2002). For example, young-of-year Smallmouth Bass are often associated with gravel and pebble substrates (George and Hadley 1979; Fore et al. 2007), but substrate use can vary considerably among streams (e.g., pebble, cobble, boulder, and bedrock; Brewer 2011). Reported depth preferences for stream-dwelling young-of-year Smallmouth Bass are also highly variable (Probst et al. 1984; Rankin 1986; Fore et al. 2007; Brewer 2011). Although nest-site fidelity is common even after males cease nest guarding (Gross and Kapuscinski 1997), there is evidence that young-of-year Smallmouth Bass can exhibit diurnal movements among stream habitats (MacCrimmon and Robbins 1981; Sabo et al. 1996). Observed variation in habitat use by young-of-year Smallmouth Bass in streams may imply a "generalist" nature or indicate the importance of multiple habitats for survival. Studies that provide better insight into the ecological characteristics of stream-dwelling young-of-year Smallmouth Bass can promote sound long-term conservation and management. This is especially important given management agencies recognize the conservation value of the different clades (Ahlert et al.1995; Malloy et al. 2000; Boxrucker et al. 2004; Quinn et al. 2004). Accordingly, our objectives were to determine young-of-year mortality of Neosho Smallmouth Bass and examine first-summer habitat use at the channel-unit scale from populations occupying two Ozark streams.

Methods

Study Area

Our study sites were a reach of Buffalo Creek in southwest Missouri and a reach of Spring Creek in northeast Oklahoma (Figure 1). Both streams are located in the Springfield Plateau of the Ozark Highlands Ecoregion (Woods et al. 2005). The underlying lithology is cherty limestone and precipitation averages 106–109 cm annually. Mean daily discharge for Buffalo Creek is 2.04 $m^3 \cdot s^{-1}$ (U.S. Geological Survey, Gage 07189100, 16 years of record, Accessed January 17, 2017) and 2.46 $m^3 \cdot s^{-1}$ (U.S. Army Corps of Engi-

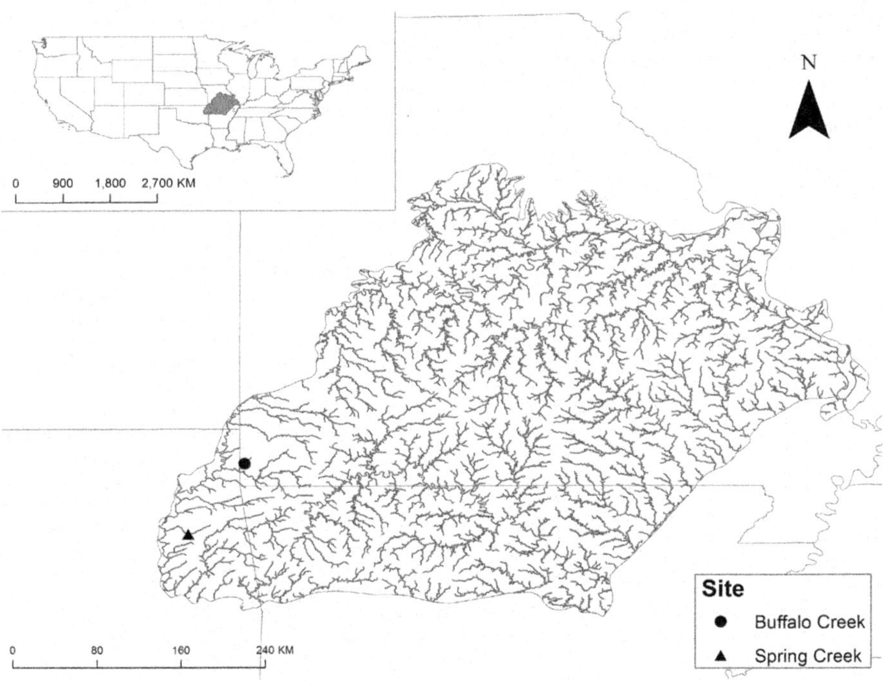

Figure 1. Location of stream reaches in the Ozark Highlands ecoregion where we conducted surveys of young-of-year Smallmouth Bass in 2013. Snorkeling surveys were conducted approximately every 2 weeks from late June through September in Buffalo Creek (black dot) and Spring Creek (black triangle) of the Ozark Highlands ecoregion (area shaded in gray).

neers, Gage LCGO2, 6 years of record, William Chatron, U.S. Army Corps of Engineers, personal communication) for Spring Creek. Both streams are influenced by springs, but Buffalo Creek has a higher groundwater gain (Y. Zhou, G. Fox, and S. Brewer, Oklahoma State University, unpublished data). Horizontal water clarity ranged 3–6 m during the study period. The most abundant piscivore in both streams is Neosho Smallmouth Bass, but there are also low densities of both Largemouth Bass *M. salmoides* and the rarely-observed Spotted Bass *M. punctulatus*.

Temperature and Discharge

Both water temperature and discharge were measured throughout the study period. Temperature loggers (HOBO Pro v2; Onset, Bourne, Massachusetts) were placed in shaded locations of a pool (0.5–0.8-m deep) in each study reach. Care was taken to ensure loggers were not placed in upwelling locations so temperatures represented average reach conditions. Daily discharges for Buffalo Creek were obtained from a nearby stream gage (U.S. Geological Survey, Gage 07189100, 16 years of record, Accessed January 17, 2017). We obtained daily discharge data for Spring Creek from the U.S. Army Corps of Engineers stream gage (U.S. Army Corps of Engineers, Gage LCGO2, 6 years of record, William Chatron, U.S. Army Corps of Engineers, personal communication).

Snorkel Surveys

Snorkel surveys for young-of-year Neosho Smallmouth Bass were conducted in Buffalo Creek and Spring Creek approximately every 2 weeks during summer 2013 (June 28–September 24). We sampled via snorkel surveys because electrofishing is biased against juvenile Smallmouth Bass (McClendon and Rabeni 1986), whereas snorkeling is more effective (i.e., provides a better estimate of abundance) when sampling Smallmouth Bass in clear-water streams (McClendon and Rabeni 1986; Brewer and Ellersieck 2011; Mollenhauer 2016). The first surveys were completed in late June following the first observed larvae swim-up period (i.e., fish were observed swimming above and adjacent to the nests) or 'black fry' in both streams. Preliminary surveys were completed at each reach to determine when larvae began to free swim from the nests. Following swim up, we conducted six abundance surveys of young-of-year Neosho Smallmouth Bass in Buffalo Creek and seven abundance surveys in Spring Creek. The last snorkel surveys were conducted on September 18 in Buffalo Creek and September 24 in Spring Creek.

Each stream reach comprised multiple channel-unit sequences (i.e., riffle-run-pool sequences) to characterize available habitat. The two reaches of Spring and Buffalo creeks (856 m and 1020-m in length, respectively) were chosen because we had permission to access these private lands where ongoing studies suggested high nest activity by Neosho Smallmouth Bass. Each site was mapped prior to sampling using the channel-unit descriptions of Rabeni and Jacobson (1993). We simplified their classification into riffle, run, pool, and backwater habitats. Briefly, we classified riffles as erosional habitats with high gradients and coarse substrates, runs as transitional habitats with moderate velocities, pools as main channel, depositional habitats with slow-moving water, and backwaters as transitional, depositional, generally standing-water habitats adjacent to the main channel. We estimated each channel-unit area by multiplying the average length by the average width. Channel-unit boundaries were marked using natural features or flags.

Three to five people conducted snorkel surveys during daylight hours (10:00–17:00) at each sampling event. All survey members were experienced snorkelers. Horizontal water clarity (see study site description) was consistent with other stream-fish snorkeling studies (e.g., Schill and Griffith 1984). The number of crew at each channel unit-site combination varied depending on water clarity, channel width, habitat complexity, and water depth (Dunham et al. 2009). For example, more people conducted surveys in larger pools with a wide channel. In this example, the snorkel lane that encompassed the pool thalweg was narrower than adjacent lanes because of the increased habitat complexity (i.e., boulders, wood). We also used more observers in shallow-water habitats (<0.2-m deep) as suggested by Dunham et al. 2009. Snorkel lanes that comprised mostly open water were generally wider, but never exceeded 6 m. We snorkeled areas ≥0.1-m deep in a slow upstream direction, avoided sudden movements, and were careful to inspect complex habitat. Each snorkeler maintained their designated lane and stayed approximately in line laterally with other surveyors. Each surveyor communicated regularly with adjacent snorkelers to minimize double counting of fish using underwater hand signals (Dunham et al. 2009).

We identified and counted fish when they were ≤2 m from the snorkeler to ensure proper identification. Sizes of Neosho Smallmouth Bass varied throughout the study period from black fry to young of year approximately 80-mm total length (TL) with tri-colored tails. In late August and September, we used dual criteria to distinguish young-of-year Neosho Smallmouth Bass from other *Micropterus* species and age classes: 1) the presence of a tri-colored tail, and 2) size approximately ≤80 mm TL. Median back-calculated length-at-age is estimated at 90-mm TL for age-1 Smallmouth Bass for North American populations (Jackson et al. 2008). Thus, we chose 80-mm TL as our cut off based both on this criterion and our observations over the growing season in these streams. It is unlikely that Smallmouth Bass could be mistaken for other Micropterus spp. given they are the dominant centrarchid in these systems. For example, we have observed only one Spotted Bass at Buffalo Creek since 2010 (sampling used both snorkeling and electrofishing; Brewer, Unpublished data). Largemouth Bass occupy both streams in low numbers, but juveniles do not have tri-colored tails like Smallmouth Bass. All surveyors practiced recognizing the fish-size cutoffs using fish silhouettes and rocks of known sizes following guidance of Dunham et al. (2009).

Analyses

We used a robust Bayesian estimator (Kruschke 2013) to assess differences in young-of-year Neosho Smallmouth Bass densities among the four channel-unit types (backwater, pool, riffle, run). We included a reach-level grouping factor in the model to account for variation in densities not associated with channel-unit type. Densities were natural-log transformed to allow model-predicted estimates to fall below zero (i.e., a true continuous variable), where we added 0.01 to all densities to allow the transformation (very small densities at channel units with observed zero counts is also biologically reasonable). We omitted density data associated with the first survey from each stream reach (i.e., June 28) because mortality had not yet stabilized (i.e., the number of fish declined considerably across all channel-unit types between the first and second surveys). The transformed data conformed well to a normal distribution assumption and we used a broad (i.e., uninformative) normal distribution for the prior on group mean parameters. A preliminary assessment indicated heterogeneous standard deviations of densities among habitat types and stream reaches. Thus, we allowed the standard deviations to vary among groups using a hierarchical prior with a broad gamma distribution. By excluding data from the first survey, the inclusion of a "day" covariate did not indicate a relationship between young-of-year Smallmouth Bass habitat use and time of year; thus, it was not included in the model. We assessed statistical differences in young-of-year densities among channel units using contrasts based on 90% highest density intervals (HDIs; Kruschke 2015). We estimated posterior distributions of model parameters using Markov chain Monte Carlo (MCMC) methods in the program JAGS (Plummer 2003) called from the statistical software R (version 3.3.0, R Core Development Team, 2016) with the package rjags (Plummer 2011). Specifically, the posterior distributions were estimated using every 20th sample from 50,000 iterations of three chains (i.e., thinning) after a 100,000 iteration burn-in phase. We confirmed model convergence by visually examining plots of MCMC chain mixture and assessing effective sample size (ESS; Kruschke 2015).

Changes in the abundance estimates over time and catch curves were used to determine survival patterns over the first summer. We anticipated abundances to be the highest following hatch and expected exponential declines over the next several weeks. We assumed some dispersal would occur during this period, but changes in the overall abundances across the reaches would relate primarily to mortality. Thus, we assumed any movements by young-of-year fish would relate to only minimal changes in overall abundance across the sample reaches. Estimated changes in abundances over this period would facilitate comparison with existing studies of the nominal species (e.g., Pflieger 1966). We developed catch curves for each sampling reach after omitting the first two-week sampling period to meet the linear assumption. The instantaneous mortality rate (Z) was obtained from the catch curve and then converted to a 2-week mortality period (A). We natural-log transformed our data, and fitted a linear regression model between the natural-log transformed abundance and Julian date. The negative value of the slope was the estimate of Z and the estimated mortality rate was calculated as $A=1-e^{-Z}$ (Miranda and Betolli, 2007). We calculated 95% confidence intervals for Z by multiplying the SE of the slope by 1.96. Our analyses were completed using the statistical software R (version 3.2.3, R Core Development Team, 2015).

Results

Temperature and Discharge

Temperature and discharge conditions were relatively stable over our study period with a few exceptions. In Spring Creek, average daily water temperature was 21.7°C in late June and 23.8°C in late September (Figure 2). Average daily water temperature of Buffalo Creek was 22.5°C at the beginning of the study period and 21.7°C by late September. Discharge in Spring Creek ranged 0.63–2.79 $m^3 \cdot s^{-1}$ throughout the study period (Figure 2). Base-flow conditions in Buffalo Creek were also relatively low during the study period (usually < 3 $m^3 \cdot s^{-1}$), but a few precipitation events in August resulted in three discharge peaks between August 4 and August 15 (Figure 2). The three peak flows were: 16.81 $m^3 \cdot s^{-1}$ (August 4), 40.50 $m^3 \cdot s^{-1}$ (August 8), and 5.89 $m^3 \cdot s^{-1}$ (August 15). Flows returned to 1.42 $m^3 \cdot s^{-1}$ (~50 cfs) on August 23.

Channel-Unit Habitat Use

Our data from Spring Creek reflected patterns of habitat use often reported for the nominal subspecies. Average young-of-year Neosho Smallmouth Bass densities were highest in pool habitat throughout the sampling period (Figure 3). We observed a slight increase in use of run habitats between June 28 and July 16 as young of year dispersed from nesting locations that were primarily located in pool habitats. Thereafter, average densities in both run and pool habitats declined, remaining rel-

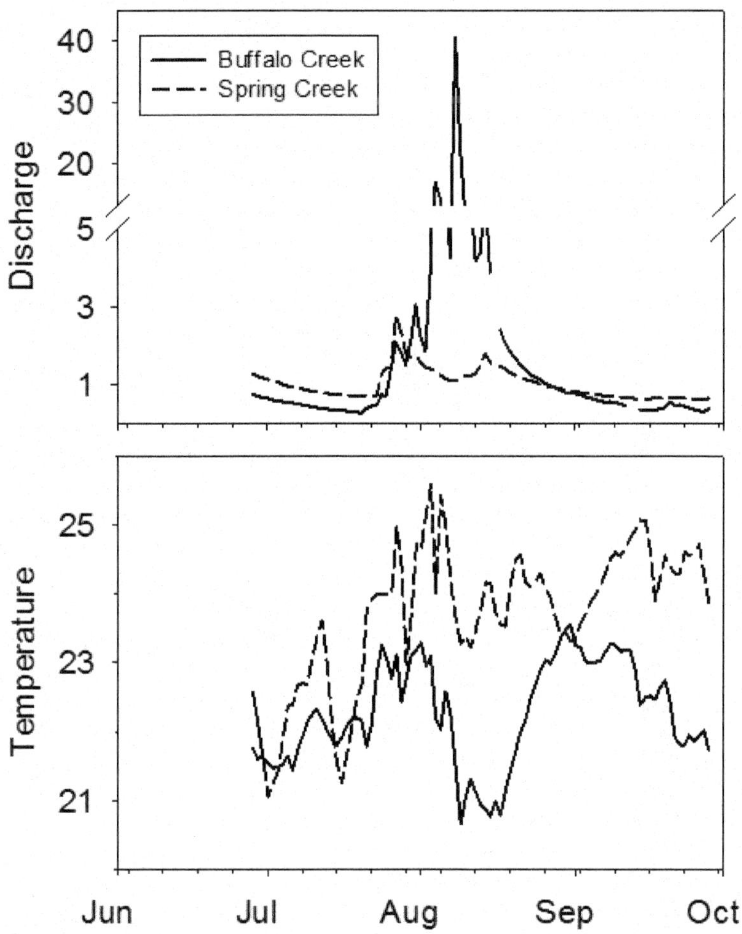

Figure 2. Daily discharge (top panel, $m^3 \cdot s^{-1}$) and water temperature (bottom panel, °C) during the 2013 sampling period (June 28 through September) in both Buffalo and Spring creeks. The sampling period is represented on the *X* axis by indicating month for simplicity.

atively unchanged from August through September. Backwater habitat was unavailable in our Spring Creek sampling reach.

Young-of-year Smallmouth Bass in Buffalo Creek used a variety of channel units during their first summer. Following hatch, average densities were high in both pool and backwater habitats (~4 and 10 fish/m^2, respectively, Figure 3). As mortality increased (Figure 4) and fish would be expected to disperse from their nests, young of year began to use run habitats and very low average densities were observed in riffles. By late July, average densities were 0.02–0.04 fish/m^2 in all available channel units. Average densities changed little during August. However, average density of fish in backwater habitats increased 10-fold during September. This trend corresponded with a decrease in average density of Smallmouth Bass in riffles, and followed an increase in discharge and decrease in temperature (i.e., the two factors co-varied) that occurred in mid to late August (Figure 2). Although fish densities were relatively

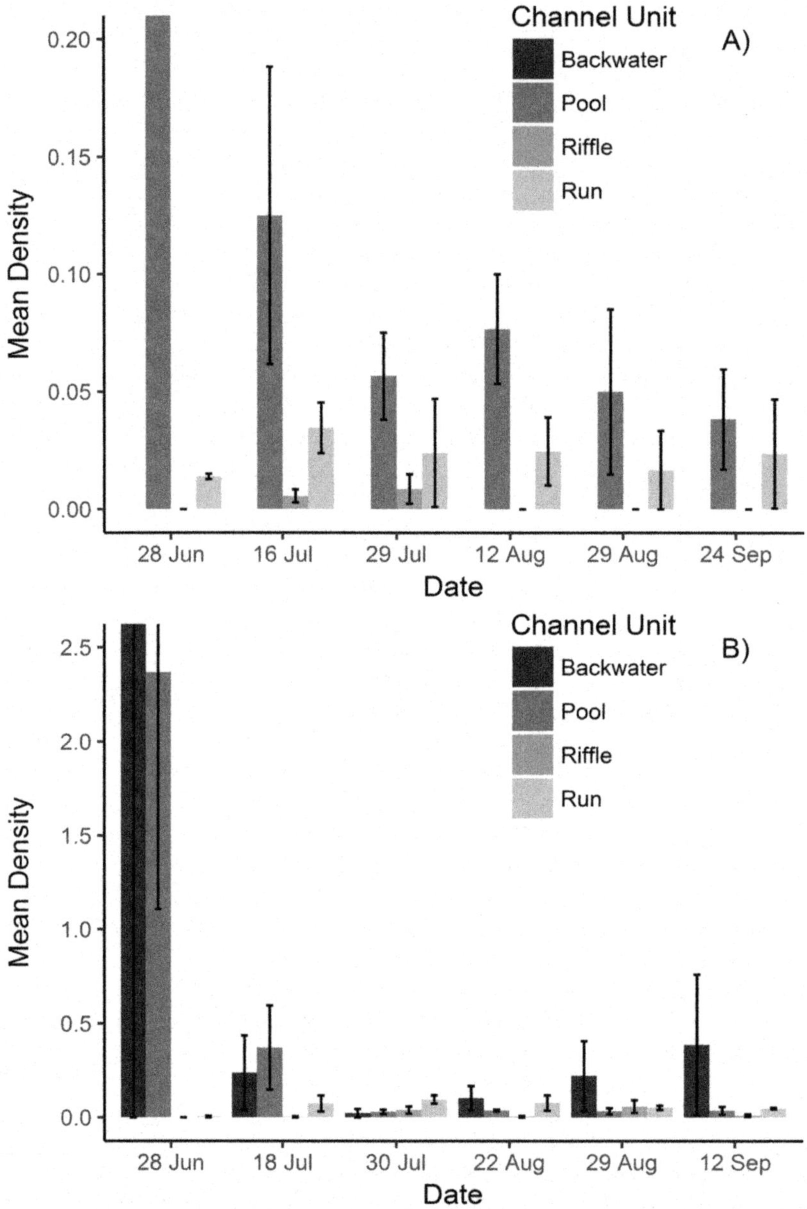

Figure 3. Average density (fish/m^2 ± SD) of young-of-year Neosho Smallmouth Bass in available channel units of Spring Creek, Oklahoma (panel A), and Buffalo Creek, Missouri (panel B). Sampling was completed via snokel surveys from late June through late September. For simplicity, data from Spring Creek sampling on September 12 were not displayed: average densities on this sample date were 0 in riffles, 0.02 in runs, and 0.04 in pools. Because of the relatively high densities in some habitats on the first sampling event, data from two channel units were truncated in the figure (average young-of-year density in pools in Spring Creek: 6.81 fish/m^2, and average young-of-year density in backwaters in Buffalo Creek: 10.44 fish/m^2).

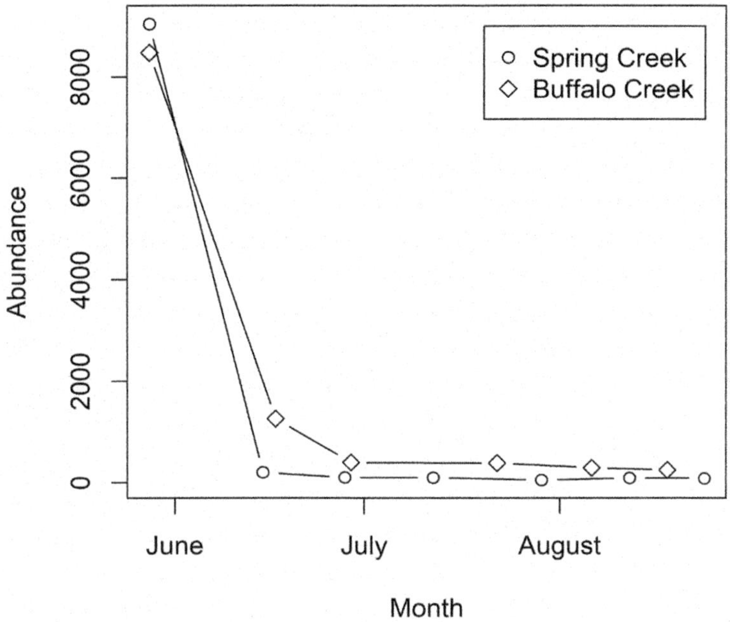

Figure 4. Abundances (number of fish in each sample reach) of young-of-year Neosho Smallmouth Bass over time (month). The Spring Creek sampling dates were: June 28, July 16, 29, August 12, 29, and September 12, 24. The Buffalo Creek sampling dates were: June 28, July 18, 30, August 22, and September 6, 18.

Table 1. Comparisons of young-of-year Smallmouth Bass densities from July 16 to September 24, 2013. We made comparisons among four channel-unit habitat types (backwater, pool, run, and riffle) in Spring Creek and Buffalo Creek using contrasts. Highest density intervals (HDIs, 90%) were calculated using a robust Bayesian estimator. Densities are reported on a natural-log scale.

Contrast	90% HDI (1)	90% HDI (2)
Backwater (1) versus pool (2)	–0.39, 1.20	0.18, 1.32
Backwater (1) versus run (2)	–0.39, 1.20	–0.01, 1.06
Pool (1) versus run (2)	0.18, 1.32	–0.01, 1.06
Backwater-pool-run (1) versus riffle (2)	–0.07, 1.19	–2.26, –1.03

high in backwaters on several occasions during our sample period, variation in densities was also quite high (Figure 3).

Considering data from the two stream reaches combined, young-of-year Smallmouth Bass densities were statistically different in riffle habitat, but not among other channel-unit types. Smallmouth Bass densities were lower in riffle habitats compared to other channel units (Table 1). The 90% HDIs did not overlap for backwater, pool, and run habitat versus riffle habitat. The 90% HDIs overlapped for backwater versus pool, backwater versus run, and pool versus run. Plots of MCMC chains confirmed adequate model convergence, and ESS was adequate for all model parameters (>10,000; Kruschke 2015).

Mortality Estimates

Abundance of swim-up Neosho Smallmouth Bass fry exceeded 8,000 at both Buffalo and Spring creeks (Figure 4). As anticipated, mortality was extremely high over the next 2-week period. In Buffalo Creek, ~85% of all young of year presumably died in the first 2 weeks, whereas ~99% of all young-of-year Neosho Smallmouth Bass were lost from Spring Creek. Abundance stabilized in both streams by the end of July, with little variation in abundances through the end of September. Instantaneous mortality in Spring Creek and Buffalo Creek were 0.011 (95% CI: 0.024–0.001) and 0.021 (95% CI: 0.035–0.007), respectively (Figure 5). Following the initial and substantial decline in young-of-year abundances (Figure 4), an average of 1.13% (95% CI: 0.13%–2.38%) and 2.11% (95% CI: 0.72%–3.48%) of the population died every 2 weeks in Spring and Buffalo creeks, respectively (Figure 5). Even though mortality was similar for the two stream reaches following the initial decline in abundance, survival throughout our study period was much lower in Spring Creek (0.56%) compared to Buffalo Creek (3.40%).

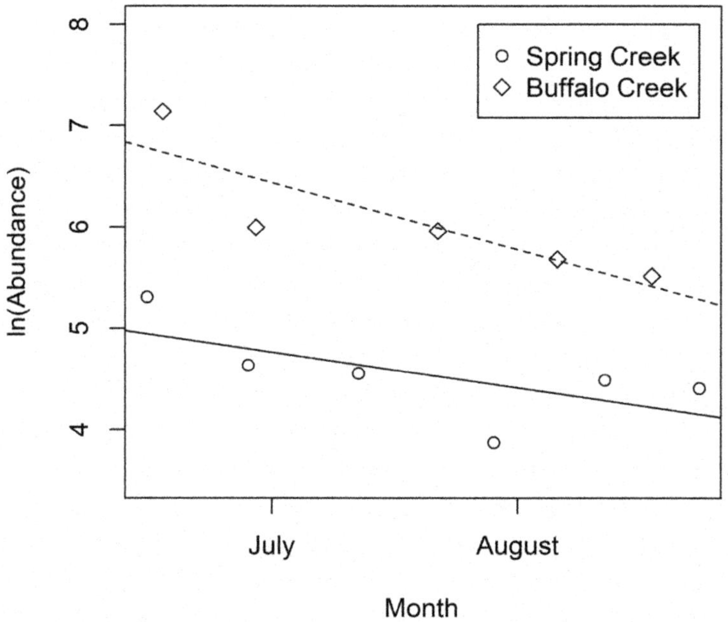

Figure 5. Catch curve of young-of-year Neosho Smallmouth Bass where the natural log (abundances) were plotted over time (shown by month) to estimate instantaneous mortality rate (Z).

Discussion

Although young-of-year Neosho Smallmouth Bass survival was low in both stream reaches, mortality was notably higher in Spring Creek than Buffalo Creek during the 2-week period following the swim-up phase. Even slight differences in annual recruitment can have a substantial effect on year-class strength in fishes (Ferreri et al. 1995; Wooton 1998; Lobón-Cerviá and Rincón 2004). Both flow (Larimore 1975; Simonson and Swenson 1990; Smith et al. 2005) and water temperature (Serns 1982; Knotek and Orth 1998) are known to influence young-of-year Smallmouth Bass survival. Flow conditions in both streams were low and declined slowly during the high-mortality period. Although stream temperatures were only slightly different during this period (21–23°C in Spring Creek versus 21–22°C in Buffalo Creek), slightly warmer temperatures in Spring Creek may have facilitated faster growth. Optimal temperatures for young-of-year growth is 29°C (Hewett and Johnson 1992), but Shuter et al. (1980) reported a positive linear relationship between daily growth rate and temperatures up to 27°C. Likewise, Coutant and Deangelis (1983) indicated the fastest growth by Smallmouth Bass fry at 25–27°C. Some faster-growing individuals via higher stream temperatures could have led to higher levels of cannibalism in Spring Creek. Smallmouth Bass fry as small as 11 mm standard length can consume fish (Pflieger 1966). We also hypothesize the higher young-of-year survival in Buffalo Creek may relate to increased habitat complexity that facilitated higher prey abundance and habitat useful for predator avoidance and higher-flow refugia.

Variation in predator densities between Spring Creek and Buffalo Creek may also have contributed to the differing survival rates of young-of-year Neosho Smallmouth Bass. Young-of-year Smallmouth Bass are highly susceptible to predation from a variety of sources (Reynolds and O'Bara 1991; Pflieger 1966; Knotek and Orth 1998). Both Longear Sunfish *Lepomis megalotis* (Pflieger 1966; Pflieger 1975; Reynolds and O'Bara 1991) and adult Smallmouth Bass (Dong and Deangelis 1998) are known to prey on young-of year Smallmouth Bass. However, tow-barge electrofishing surveys in 2014 revealed low Longear Sunfish densities in both streams and densities of age ≥1 Smallmouth Bass were also comparable (Mollenhauer, Unpublished data). Crayfishes are also known predators of young-of-year Smallmouth Bass (Dorn and Mittelbach 2004; Brewer and Orth 2015) and were a plausible source of variation in survival between the streams. Anecdotal observations suggest crayfish densities in Spring Creek may be higher than Buffalo Creek, but we do not have quantitative data to support this notion.

Our study demonstrates the importance of multiple habitat types for young-of-year Neosho Smallmouth Bass, and how habitat deemed unimportant to adult fishes (i.e., backwater habitat) may be fundamental to early life stages. Perceived habitat use by young-of-year stream fishes is sometimes based on a single survey (e.g., Muhlfeld et al. 2001, Fore et al. 2007, Schager and Burkhardt-Holm 2007), which may overemphasize the importance of certain habitats and provides minimal information on how dispersal relates to trends in survival among streams. Dispersal can transition young-of-year fishes to a variety of habitats that are equally important for survival (Moore and Thorp 2008). By conducting multiple surveys over time, we show how Neosho Smallmouth Bass shift their use of habitats over time. Interestingly, young-of-year fish in Buffalo Creek had very high densities in backwater and pool habitats immediately following spawning. Although overall densities declined in all channel unit types through July and the first part of August, average densities of fish using backwaters increased in late August and September. Adult Smallmouth Bass are typically not associated with backwaters (Brewer and Orth 2015; but see Brewer 2013), but our findings suggest these habitats may be important for juvenile life stages at certain times (i.e., hatch and some periods of juvenile rearing), and management strategies would benefit from this consideration. Similarly, Moore and Gregory (1988) found backwaters provided areas of flow refuge and increased foraging opportunities for young-of-year Cutthroat Trout *Oncorhynchus clarki*, although these habitats were seldom used by adult fish. Fish are presumed to choose habitat that balances foraging opportunities and predation risk (Wooton 1998).

We hypothesize that the use of backwater habitats in late August and September following the higher-flow events may be due to increased prey availability in these habitats, particularly near the confluence with the main channel. We found a 10-fold increase in young-of-year densities following higher-flow events in August, but the use of these habitats continued after low-flow conditions returned. Backwater habitats can provide refuge for young-of-year fishes during high-flow events (Bolland et al. 2012). Other studies have also related backwaters to increased survival of young-of-year stream fishes (e.g., Kwak 1988; Bowen et al. 1998; Freeman et al. 2001; Bowen et al. 2003). We also observed that young-of-year Smallmouth Bass tended to occupy backwater areas near the confluence with the main channel (i.e., the 'mouth' of the backwater). These confluence areas may be ideal foraging areas, particularly following higher flows, when prey items may increase due to drifting organisms in the main channel coming in contact with much lower velocity areas. A sufficient invertebrate forage base is essential for young-of-year Smallmouth Bass growth and development (Easton and Orth 1992), and backwaters are typically highly productive habitats with greater zooplankton and macroinvertebrate densities than other stream areas (Ward and Stanford 1995; Allan and Castillo 2007). These confluence areas would be ideal for drift feeding, which is a known foraging strategy by sub-adult Smallmouth Bass (Simonson and Swenson 1990). In addition, juvenile Smallmouth Bass are able to make movements within and between channel units (A. Miller and S. Brewer, Oklahoma Cooperative Fish and Wildlife Research Unit, Oklahoma State University, unpublished data), thus, regular movements between backwaters and the main channel may also occur.

Although young-of-year Smallmouth Bass used backwater habitats in Buffalo Creek, the lack of backwater habitats in the Spring Creek reach and highly variable fish densities limited the use of meaningful statistical comparisons among the channel-unit types. Although biologically meaningful, we did not find that young-of-year densities were statistically different in backwater habitats despite the observed 10-fold increase on one sampling occasion. Backwater habitats were not available in the Spring Creek reach, which increased model uncertainty when estimating the group means (i.e., with only two reaches, there was considerable uncertainty when estimating densities in backwater habitats in Spring Creek). Further, the highly variable fish densities in both backwater and pool habitats suggest other factors may also be associated with variation in young-of-year habitat use. Classifying channel units, though convenient for investigators, lacks important details such as spatial proximity to other habitat features and physicochemical factors (e.g., water depth and temperature may vary among pools) that may be important to fish. Measuring additional habitat elements would necessitate increased sampling effort, which is already high with repeated sampling throughout the juvenile rearing season.

Our results may have been somewhat influenced by sampling artifacts. Water clarity and time of sampling are primary factors related to variability in stream-fish snorkeling efficiency (Dunham et al. 2009), but visibility was ideal in both streams across the study period and all surveys were completed during daylight hours. However, small fish that occupy interstitial spaces are difficult to detect (Mullner et al. 1998; Wildman and Neumann 2003) and clustering by fish may result in undercounting (Hillman et al. 1992). Thus, the high abundances of very small fish (<30 mm-TL) in late June may be underestimated, and snorkeling efficiency likely increased as young-of-year abundance decreased and fish size increased. Nevertheless, we believe the observed higher mortality in Spring Creek in late June and early July was accurate, although overall survival is likely overestimated in both streams. Additionally, constant capture probability (i.e., the proportion of available individuals identified) was more important than perfect capture probability for comparable results between the reaches. We feel that constant capture probability was a reasonable assumption given the relatively short duration of the study, the similar environmental conditions between the reaches, and our standardization of the number of crew and size of designated snorkel lanes among different habitats. Our young-of-year abundance and density estimates do not account for possible movements from adjacent stream reaches. Young-of-year Smallmouth Bass can have a propensity to move downstream in some systems (Humston et al. 2010). However, movement during the swim-up phase was likely inconsequential because young-of-year Smallmouth Bass tend to remain near the nest during that period (Gross and Kapuscinski 1997). Further, a recent tagging study of larger juvenile Neosho Smallmouth Bass revealed no directional movement tendency (A. Miller and S. Brewer, unpublished data). Additionally, we do not think flows were high enough to displace young-of-year fish in June or July (i.e., we did not observe repeated spawning events). Therefore, movement likely did not confound our survival estimates early in the season when we observed the variable mortality rates between the two streams.

Because of the limited human modifications in Spring Creek and Buffalo Creek (i.e., no dams or channelization, primarily forested areas), these streams are ideal for examining the effects of abiotic and biotic factors on young-of-year Neosho Smallmouth Bass survival. Our study revealed both high and differential young-of-year mortality during a relatively short time period immediately following the swim-up phase, which is similar to findings for the nominal species. Specifically, Pflieger (1966) noted a substantial decrease in Smallmouth Bass fry abundances within the first few days of leaving the nest. Likewise, Knotek and Orth (1998) indicated mortality was highest during the swim-up phase when young-of-year Smallmouth Bass have limited abilities to escape predation. Future research aimed at understanding factors driving young-of-year Smallmouth Bass mortality during this critical time period would result in a greater ability to predict year-class strength based on prevailing biotic and abiotic conditions. It would also be beneficial if future studies examined predation, food availability, and survival under more variable thermal and flow conditions. In addition, future efforts to understand young-of-year Smallmouth Bass dispersal and habitat use across multiple streams (including very small streams thought to be unimportant to adults) and including factors at coarse spatial scales (e.g., groundwater, land use) would allow for more relevant management actions. For example, we may find opportunities to protect and enhance Smallmouth Bass spawning and rearing habitats or identify important movement corridors that provide areas of refuge for young-of-year fishes. Currently, our understanding of movements by young-of-year Smallmouth Bass, particularly the Neosho subspecies, is limited and habitat use across coarse spatial and temporal scales remains largely unknown.

Acknowledgments

This research is a contribution of the Oklahoma Cooperative Fish and Wildlife Research Unit (U.S. Geological Survey, Oklahoma Department of Wildlife Conservation, Oklahoma State University, and Wildlife Management Institute cooperating). Any use of trade, firm, or product names is for descriptive purposes and does not imply endorsement by the U.S. Government. We thank Dan Shoup, Mike Siepker, and three anonymous reviewers for their thoughtful comments that improved the overall quality of our manuscript.

References

Ahlert, J., S. Barkley, L. Claybrok, R. Fourt, J. Hogue, B. Howell, B. Limbird, R. Moore, M. Oliver, T. Penniston, C. Perrin, L. Red, B. Wagner, and S. Wooldridge. 1995. The Arkansas Smallmouth Bass management plan. Arkansas Game and Fish Commission, Little Rock.

Allan, J. D. and M. M. Castillo. 2007. Stream ecology: structure and function of running waters, 2nd edition. Springer, Dordrecht, The Netherlands.

Bolland, J. D., A. D. Nunn, M. C. Lucas, and I. G. Cowx. 2012. The importance of variable lateral connectivity between artificial floodplain waterbodies and river channels. River Research and Applications 28:1189–1199.

Bowen, Z. H., M. C. Freeman, and K. D. Bovee. 1998. Evaluation of generalized habitat criteria for assessing impacts of altered flow regimes on warmwater fishes. Transactions of the American Fisheries Society 127:455–468.

Bowen, Z. H., K. D. Bovee, and T. J. Waddle. 2003. Effects of flow regulation on shallow-water habitat dynamics and floodplain connectivity. Transactions of the American Fisheries Society 132:809–823.

Boxrucker, J., A. A. Echelle, and R.A. Van Den Bussche. 2004. Determining the degree of hybridization in the Smallmouth Bass population of Broken Bow Reservoir and the Mountain Fork River. Oklahoma Department of Wildlife Conservation, Final Report F-50-R, Project 19, Oklahoma City.

Brewer, S. K. 2011. Patterns in young-of-year Smallmouth Bass microhabitat use in multiple stream segments with contrasting land uses. Fisheries Management and Ecology 18:506–512.

Brewer, S. K. 2013. Groundwater influence on the distribution and abundance of riverine Smallmouth Bass *Micropterus dolomieu* in pasture landscapes of the Midwestern USA. River Research and Applications 29:269–278.

Brewer, S. K., C. F. Rabeni, S. P Sowa and G. Annis. 2007. Natural-occurring landscape and in-channel factors affecting the distribution and relative abundance of riverine Smallmouth Bass in Missouri. North American Journal of Fisheries Management 27:326–341.

Brewer, S. K., and D. J. Orth. 2015. Smallmouth Bass *Micropterus dolomieu* Lacepède, 1802. Pages 9–26 *in* M. D. Tringali, J. M. Long, T. W. Birdsong, and M. S. Allen, editors. Black bass diversity: multidisciplinary science for conservation. American Fisheries Society, Symposium 82, Bethesda, Maryland.

Brewer, S. K., and J. M. Long. 2015. Biology and ecology of Neosho Smallmouth Bass and the genetically distinct Ouachita lineage. Pages 281–286 *in* M. D. Tringali, J. M. Long, T. W. Birdsong, and M. S. Allen, editors. Black bass diversity: multidisciplinary science for conservation. American Fisheries Society, Symposium 82, Bethesda, Maryland.

Brewer, S. K., and M. R. Ellersieck. 2011. Evaluating two observational sampling techniques for determining the distribution and detection probability of age-0 Smallmouth Bass in clear, warmwater streams. North American Journal of Fisheries Management 31:894–904.

Coutant, C. C., and D. L. Deangelis. 1983. Comparative temperature-dependent growth of Largemouth and Smallmouth Bass fry. Transactions of the American Fisheries Society 112:416–423.

Curry, R. A., S. L. Currie, S. K. Arndt, and A. T. Bielak. 2005. Winter survival of age-0 Smallmouth Bass, *Micropterus dolomieu*, in north eastern lakes. Environmental Biology of Fishes 72:111–122.

Dauwalter, D. C., and W. L. Fisher. 2007. Spawning chronology, nest site selection and nest success of Smallmouth Bass during benign streamflow conditions. The American Midland Naturalist 158:60–78.

Dong, Q., and D. L. Deangelis. 1998. Consequences of cannibalism and competition for food in a Smallmouth Bass population: an individual-based modeling study. Transactions of the American Fisheries Society 127:174–191.

Dorn, N. J., and G. G. Mittelbach. 2004. Effects of a native crayfish (*Orconectes virilis*) on the reproductive success and nesting behavior of sunfish (*Lepomis* spp.). Canadian Journal of Fisheries and Aquatic Sciences 61:2135–2143.

Dunham, J. B., A. E. Rosenberger, R. F. Thurow, C. A. Dolloff, and P. J. Howell. 2009. Coldwater fish in wadeable streams. Pages 119–138 *in* S. A. Bonar, W. A. Hubert, and D. W. Willis, editors. Standard methods for sampling North American freshwater fishes. American Fisheries Society, Bethesda, Maryland.

Easton, R. S., and D. J. Orth. 1992. Ontogenetic diet shifts of age-0 Smallmouth Bass (*Micropterus dolomieu* Lacepède) in the New River, West Virginia, USA. Ecology of Freshwater Fish 1:86–98.

Ferreri, C., W. W. Taylor, and D. B. Hayes. 1995. Evaluation of age-0 survival and its effect on Lake Trout rehabilitation in the Michigan waters of Lake Superior. Journal of Great Lakes Research 1:218–224.

Fore, J. D., D. C. Dauwalter, and W. L. Fisher. 2007. Microhabitat use by Smallmouth Bass in an Ozark stream. Journal of Freshwater Ecology 22:189–199.

Freeman, M. C., Z. H. Bowen, K. D. Bovee, and E. R. Irwin. 2001. Flow and habitat effects on juvenile fish abundance in natural and altered flow regimes. Ecological Applications 11:179–190.

George, E. L., and W. F. Hadley. 1979. Food and habitat partitioning between Rock Bass (*Ambloplites rupestris*) and Smallmouth Bass (*Micropterus dolomieui*) young of the year. Transactions of the American Fisheries Society 108:253–261.

Gross, M. L., and A. R. Kapuscinski. 1997. Reproductive success of Smallmouth Bass estimated and evaluated from family-specific DNA fingerprints. Ecology 78:1424–1430.

Hewett, S. W., and B. L. Johnson. 1992. Fish bioenergetics model 2. Technical report WIS-SG-92-250, University of Wisconsin Sea Grant Institute: Madison, Wisconsin.

Hillman, T. W., J. W. Mullan, and J. S. Griffith. 1992. Accuracy of underwater counts of juvenile Chinook Salmon, Coho Salmon, and steelhead. North American Journal of Fisheries Management 12:598–603.

Humston, R., B. M. Priest, W. C. Hamilton, and P. E. Bugas, Jr. 2010. Dispersal between tributary and main-stem rivers by juvenile Smallmouth Bass evaluated using otolith microchemistry. Transactions of the American Fisheries Society 139:171–184.

Jackson, Z. J., M. C. Quist, and J. G. Larscheid. 2008. Growth standards for nine North American fish species. Fisheries Management and Ecology 15:107–118.

Knotek, W. L., and D. J. Orth. 1998. Survival for specific life intervals of Smallmouth Bass, *Micropterus dolomieu*, during parental care. Environmental Biology of Fishes 51:285–296.

Kruschke, J. K. 2013. Bayesian estimation supersedes the t test. Journal of Experimental Psychology: General 142:573–603.

Kruschke, J. K. 2015. Doing Bayesian data analysis: a tutorial with R, JAGS, and Stan, 2nd Edition. Elsevier, San Diego, California.

Kwak, T. J. 1988. Lateral movement and use of floodplain habitat by fishes of the Kankakee River, Illinois. The American Midland Naturalist 120:241–249.

Larimore, R. W. 1975. Visual and tactile orientation of Smallmouth Bass fry under floodwater conditions. Pages 323–332 *in* R. H. Stroud and H. Clepper, editors. Black bass biology and management. Sport Fishing Institute, Washington, D.C.

Lobón-Cerviá, J., and P. A. Rincón. 2004. Environmental determinants of recruitment and their influence on the population dynamics of stream-living Brown Trout *Salmo trutta*. Oikos 105:641–646.

Lyons, J. 1997. Influence of winter starvation on the distribution of Smallmouth Bass among Wisconsin streams: a bioenergetics modeling assessment. Transactions of the American Fisheries Society 126:157–162.

MacCrimmon, H. R., and W. H. Robbins. 1981. Influence of temperature, water current, illumination, and time on activity and substrate selection in juvenile Smallmouth Bass (*Micropterus dolomieui*). Canadian Journal of Zoology 59:2322–2330.

Malloy, T. P., A. A. Echelle, R. A. Van Den Bussche, W. L. Fisher. 2000. Genetic structure and age and growth of Smallmouth Bass in two Oklahoma reservoirs. Oklahoma Department of Wildlife Conservation, Final Report F-41-R, Project 21, Oklahoma City.

McClendon, D. D. and C. F. Rabeni. 1986. Sampling stream centrarchids: comparing electrofishing and underwater observation. Proceedings of the Annual Conference of Southeastern Association of Fisheries and Wildlife Agencies 40:92–101.

Miranda L. E. and P. W. Betolli. 2007. Mortality. Pages 229–277 *in* C. Guy and M. Brown, editors. Analysis and Interpretation of Freshwater Fisheries Data. American Fisheries Society, Bethesda, Maryland.

Mollenhauer, R. 2016. Sampling the stream landscape: factors influencing detection and distribution of stream fishes. Doctoral dissertation. Oklahoma State University, Stillwater.

Moore, K. M. S., and S. V. Gregory. 1988. Summer habitat utilization and ecology of Cutthroat Trout fry (*Salmo clarki*) in Cascade Mountain streams. Canadian Journal of Fisheries and Aquatic Sciences 45:1921–1930.

Moore, S. L., and J. H. Thorp. 2008. Coping with hydrogeomorphic variations in a prairie river: resiliency in young-of-the-year fishes. River Research and Applications 24:267–278.

Muhlfeld, C. C., D. H. Bennett, and B. Marotz. 2001. Summer habitat use by Columbia River Redband Trout in the Kootenai River drainage, Montana. North American Journal of Fisheries Management 21:223–235.

Mullner, S. A., W. A. Hubert, and T. A. Wesche. 1998. Snorkeling as an alternative to depletion electrofishing for estimating abundance and length-class frequencies of trout in small streams. North American Journal of Fisheries Management 18:947–953.

Pert E. J., D. J. Orth, and M. J. Sabo. 2002. Lotic-dwelling age-0 Smallmouth Bass as both resource specialists and generalists: reconciling disparate literature reports. Pages 185–189 *in* D. P. Phillip and M. S. Ridgway, editors. Black bass: ecology, conservation, and management. American Fisheries Society, Symposium 31, Bethesda, Maryland.

Pflieger, W. L. 1966. Reproduction of the Smallmouth Bass (*Micropterus dolomieui*) in a small Ozark stream. The American Midland Naturalist 76:410–418.

Pflieger, W. L. 1975. Reproduction and survival of the Smallmouth Bass in Courtois Creek. Pages 231–239 *in* R. H. Stroud and H. Clepper, editors. Black bass biology and management. Sport Fishing Institute, Washington, D.C.

Plummer M. 2003. JAGS: a program for analysis of Bayesian graphical models using Gibbs sampling. Proceedings of the Third International Workshop on Distributed Statistical Computing (DSC 2003), Vienna, Austria. Available: http://mcmc-jags.sourceforge.net/. (August 2017).

Plummer M. 2011. rjags: Bayesian graphical models using MCMC. R package version 2.2.0-4. Available: http://cran.rproject.org/web/packages/rjags/. (August 2017).

Probst, W. E., C. F. Rabeni, W. G. Covington, and R. E. Marteney. 1984. Resource use by stream-dwelling Rock Bass and Smallmouth Bass. Transactions of the American Fisheries Society 113:283–294.

Quinn, J., B. Wagner, and S. Filipek. 2004. Characteristics of Black bass populations in the rivers and streams of Arkansas. Arkansas Game and Fish Commission, Stream Fisheries Program Report STP2004–01, Little Rock.

R Development Core Team. 2015. R: a language and environment for statistical computing. R Foundation for Statistical Computing, Vienna. Available: http://www.R-project.org/. (October 2017).

R Development Core Team. 2016. R: a language and environment for statistical computing. R Foundation for Statistical Computing, Vienna. Available: http://www.R-project.org/. (August 2017).

Rabeni, C. F., and R. B. Jacobson. 1993. Geomorphic and hydraulic influences on the abundance and distribution of stream centrarchids in Ozark USA streams. Polskie Archiwun Hydrbiologii 40:87–99.

Rankin, E. T. 1986. Habitat selection by Smallmouth Bass in response to physical characteristics in a natural stream. Transactions of the American Fisheries Society 115:322–334.

Reynolds, C. R., and C. J. O'Bara. 1991. Reproductive ecology and spawning habitat of Smallmouth Bass in two small streams of the Tennessee River system. Pages 61–65 *in* D. C. Jackson, editor. The First International Smallmouth Bass Symposium. Mississippi Agricultural and Forestry and Experiment Station, Mississippi State University.

Sabo, M. J., D. J. Orth, and E. J. Pert. 1996. Effect of stream microhabitat characteristics on rate of net energy gain by juvenile Smallmouth Bass, *Micropterus dolomieu*. Environmental Biology of Fishes 46:393–403.

Schager, E., A., and P. Burkhardt-Holm. 2007. Status of young-of-the-year Brown Trout (*Salmo trutta fario*) in Swiss streams: factors influencing YOY trout recruitment. Aquatic Sciences 69:41–50.

Schill, D. J. and J. S. Griffith. 1984. Use of underwater observations to estimate Cutthroat Trout abundance in the Yellowstone River. North American Journal of Fisheries Management 4:479–487.

Shuter, B. J., J. A. Maclean, F. E. J. Fry, and H. A. Regier. 1980. Stochastic simulation of temperature effects on first-year survival of Smallmouth Bass. Transactions of the American Fisheries Society 109:1–34.

Serns, S. L. 1982. Relation of temperature and population density to first-year recruitment and growth of Smallmouth Bass in a Wisconsin lake. Transactions of the American Fisheries Society 111:570–574.

Simonson, T. D., and W. A. Swenson. 1990. Critical stream velocities for young-of-year Smallmouth Bass in relation to habitat use. Transactions of the American Fisheries Society 119:902–909.

Smith, S. M., J. S. Odenkirk, and S. J. Reeser. 2005. Smallmouth Bass recruitment variability and its relation to stream discharge in three Virginia rivers. North American Journal of Fisheries Management 25:1112–1121.

Stark, W. J., and A. A. Echelle. 1998. Genetic structure and systematics of Smallmouth Bass, with emphasis on Interior Highlands populations. Transactions of the American Fisheries Society 127:393–416.

Ward, J. V., and J. A. Stanford. 1995. Ecological connectivity in alluvial river ecosystems and its disruption by flow regulation. Regulated Rivers: Research and Management 11:105–119.

Wildman, T. L. and R. M. Neumann. 2003. Comparison of snorkeling and electrofishing for estimating abundance and size structure of Brook Trout and Brown Trout in two southern New England streams. Fisheries Research 60:131–139.

Woods A. J., J. M. Omernik, D. R. Butler, J. G. Ford, J. E. Henley, B. W. Hoagland, D. S. Arndt, and B.C. Moran. 2005. Ecoregions of Oklahoma (color poster with map, descriptive text, summary tables, and photographs). U.S. Geological Survey, Reston, Virginia.

Wooton, R. J. 1998. Ecology of teleost fishes, 2nd edition. Kluwer, Dordrecht, The Netherlands.

American Fisheries Society Symposium 87:39–53, 2019

Local and Landscape Determinants of Smallmouth Bass Habitat Suitability in Michigan Streams

Jan-Michael Hessenauer*
Lake St. Clair Fisheries Research Station
Michigan Department of Natural Resources, 33135 South River Road
Harrison Township, Michigan 48045, USA

Kevin Wehrly and Danielle Forsyth Kilijanczyk
Institute for Fisheries Research, Michigan Department of Natural Resources
400 North Ingalls, G250, Ann Arbor, Michigan 48109, USA

Todd Wills
Lake St. Clair Fisheries Research Station
Michigan Department of Natural Resources, 33135 South River Road
Harrison Township, Michigan 48045, USA

Troy Zorn
Marquette Fisheries Research Station, Michigan Department of Natural Resources, 484 Cherry Creek Road, Marquette, Michigan 49855, USA

Abstract.—Black bass *Micropterus* spp. are among the most popular sportfish species in North America, however, managers currently have insufficient information about patterns and drivers of their distribution. Tradeoffs exist with the spatial scale at which to collect information associated with the distribution of species. Local-scale data are expensive to collect, but are likely directly associated with occupancy of a particular site. Landscape-scale data are increasingly available, are easily scaled and linked to site occupancy, but management often does not occur at this scale. We utilized site-specific habitat data collected as part of a statewide stream sampling program from 2002 to 2013 to generate presence–absence data for Smallmouth Bass *M. dolomieu*. We used site specific local-scale, landscape-scale, and a combination of local- and

*Corresponding author: HessenauerJ1@michigan.gov

landscape-scale data to create random forest classification tree models of Smallmouth Bass presence and absence. All three models had similar total error rates ranging from 13.9% to 16.6%. Overall rates of success and error did not differ among the three models ($P = 0.95$). The model using only landscape-scale data were used to predict the presence or absence of Smallmouth Bass for all Michigan stream segments and had an error rate of 17% based on an independent data validation. These data suggest that our approach has utility for predicting Smallmouth Bass occurrence in Michigan streams by identifying important habitat features associated with Smallmouth Bass occurrence. This approach could be extended to understand the distribution of other black bass species in locations where distributional data may be more limited.

Introduction

Black bass *Micropterus* spp. are among the most popular sportfish species in North America (USFWS 2011). Future management and research priorities for these species require information about their current spatial and temporal distributions, and the critical habitat features that drive these distributions. This is especially important as black bass populations face increasing pressure from land use, habitat degradation, climate change, establishment and spread of exotic species, and angling pressure. Such changes have already been documented to some extent. For example, Smallmouth Bass *M. dolomieu* are known to be sensitive to anthropogenic changes to fluvial systems which may increase stream temperatures and ultimately favor competitors such as Largemouth Bass *M. salmoides* (e.g., Sowa and Rabeni 1995).

Traditionally, most efforts to identify critical habitat were site focused (e.g., Todd and Rabeni 1989), and in practice most fisheries management actions still occur at the local-scale. Site-specific habitat or food availability are undoubtedly critical components of site occupancy at the local scale. For example, coarse woody materials have been identified as critical habitat features for stream fishes (e.g., Angermeier and Karr 1984; Harmon et al. 1986; Everett and Ruiz 1993) and a perceived lack of coarse woody materials at a local site may prompt management action (Roni et al. 2015). In contrast, a growing number of studies have assessed fish-habitat relationships from a landscape perspective (e.g., Wiley et al. 1997; Zorn et al. 2002; Zorn et al. 2004; Steen et al. 2008; Lyons et al. 2010). The now ubiquity of GIS-based datasets provides an abundance of landscape-scale data that have been used to predict stream fish distributions (e.g., Wiley et al. 1997; Zorn et al. 2002; Zorn et al. 2004; Steen et al. 2008). These systems are useful because they are easily scaled up or down, and capture critical landscape-scale processes that may be missed if only site-specific data are utilized (e.g., Wiley et al. 1997; Fausch et al. 2002; Allan 2004).

Tradeoffs exist for fisheries managers and researchers regarding the spatial scale at which habitat is viewed. For example, local approaches may be critical for identifying habitat features to protect and restore, but they are labor intensive especially at scales useful for predicting species occurrence across broad spatial extents. Landscape-based approaches, on the other hand, provide a means to identify and characterize habitat suitability across large spatial extents, but potentially at the expense of reduced accuracy. It is well established that the distribution of fishes are influenced by habitat features operating at both local and landscape scales (e.g., Tonn 1990; Wiley et al. 1997), but relatively few studies have utilized local- and landscape-scale data to predict fish occurrence and abundance (e.g., Brewer et al. 2007). The goals of this study were to 1) identify important habitat features associated with Smallmouth Bass occurrence and 2) estimate the current distribution of Smallmouth Bass in Michigan at the stream valley segment (see below) scale. While Smallmouth Bass distributions are broadly defined for Michigan (e.g., Bailey et al. 2004), the generation of fine-scale distributions provides managers with predictions of Smallmouth Bass presence or absence in streams that have never been sampled. To achieve these goals, we used habitat data collected at individual survey sites and landscape-scale data to develop predictive models of Smallmouth Bass presence or absence (PA). The diversity of data collected provided the opportunity to compare the efficacy of modeling Smallmouth Bass occurrence utilizing locally-collected data only, using landscape-scale data only, and a combination of the two data types. This approach is particularly important because of the aforementioned tradeoffs faced by fisheries managers and researchers regarding the spatial scales of data collection. Therefore, we sought to specifically examine the accuracy of models of Smallmouth Bass distributions in the State of Michigan using data collected at multiple spatial scales.

Methods

This study used data from 212 stream surveys conducted as part of the Michigan Stream Status and Trends Program (MSSTP). The MSSTP is a statewide inventory and monitoring program of Michigan's stream aquatic resources (Hayes et al. 2003). The MSSTP program methods are well described by Wills et al. (2015) and are summarized briefly here. The MSSTP defines streams as flowing waters of all sizes (Wills et al. 2015), including brooks, creeks, rivers, etc. A stratified random sampling approach was employed to sample river valley segments (Baker 2006; Seelbach et al. 2006), which represent contiguous stream segments of homogenous habitat (Wills et al. 2015). These segments are stratified by local fisheries management unit, catchment area, and temperature class, and are randomly sampled without replacement, generating a representative sample of stream types and habitats throughout the State of Michigan (Wills et al. 2015).

Stream surveys were conducted by Michigan Department of Natural Resources personnel in July and August during low flow, and consisted of both fish and habitat sampling (Wills et al. 2015). Fish sampling was conducted by electrofishing a single pass through the stream with backpack, tow-barge, or boat electrofishing units, depending on stream size. Electrofishing station length varied by site but was typically between 150 and 460 m, or roughly equivalent to 35 stream widths (Lyons 1992). All fish greater than 2.5 cm in total length were captured, identified and measured for total length.

Habitat sampling was conducted separately, but typically within a few days of fish sampling. Transects were placed throughout the fish sampling reach and stream depth, substrate, and habitat type (e.g., riffle, run, pool) were quantified along evenly spaced points. Bank conditions and riparian vegetation were assessed where each transect intersected the shore. Large woody materials and other types of instream structure were counted throughout the entire sampling station. For nonwadeable streams, protocols were modified and data collected by boat. Wills et al. (2015) provide a summary and description of statewide distribution of habitat data collected by the MSSTP.

In addition to data collected at localized field sites by the MSSTP we used landscape-scale data generated by the Michigan Department of Natural Resources, Institute for Fisheries Research (unpublished data). We used the 1:24,000 National Hydrography Dataset (NHD; USGS 2013) as the spatial framework for this study. Landscape descriptors of surficial geology texture (Farrand and Bell 1982), soil permeability (NRCS 2013), and potential groundwater movement (Baker et al. 2003) were summarized for the watershed of each segment. Stream size was characterized using watershed area. Landform was characterized as the elevation of the outlet of each segment and by calculating the gradient of each segment and segment watershed using data from the National Elevation Dataset (NED; USGS 2013). We characterized flow regime (mean annual discharge, April 10% exceedance yield, and August 90% exceedance yield) in each segment using landscape-based flow models developed by Seelbach et al. (2011). Finally, we characterized July mean water temperature for each segment using a landscape-based temperature model developed by Wehrly et al. (2009).

We used a classification tree (e.g., Breiman et al. 1984; De'ath and Fabricius 2000) approach to assess the association of local and landscape habitat variables with Smallmouth Bass presence or absence (PA). Classification trees seek to divide observations into groups based on predictor variables which minimize within group variance and maximize among group variance. This approach is ideal because it is capable of handling large numbers of variables, including continuous and categorical variables, and does not require distributional assumptions about the data (Breiman et al. 1984; De'ath and Fabricius 2000). We implemented the random forest approach (Breiman 2001), using the program R (R Core Team 2015) package randomForest (Liaw and Wiener 2002).

Three separate random forest models were created: one using only site-specific habitat data collected by the MSSTP from 2002 to 2013, a second model using only landscape-scale data estimated for each stream reach, and a third model using a combination of site-specific habitat data and the landscape-scale data (Table 1). The random forest model as implemented in R does not accept missing data, therefore surveys with missing observations were excluded, resulting in slightly different sample sizes among each of the three models (Table 2). To ease interpretation and improve model accuracy we sought to reduce the total number of variables considered by the site-specific model by conducting a preliminary filter of variables. This filter compared the mean values of each variable for sites where Smallmouth Bass were present or absent. If the difference between the mean presence and absence values for a particular variable was greater than or equal to 10% of the observed range of the variable, then the variable was considered potentially important and included in further analysis. The combined landscape- and site-specific dataset used all variables identified as potentially important from the site-specific model and all landscape variables (Table 1). For each of the three datasets we generated 10,000 random trees. Each individual tree was based on a random two-thirds subset of the individual dataset, and used a random subset of the predictor variables to divide the data (Breiman 2001; Liaw and Wiener 2002). The accuracy of each tree was assessed by its ability to assign the one-third random subset of data not used for model construction (defined as out-of-bag OOB error rate). The relative importance of each predictor variable was assessed by comparing the accuracy of trees containing and not containing the variable across all 10,000 trees. Finally, to compare error and accuracy among the three random forest models, we used predicted and observed PA data to create four groups:

Table 1. Variables included in random forest models for Smallmouth Bass presence or absence. Variables are listed by their order of importance in their primary model (local or landscape). All variables were used in a combined model.

Variable	Model	Definition
S_T_size	Local	Michigan Status and Trends Program size strata (Wills et al. 2015)
S_T_temp	Local	Michigan Status and Trends Program temperature strata (Wills et al. 2015)
Discharge_cfs	Local	Measured discharge
Depth.ft.	Local	Measured stream average depth
totalareastr	Local	Total area of instream structure
riffle.	Local	Percent of stream habitat classified as riffle
Avg_Undercut_Depth	Local	Average depth of stream undercuts
Pct_Sand	Local	Percent of stream substrate classified as sand
AvgPctRootedPlants	Local	Average percent cover of rooted plants
areaBD	Local	Area of streambed covered by boulders
BC2.	Local	Percent of shoreline points with fair bank condition (25–50% bare soil; Wills et al. 2015)
Gear_Type	Local	Gear used to sample streams: backpack, barge or boat electrofishing (Wills et al. 2015)
LRG_dec	Local	Percent of transects with large deciduous trees DBH > 15 cm (Wills et al. 2015)
totdbh18	Local	Amount of instream woody debris with DBH > 45 cm
YARD	Local	Percent of transects with yards as dominant vegetation type
PCT_Scobble	Local	Percent of small cobble (6-13 cm) diameter
SML_dec	Local	Percent of transects with large deciduous trees DBH < 15 cm (Wills et al. 2015)
AvgWidth	Local	Average stream width in meters
DA_KM2	Landscape	Drainage area
QANN50_CFS	Landscape	Predicted mean annual flow
JULYSTEMP	Landscape	Predicted mean July stream temperature
ELEVATION	Landscape	Stream reach elevation
QAUG90_YLD	Landscape	Predicted base flow yield
W_AIRTEMP	Landscape	Mean July air temperature in local catchment
WT_FINE	Landscape	Percent fine surficial geology in the cumulative catchment
QAPR10_YLD	Landscape	Predicted high flow yield
WT_PRECIP	Landscape	Average annual precipitation in the cumulative catchment
REGION	Landscape	Region of the state of Michigan: Upper Peninsula, Northern Lower Peninsula, Southern Lower Peninsula
WT_SOIL	Landscape	Soil permeability in the cumulative catchment
W_DARCY	Landscape	Average groundwater movement in the local catchment
GRADIENT	Landscape	Reach slope
WT_SLOPE	Landscape	Slope of the cumulative catchment
GLAKES	Landscape	Stream Great Lakes access (Y/N)

Table 2. Contingency table used to compare the number of surveys assigned correctly and incorrectly among the Site-Specific, Landscape, and Combined random forest models. The four assignment groups represent whether the model predicted presence (P) or absence (A) compared with what was observed in the field.

Model/Survey	Site-Specific	Model Landscape	Combined
P/P	18	29	18
P/A	16	24	16
A/A	118	142	115
A/P	6	10	7
Total	158	205	156

A/A: model predicts absence, Smallmouth Bass absent from survey; P/A: model predict presence, Smallmouth Bass absent from survey; A/P: model predicts absence, Smallmouth Bass present in survey; P/P: model predicts presence, Smallmouth Bass present in survey. The proportion of surveys assigned among the four groups, i.e., model accuracy and error, were compared using a 4 × 3 contingency table (Table 2) with significance tested against the chi-square distribution.

The final model was selected based on statistical comparison among the models, class-specific error rates, and the ease of statewide extrapolation of model results. In addition to the internal model validation that occurs as part of the random forest procedure, the final model was further independently validated using Smallmouth Bass PA data collected by 24 MSSTP surveys conducted from 2014 to 2016. For this validation, Smallmouth Bass PA predicted by the random forest model was compared with observed PA during MSSTP surveys. Finally, the model was used to predict the PA of all Michigan stream valley segments.

Results

Smallmouth Bass were detected by 54 of the 212 stream surveys conducted as part of the MSSTP from 2002 to 2013 (Figure 1). Smallmouth Bass were most commonly detected (24% of detections) in streams classified as medium-sized (catchment area of 103.6–463.6 km^2), and cool (mean July temperatures 19–22°C).

Overall, the three random forest models did not differ in terms of accuracy and error rates (chi-square = 1.64, df = 6, P = 0.95). Out of bag error rates ranged from 13.9% to 16.6% (Table 3). Across the three models, class-specific error rates were similar. Streams not predicted to have Smallmouth Bass had an OOB error rates of between 5% and 7%; however, streams that were predicted to have Smallmouth Bass had an OOB error rate ranging from 45% (landscape data only) to 47% (site-specific model and combined site-specific and landscape model).

Variables associated with predicted PA of Smallmouth Bass were similar among the three random forest models. In all three models variables associated with stream size had the highest importance (Figures 2 and 3, 4). For the site-specific data model stream average width and the assigned size strata were the most influential variables; followed by measured discharge, assigned temperature strata and stream depth (Figure 2). For the landscape data only model drainage area, predicted mean July stream temperature, and predicted mean annual discharge were the most influential variables (Figure 3). Finally,

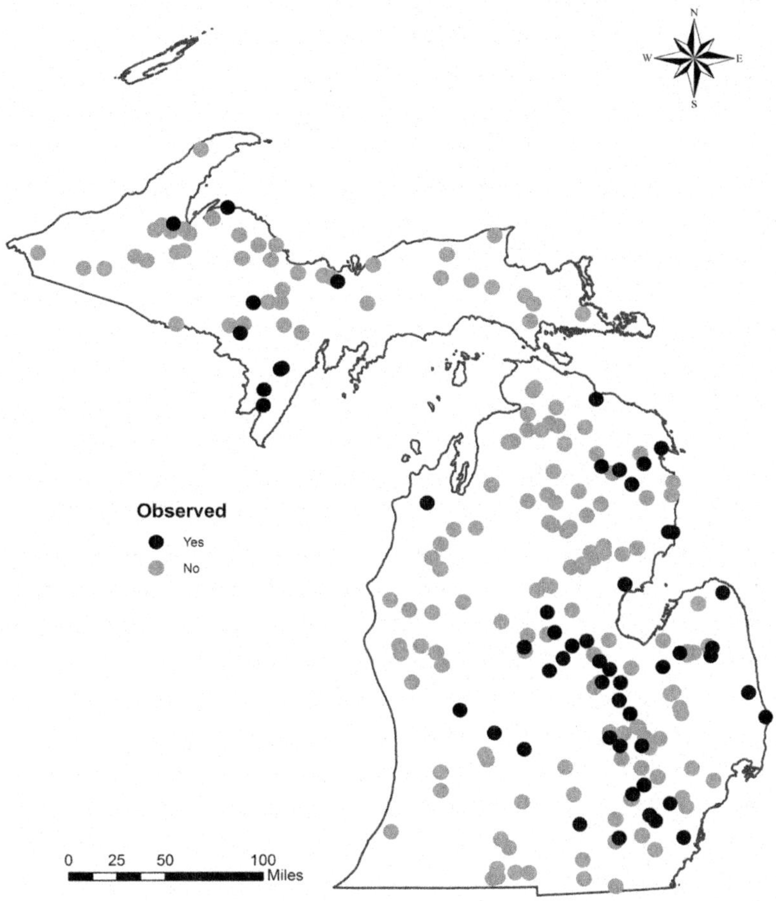

Figure 1. Sampling locations of the Michigan Stream Status and Trends Program from 2002–2013. Observations of Smallmouth Bass are indicated by the black circles, whereas surveys where Smallmouth Bass were not observed are indicated by the grey circles.

Table 3. Comparison of random forest model out-of-bag (OOB) error rates for the site specific, landscape-only and combined sites specific and landscape (combined) models. Error rates are represented as mean values, for sites where Smallmouth Bass were predicted to be present (Present), sites where Smallmouth Bass were predicted to be absent (Absent), and the number of surveys used in each model (N).

	Error Rate (%)			
Model	Mean	Present	Absent	N
Site-specific	13.9	47	5	158
Landscape-only	16.6	45	7	205
Combined	14.7	47	6	156

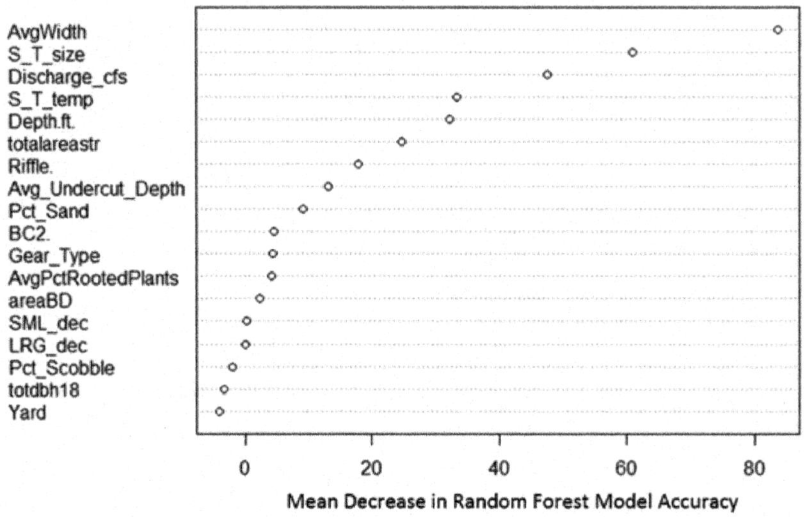

Figure 2. Measures of variable importance as assessed by the mean decrease in random forest model accuracy with and without the variable of interest for local stream reach data measured by the Michigan Stream Status and Trends Program. Variables along *Y*-axis are defined in Table 1.

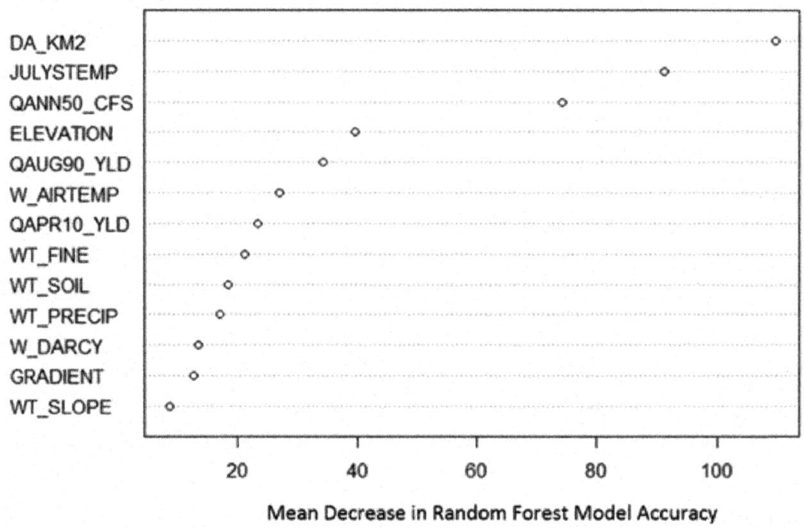

Figure 3. Measures of variable importance as assessed by the mean decrease in random forest model accuracy with and without the variable of interest for landscape-scale stream reach data estimated by the Michigan Department of Natural Resources Institute for Fisheries Research. Variables along *Y*-axis are defined in Table 1.

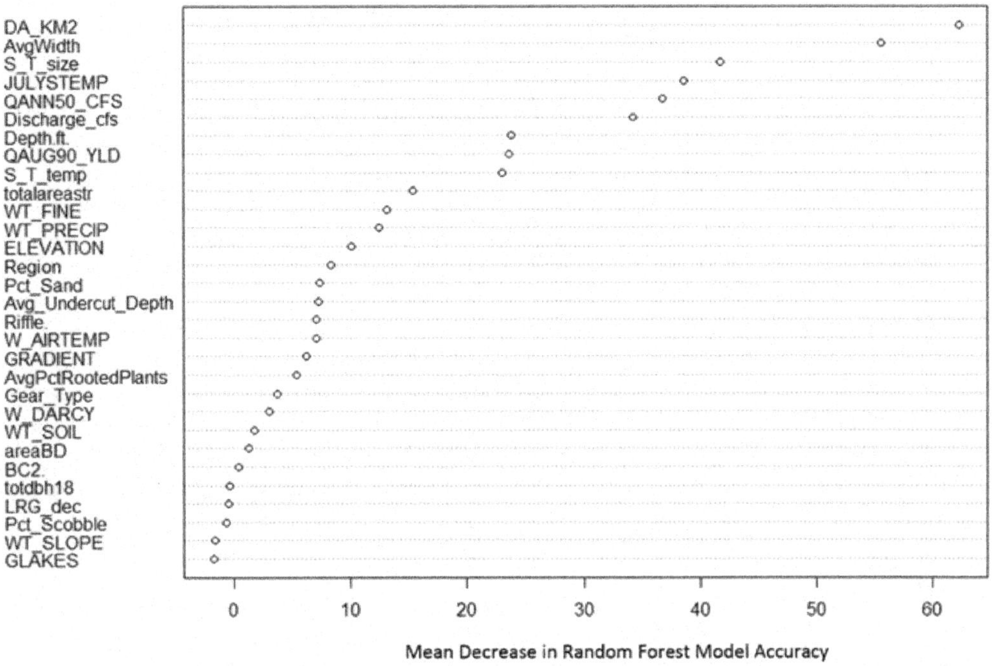

Figure 4. Measures of variable importance as assessed by the mean decrease in random forest model accuracy with and without the variable of interest for landscape-scale and local stream reach data. Variables along *Y*-axis are defined in Table 1.

the model constructed with both site specific and landscape data identified drainage area, average width, stream size strata, predicted mean July stream temperature, and predicted mean annual flow as the most influential variables (Figure 4).

Given the overall similarity in accuracy and error rates among the three models it is likely that similar results could have been obtained with all three models. We selected the landscape only model to predict the PA of Smallmouth Bass for all statewide stream segments, because the nature of data collection (remote sensing and GIS based tools) allowed for easy extrapolation to all statewide stream valley segments. Additionally when site-specific and landscape data were utilized in the same model three of the five most influential variables (drainage area, predicted mean July stream temperature, and predicted mean annual flow) were landscape-based and the two site-specific variables (average stream width and stream size strata) were both associated with stream size. While the selection of the model based only on landscape data results in the exclusion of site-specific data, this approach has several advantages. First, while the landscape model had a slightly higher OOB error rate than the site-specific data model (16.6% versus 13.9%), the landscape model had a lower OOB error rate for streams that were predicted to have Smallmouth Bass present than the site specific data model (45% versus 47%). Second, because landscape-scale data are calculated for each stream segment based on statewide models fewer missing data points exist and therefore more surveys are included in the model. Finally, given that the landscape data were calculated from statewide models, predictions based on this model are easily applied statewide at the same scale as the model data (1:24,000).

The landscape model successfully predicted the presence or absence of Smallmouth

Bass for 20 of the 24 validation surveys, for a total overall error rate of 17%. Smallmouth Bass were not detected at 15 validation sites, and all 15 sites were predicted to not contain Smallmouth Bass (class specific error rate of 0%). Smallmouth Bass were detected at nine validation sites, of which five were predicted to contain Smallmouth Bass (class specific error rate of 44%). Model error rates for the independent validation closely mirrored the random forest internal model validation, suggesting that the landscape only model has broad applicability for predicting Smallmouth Bass occupancy of Michigan streams. Based on the success of the landscape only model, predictions of Smallmouth Bass presence and absence were made for all statewide stream reaches (Figure 5).

Discussion

Random forest modeling of Smallmouth Bass PA based on site-specific and landscape-scale data resulted in relatively accurate predictions of Smallmouth Bass PA in Michigan streams. Importantly, for the management of Smallmouth Bass, we found that landscape-scale data provided a surrogate for site-specific habitat data, with minimal loss of accuracy. This is not surprising given causal relationships between landscape-scale attributes and key site-level habitat variables that fish directly interact with (Zorn and Wiley 2006). Indeed, the predicted distribution of PA (Figure 5), closely mirrored the observed PA of Smallmouth Bass based on field sampling (Figure 1), and compares

Figure 5. Predicted Smallmouth Bass distribution for Michigan stream reaches based on random forest modeling using landscape-scale data.

favorably with the historical distribution of Smallmouth Bass in Michigan streams (Bailey et al. 2004). Both the site-specific and landscape modeling identified stream size as an important predictor of Smallmouth Bass presence and absence. Correlates of stream size have been identified as important variables for Smallmouth Bass presence and abundance in Missouri (Brewer et al. 2007) and Wisconsin (Lyons et al. 2010). The presence-absence predictions we generated for Michigan stream Smallmouth Bass provide a valuable bench mark against which research and management actions may be weighed, and changes in Smallmouth Bass distributions may be assessed.

Fisheries managers are often faced with managing waters where little or no information about species occurrence exists. Constituents often want to know where a sportfish is likely to occur or not occur; or why a specific waterbody is not (or is) stocked with sportfish. In water rich states like Michigan most individual stream reaches have either not been surveyed recently, or have never been surveyed at all. The combination of a systematic statewide random sampling program, landscape scale datasets, and random forest modeling can help answer these types of common constituent questions by generating species presence and absence predictions for all statewide stream reaches. Likewise, random forest models may be valuable to managers trying to decide how to manage specific waterbodies. For example, reducing or eliminating stocking prescriptions for waterbodies with little opportunity for persistence. Additionally, the predicted presence and absence of Smallmouth Bass generated for each of Michigan's stream segments in this study represents a hypothesis that may be tested by additional field sampling. Finding no Smallmouth Bass in a stream where Smallmouth Bass are predicted to be present may indicate local disturbances for a particular stream segment, and provide an opportunity for restoration. Conversely, since the presence or absence of a species at a site can be determined by a single fish, investigators looking at mis-predicted sites should consider connectivity and fragmentation by assessing the types and proximity of habitats that each site is connected to. Finally, landscape scale predictions of species presence and absence are valuable for driving research programs investigating a broad variety of questions with management implications. For example, stretches of unsuitable habitat, where a species is not predicted to exist, may separate patches of suitable habitat where a species is predicted to exist. Such "gaps" may form the basis of cryptic population boundaries driving population genetic structure and metapopulation dynamics (e.g., Levin 1992). Cryptic population boundaries and metapopulation dynamics may influence the ability of particular populations to withstand exploitation and thus have important management implications.

Our random forest model of Smallmouth Bass presence and absence performed similarly to models created by Steen et al. (2008) and Lyons et al. (2010). Steen et al. (2008) generated predictions of presence or absence for 93 Michigan stream fishes based on a variety of data sources collected from 1980 to 2000. Lyons et al. (2010) estimated predictions for 50 stream fishes in Wisconsin. Similar to our model, Steen et al. (2008) averaged 73.3% overall accuracy and Lyons et al. (2010) estimated 75.3% accuracy on their validation streams (we averaged 83%). Both Steen et al. (2008) and Lyons et al. (2010) had higher accuracy (85% and 97%, respectively) for streams predicted not to contain Smallmouth Bass (our model: 93%), and lower accuracy (Steen et al. 2008: 62%; Lyons et al. 2010: 54.1%) for streams predicted to contain Smallmouth Bass (our model: 55%). However, our study considered local reach-specific habitat variables in addition to landscape scale remotely detected variables.

For Smallmouth Bass, a model based only on landscape features was about as accurate as a model based on a combination of landscape and local features (Table 3). However, for other stream fishes this may not be the case. For example, Zorn et al. (2004) found that spatial patterns in abundance of 68 common fish species in streams of Michigan's Lower Peninsula were most often explained by a combination of variables measured at the catchment, reach, and site scales (i.e., catchment area, July mean temperature, reach gradient, substrate conditions, total phosphorus). The combination of the MSSTP sampling for species presence and local habitat variables, and the modeling approach we used, could be easily expanded to incorporate other species utilizing both local and landscape variables as recommended by Wiley et al. (1997).

Our study revealed a dichotomy in model accuracy for predicted Smallmouth Bass PA in stream reaches, which has also been observed by similar efforts (Steen et al. 2008; Lyons et al. 2010); such that reaches predicted to contain Smallmouth Bass had substantially lower accuracy than reaches predicted to be absent of Smallmouth Bass. This result was consistent across the three models we ran (Table 3). There are several likely causes of this discrepancy, which we feel warrant additional research. First, the MSSTP protocols used for Smallmouth Bass detection are conducted using a single pass of electrofishing in each stream, using different electrofishing gears depending on stream size (Wills et al. 2015). The probability of detecting Smallmouth Bass using these methods for a given survey or under particular habitat conditions is unknown. Detection probability is a crucial element of occupancy modeling, and likely varies across space and time (e.g., MacKenzie et al. 2002; Gu and Swihart 2004). Low or variable detection probability of Smallmouth Bass may contribute to the relatively high rate of error for streams where random forest modeling predicts that Smallmouth Bass are present, i.e., they are in fact present but are not detected by MSSTP. Similarly, low detection probability may result in an underestimation of errors for streams where random forest modeling predicts Smallmouth Bass are absent. Future work to estimate detection probability of Smallmouth Bass and other species of interest under the variety of gear types and habitat conditions surveyed is of high importance to the MSSTP. Finally, despite the broad spatial coverage in sampling locations achieved by the MSSTP (Figure 1), total sample size, especially where Smallmouth Bass were detected, remains relatively small (212 total surveys 2002–2013, with 54 detections of Smallmouth Bass). The MSSTP is an ongoing program, so future modeling efforts will benefit from increased sample size, possibly resulting in reduced class-specific error rates.

In addition to abiotic factors, biological factors such as predator, prey, and competitor abundances may improve the accuracy of Smallmouth Bass presence over models such as ours that only incorporate abiotic habitat variables. This may be particularly important as interactions among biotic variables are also likely to influence species distributions. For example, focusing on 100 sites in Michigan streams with Smallmouth Bass, Zorn and Wiley (2010) found that Smallmouth Bass densities at sites were slightly more sensitive to biological variables than habitat variables. By collecting fish community data, the MSSTP provides some opportunity for the expansion of modeling efforts, yet key factors such as invertebrate community characteristics are not collected (Wills et al. 2015). Incorporating these factors at broad spatial scales remains challenging. Developing robust predictive models of stream fish presence/absence may also relate to an inability to more heavily weigh conditions of sites where a species thrives, relative to sites where conditions are barely suitable or the species occurrence is transient. This contrasts with abundance-based models where conditions at better locations for a species are weighted more heavily in model development (e.g., Steen et al. 2008; Zorn et al. 2012).

Ecosystem engineers (e.g., North American Beavers *Castor canadensis*) and human modifications to the landscape may also disrupt expected distribution patterns of warmwater and coldwater stream fishes within drainage networks (e.g., Larson and Moore 1985; Stoks and McPeek 2003; Burchsted et al. 2010). For example, predicted mean July water temperature, an important determinant of fish species distributions and assemblage structure in Michigan (Wehrly et al. 2003; Zorn et al. 2004; Steen et al. 2008), can be altered by main-stem lakes, impoundments, and agricultural and urban land uses (Wehrly et al. 2006) leading to changes in species PA patterns. For example, human modification of Missouri Ozark Border region streams caused increased stream temperatures and resulted in declines of Smallmouth Bass due to their competitive disadvantage with Largemouth Bass at higher water temperatures (Sowa and Rabeni 1995).

The combination of site specific fish community and habitat sampling, and landscape-scale habitat variables facilitated random forest modeling of Smallmouth Bass PA at a state wide scale. Our approach is easily expanded to other species and may reveal drivers of species distributions at multiple spatial scales. In the absence of other data needed for management decisions, these models provide a valuable estimate of species PA. These data are particularly useful for addressing inquiries from constituents about stream reaches where little or no survey data exists. Additionally, such models can provide hypotheses about population boundaries associated with unsuitable habitat acting as cryptic barriers to population dispersal. Finally, our study provides a valuable baseline of habitat conditions against which current and future changes in species distributions can be assessed. This is increasingly important as scientists and managers attempt to quantify and manage the effects of large scale change to the landscape.

Acknowledgments

We thank the many dedicated fisheries professional across the State of Michigan for their efforts to collect stream fish data as part of the Michigan Stream Status and Trends Program. We also sincerely thank three anonymous reviewers and the editors whose comments improved earlier versions of this manuscript.

References

Allan, J. D. 2004. Landscapes and riverscapes: the influence of land use on stream ecosystems. Annual Review of Ecology, Evolution, and Systematics 35:257–284.

Angermeier, P. L., and J. R. Karr. 1984. Relationship between woody debris and fish habitat in a small warmwater stream. Transactions of the American Fisheries Society 113:716–726.

Bailey, R. M., W. C. Latta, and G. R. Smith. 2004. An atlas of Michigan fishes with keys and illustrations for their identification. University of Michigan Museum of Zoology, Ann Arbor.

Baker, E. A. 2006. A landscape-based ecological classification system for river valley segments in Michigan's Upper Peninsula. Michigan Department of Natural Resources, Fisheries Research Report 2085. Ann Arbor, MI, USA.

Baker, M. E., M. J. Wiley, P. W. Seelbach, and M. L. Carlson. 2003. A GIS model of subsurface water potential for aquatic resource inventory, assessment, and environmental management. Environmental Management 32:706–719.

Breiman, L., J. H. Friedman, R. A. Olshen, and C. J. Stone. 1984. Classification and regression trees. Wadsworth and Brooks, Monterey, California.

Breiman, L. 2001. Random forests. Machine Learning 45:5–32.

Brewer, S. K., C. F. Rabeni, S. P. Sowa, and G. Annis. 2007. Natural landscape and stream segment attributes influencing the distribution and relative abundance of riverine Smallmouth Bass in Missouri. North American Journal of Fisheries Management 27:326–341.

Burchsted, D., M. Daniels, R. Thorson, and J. Vokoun. 2010. The river discontinuum: applying beaver modifications to baseline conditions for restoration of forested headwaters. BioScience 60:908–922.

De'ath, G., and K. E. Fabricius. 2000. Classification and regression trees: a powerful yet simple technique for ecological data analysis. Ecology 81:3178–3192.

Everett, R. A., and G. M. Ruiz. 1993. Coarse woody debris as a refuge from predation in aquatic communities. Oecologia 93:475–486.

Farrand, W. R., and D. L. Bell. 1982. Quaternary geology of Michigan. State of Michigan Department of Natural Resources, Geology Survey Map, Lansing.

Fausch, K. D., C. E. Torgersen, C. V. Baxter, and H. W. Li. 2002. Landscape to riverscapes: bridging the gap between research and conservation of stream fish. BioScience 52:483–498.

Gu, W., and R. K. Swihart. 2004. Absent or undetected? Effects of non-detection of species occurrence on wildlife-habitat models. Biological Conservation 116:195–203.

Harmon, M. E., J. F. Franklin, F. J. Swanson, P. Sollins, S. V. Gregory, J. D. Lattin, N. H. Anderson, S. P. Cline, N. G. Aumen, J. R. Sedell, G. W. Lienkaemper, K. Cromack, Jr., K. W. Cummins. 1986. Ecology of coarse woody debris in temperate ecosystems. Advances in Ecological Research 15:133–302.

Hayes, D., E. Baker, R. Bednarz, D. Borgeson, Jr., J. Braunscheidel, J. Breck, M. Bremigan, A. Harrington, R. Hay, R. Lockwood, A. Nuhfer, J. Schneider, P. Seelbach, J. Waybrant, and T. Zorn. 2003. Developing a standardized sampling program: the Michigan experience. Fisheries 28:18–25.

Larson, G. L., and S. E. Moore. 1985. Encroachment of exotic Rainbow Trout into stream populations of native Brook Trout in the southern Appalachian Mountains. Transactions of the American Fisheries Society 114:195–203.

Levin, S. A. 1992. The problem of pattern and scale in ecology. Ecology 73:1943–1967.

Lyons, J. 1992. The length of stream to sample with a towed electrofishing unit when fish species richness is estimated. North American Journal of Fisheries Management 12:198–203.

Lyons, J., J. S. Stewart, and M. Mitro. 2010. Predicting effects of climate warming on the distribution of 50 stream fishes in Wisconsin, U.S.A. Journal of Fish Biology 77:1867–1898.

Liaw, A., and M. Wiener. 2002. Classification and regression by randomForest. R News 2:18–22.

MacKenzie, D. I., J. D. Nichols, G. B. Lachman, S. Droege, J. A. Royle, and C. A. Langtimm. 2002. Estimating site occupancy rates when detection probabilities are less than one. Ecology 83:2248–2255.

R Core Team. 2015. R: a language and environment for statistical computing. R Foundation for Statistical Computing, Vienna, Austria, https://www.R-project.org.

Roni, P., T. Beechie, G. Pess, and K. Hanson. 2015. Wood placement in river restoration: fact, fiction, and future direction. Canadian Journal of Fisheries and Aquatic Sciences 72:1–13.

Seelbach, P. W., M. J. Wiley, M. E. Baker, and K. E. Wehrly. 2006. Initial classification of river valley segments across Michigan's lower peninsula. Pages 25–48 *in* R. M. Hughes, L. Wang, and P. W. Seelbach, editors. Influences of landscapes on stream habitats and biological assemblages. American Fisheries Society, Symposium 48, Bethesda, Maryland.

Seelbach, P. W., L. C. Hinz, M. J. Wiley, and A. R. Cooper. 2011. Using multiple linear regression to estimate flow regimes for all rivers across Illinois, Michigan, and Wisconsin. Michigan Department of Natural Resources, Fisheries Research Report 2095, Lansing.

Sowa, S. P., and C. F. Rabeni. 1995. Regional evaluation of the relation of habitat to distribution and abundance of Smallmouth Bass and Largemouth Bass in Missouri streams.

Transactions of the American Fisheries Society 124:240–251.

Steen, P. J., T. G. Zorn, P. W. Seelbach, J. S. Schaeffer. 2008. Classification tree models for predicting distributions of Michigan stream fish from landscape variables. Transactions of the American Fisheries Society 137:976–996.

Stoks, R., and M. A. McPeek. 2003. Predators and life histories shape Lestes damselfly assemblages along a freshwater habitat gradient. Ecology 84:1576–1587.

Todd, B. L., and C. F. Rabeni. 1989. Movement and habitat use by stream-dwelling Smallmouth Bass. Transactions of the American Fisheries Society 118:229–242.

Tonn, W. M. 1990. Climate change and fish communities: a conceptual framework. Transactions of the American Fisheries Society 119:337–352.

NRCS (Natural Resources Conservation Service), United States Department of Agriculture. 2013. Web Soil Survey. Natural Resources Conservation Service, Washington, D. C. Available: http://websoilsurvey.nrcs.usda.gov/. (August 2018).

USFWS (U.S. Fish and Wildlife Service) and U.S. Census Bureau. 2011. National survey of fishing, hunting, and wildlife-associated recreation. USFWS, Washington, D.C.

USGS (U.S. Geological Survey). 2013. National Hydrography Geodatabase: The National Map viewer. United States Geological Survey, Washington, D.C. Available: http://viewer.nationalmap.gov/advanced-viewer/. (August 2018).

Wehrly, K. E. M. J. Wiley, and P. W. Seelbach. 2003. Classifying regional variation in thermal regime based on stream fish community patterns. Transactions of the American Fisheries Society 132:18–38.

Wehrly, K. E., M. J. Wiley, and P.W. Seelbach. 2006. Influence of landscape features on summer water temperatures in lower Michigan streams. Pages 113–127 *in* R. M. Hughes, L. Wang, and P. W. Seelbach, editors. Influences of landscapes on stream habitats and biological assemblages. American Fisheries Society, Symposium 48, Bethesda, Maryland.

Wehrly, K. E., T. Brenden, and L. Wang. 2009. A comparison of statistical approaches for predicting stream temperatures across heterogeneous landscapes. Journal of the American Water Resources Association 45:986–997.

Wiley, M. J., S. L. Kohler, and P. W. Seelbach. 1997. Reconciling landscape and local views of aquatic communities: lessons from Michigan trout streams. Freshwater Biology 37:133–148.

Wills, T. C., T. G. Zorn, D. B. Hayes, and K. E. Wehrly. 2015. Status and trends of Michigan stream resources, 2002–2007. Michigan Department of Natural Resources, Fisheries Report 09, Lansin.

Zorn, T. G., P. W. Seelbach, and M. J. Wiley. 2002. Distributions of stream fishes and their relationship to stream size and hydrology in Michigan's Lower Peninsula. Transactions of the American Fisheries Society 131:70–85.

Zorn, T. G., P. W. Seelbach, and M. J. Wiley. 2004. Utility of species-specific, multiple linear regression models for prediction of fish assemblages in rivers of Michigan's Lower Peninsula. Michigan Department of Natural Resources, Fisheries Research Report 2072, Ann Arbor.

Zorn, T. G., and M. J. Wiley. 2006. Influence of landscape characteristics on local habitat and fish biomass in streams of Michigan's Lower Peninsula. Pages 375–394 *in* R. M. Hughes, L. Wang, and P. W. Seelbach, editors. Influences of landscapes on stream habitats and biological assemblages. American Fisheries Society, Symposium 48, Bethesda, Maryland.

Zorn, T. G., and M. J. Wiley. 2010. Influence of sampling extent on the relative importance of biotic and abiotic factors in explaining variation in stream fish density. Pages 487–502 *in* K. Gido and D. Jackson, editors. Community ecology of stream fishes: concepts, approaches, and techniques. American Fisheries Society, Symposium 73, Bethesda, Maryland.

Zorn, T. G., P. W. Seelbach, and E. S. Rutherford. 2012. A regional-scale habitat suitability model to assess the effects of flow reduction on fish assemblages in Michigan streams. Journal of the American Water Resources Association 48:871–895.

American Fisheries Society Symposium 87:55–73, 2019

Trophic Ecology of Smallmouth Bass in Hierarchical Stream Networks

OLIVER NETTERE AND E. WILLIAM HAMILTON
Department of Biology, Washington and Lee University
204 West Washington Street, Lexington, Virginia 24450, USA

RYAN J. WOODLAND
Chesapeake Biological Laboratory, Center for Environmental Science
University of Maryland, 146 Williams Street
Post Office Box 38, Solomons, Maryland 20688, USA

ROBERT HUMSTON*
Department of Biology, Washington and Lee University
204 West Washington Street, Lexington, Virginia 24450, USA

Abstract.—Prey selection and diet are highly plastic and can vary with temporal and spatial differences in competition or prey availability. This study investigated the possibility that the trophic position of Smallmouth Bass *Micropterus dolomieu* might change in response to systematic, hierarchical variation in community structure in stream networks. We hypothesized that a shift toward increased insectivory and decreased piscivory would be observed in smaller streams, resulting in a lower trophic position of Smallmouth Bass and reflecting differences in community structure and prey availability. We applied a combination of diet analyses and stable isotope methods to compare prey selection and trophic position of Smallmouth Bass across a range of stream sizes. Stable isotope analyses indicated that Smallmouth Bass trophic position was slightly elevated in smaller watersheds, contradicting our initial hypothesis. However, differences in average trophic position in watershed size categories were small (ranging from 3.6 to 3.8) and of limited ecological significance. Isotopic niche width did not vary among stream size categories, and gut content analyses revealed no differences in frequency of occurrence of fish, crayfish, or insects (larvae and adults). Collectively these results indicate that trophic position, and perhaps trophic niche, of Smallmouth Bass are consistent across hierarchical variation in stream size and habitat.

*Corresponding author: HumstonR@wlu.edu

Introduction

Smallmouth Bass *Micropterus dolomieu* were introduced into Virginia and other Atlantic slope rivers in the late 1800s and have since become well-established in a variety of aquatic systems in the state to support a flagship recreational fishery (Jenkins and Burkhead 1994). Since their introduction, they have continued to disperse throughout river networks and have expanded into upper watersheds. Smallmouth Bass are classified as a hardy cool/warmwater species, capable of inhabiting a range of conditions (Brown et al. 2009). Smallmouth Bass are known predators and competitors of native species in the region and are considered strong invaders of new systems (Gard 2004; Brown et al. 2009). Smallmouth Bass have become invasive in a variety of other systems where they have had a quantifiable deleterious effect on native fish species, including salmonids (Whittier and Kincaid 1999; MacRae and Jackson 2001; Fritts and Pearsons 2004; Carey et al. 2011).

Smallmouth Bass are expanding their range as climate change alters thermal regimes of aquatic ecosystems, with consequences for existing fish assemblages (Shuter and Post 1990; Jackson and Mandrak 2002; Vander Zanden et al. 2004; Chu et al. 2005; Sharma et al. 2007; Loppnow et al. 2013). This manifests as poleward expansion (Sharma et al. 2007) as well as upstream expansion into current coldwater habitats (Lawrence et al. 2012, 2014), as upstream distribution is likewise limited by thermal restrictions on young-of-year growth (Shuter and Post 1990; Lawrence et al. 2012). It is therefore likely that Smallmouth Bass will continue to expand their range in Virginia into low-order streams where they had been previously restricted by thermal regimes. Recent research indicates that many fish species are plastic in their thermal requirements and therefore have the capacity to compensate for shifts in thermal regimes, which could make coldwater fish species distribution more resilient to climate change effects (Stitt et al. 2014; Farrell and Franklin 2016). Likewise, Myers et al. (2018) note that even small changes in coldwater stream temperatures could substantially shift assemblage production to favor coolwater species. Thermal plasticity of coldwater fish species would therefore lead to increased sympatry among species presently considered cool- and coldwater adapted.

Proactive management and conservation actions require some understanding of what effects Smallmouth Bass may have on cool- and coldwater aquatic communities. Characterizing trophic ecology is central to identifying likely interspecific interactions and predicting the effects of such biogeographic changes. Much of the concern surrounding expansion of Smallmouth Bass populations has focused on effects on other top predators, particularly native salmonids, via direct predation (Petersen and Kitchell 2001; Fritts and Pearsons 2004; Lawrence et al. 2014) or through competition for prey (Vander Zanden et al. 1999, 2004). Other studies have demonstrated effects on prey fish assemblages through predation (Jackson and Mandrak 2002; Johnson et al. 2008) and changes in habitat selection behavior (Schlosser 1987). Though the feeding ecology and trophic dynamics of Smallmouth Bass are relatively well studied, to our knowledge there has been no systematic assessment of how trophic ecology might vary across gradients of stream order and corresponding community ecology in lotic systems. This knowledge could help anticipate the trophic effects of Smallmouth Bass as they increasingly populate the upper reaches of watersheds.

The River Continuum Concept suggests that the physical characteristics of streams structure macroinvertebrate and fish communities in a hierarchical manner (Vannote et al. 1980). Diversity of macroinvertebrate species and feeding guilds varies inversely with stream size, while fish species diversi-

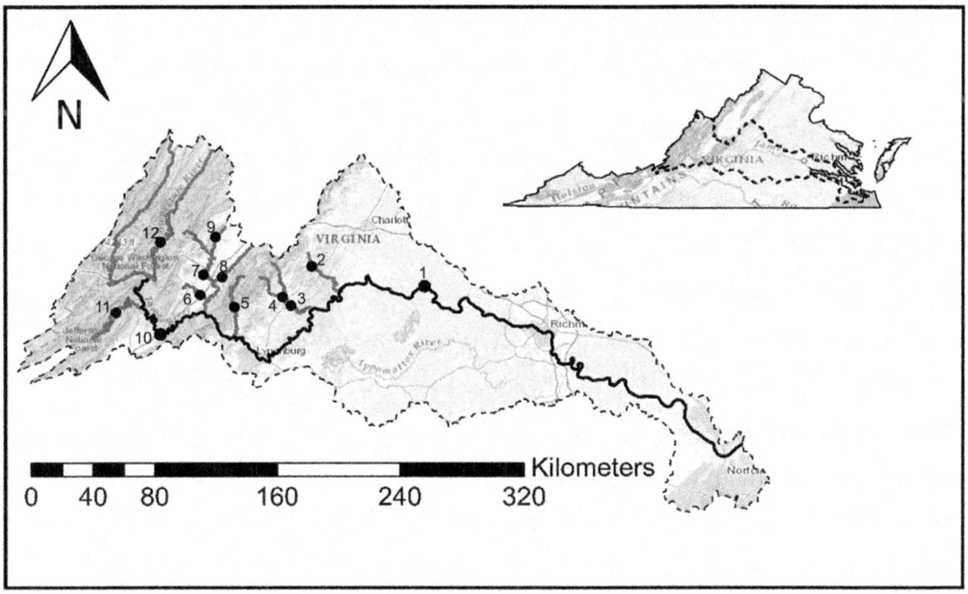

Figure 1. Study area map with collection locations numbered as follows: (1) middle James River; (2) Rockfish River; (3) Tye River; (4) Piney River; (5) Pedlar River; (6) Buffalo Creek; (7) Maury River; (8) South River; (9) Walkers Creek; (10) upper James River; (11) Craig Creek; (12) Cowpasture River.

ty is positively correlated to stream size. As opportunistic predators responding to prey availability, Smallmouth Bass could exhibit reduced levels of piscivory and increased reliance on insectivory in low-order tributaries. This would cause Smallmouth Bass to occupy a lower trophic position in the smaller streams which some studies suggest is possible (Johnson and Dropkin 1993, 1995) and would be consistent with documented scaling of food chain length with watershed size (Sabo et al. 2010).

The objective of this study was to employ complementary stable isotope analysis and gut content methods to assess Smallmouth Bass trophic position across a range of stream sizes in a hierarchical river network (James River Basin, Virginia, Figure 1). We hypothesized that Smallmouth Bass diet would shift to increased insectivory and decreased piscivory in smaller streams, and this would be reflected in both decreased trophic position calculated via stable nitrogen isotope analysis and a decreased frequency of fish prey observed in gut contents. These two measures integrate feeding behavior over timescales ranging from 2 to 3 months (stable isotopes) to hours or days (gut contents), thus providing complementary inference on trophic ecology.

Methods

Study Area and Sampling

The James River is formed by the confluence of the Jackson and Cowpasture Rivers in the Ridge and Valley physiographic province of western Virginia. It is a tributary of the Chesapeake Bay that reaches roughly 7th order (Strahler) in size before its estuary confluence (Benke and Cushing 2005). It is locally divided into three sections: the tidal-influenced 'lower' section below the fall line of the Coastal Plain (near Richmond, VA); the 'middle' section stretching roughly from the Blue Ridge through the Piedmont provinces; and the higher-gradient 'upper' section flowing through the Appalachian (Ridge and Valley) and Blue Ridge mountains. Smallmouth Bass were introduced to the basin in the mid 1800s and now inhabit all the major tributaries of the upper and middle basin (Jenkins and Burkhead 1994).

Sample site locations were chosen in the James River watershed with the goal of maximizing the range of watershed sizes (area) represented (Figure 1). We sampled 12 sites in 11 different rivers (Table 1) ranging from the mainstem of the "middle" James River above the fall line (15,132 km^2 watershed area above sampling site) to low-order streams in the upper basin's headwater reaches (e.g., Walker's Creek, 39 km^2). Tributaries of the mainstem James River were sampled a minimum of 8-km upstream from their confluence. We collected Smallmouth Bass primarily by boat or backpack electrofishing. However, in locations that were too shallow for boat access but too large for effective sampling with backpack electrofishers, hook-and-line capture was used to augment collections. Only fish greater than 100 mm TL were collected, following the suggestion by Mittlebach and Persson (1998) that Smallmouth Bass switch to piscivory at 40–100 mm TL. Thus all fish in our sample were above the minimum size associated with piscivory. Fish collections followed IACUC-approved guidelines (protocol RH-0618a). Macroinvertebrates including insect larvae, snails, and crayfish were collected at each site by hand and using D-frame nets, transported on ice to the laboratory, and stored frozen at –20 C. Samples of dorsal muscle tissue roughly 1–2 cm^3 were collected from each Smallmouth Bass and stored frozen in clean polyethelyne vials. Guts were removed from each fish and their contents identified; fish prey were identified to the lowest taxonomic order possible, while macroinvertebrates and crayfish were identified to family level. We recorded the number of each prey item in individual guts.

Stable Isotope Analyses

Smallmouth Bass muscle tissue and invertebrate samples were dried at 60°C for 48 h and then crushed to a fine powder by mortar and pestle. Invertebrate samples were grouped to lowest taxonomic level identified and combined for each location. Approximately 1 mg of each homogenized sample was weighed out by microbalance (Sartorious, Göttingen, Germany) and sealed in tin capsules for isotopic analysis. Two samples were analyzed from each individual fish or from bulk invertebrate tissue and average $\delta^{15}N$ and $\delta^{13}C$ values from duplicate samples were used for data analysis. Stable isotope ratios for the powdered samples were determined using a Costech ECS 4010 elemental analyzer (Valencia, California) coupled to a Thermo Electron Delta Plus stable isotope ratio mass spectrometer (Bremen, Germany) at Washington and Lee University (Lexington, Virginia). Stable isotope values were expressed in delta (δ) notation as per mille (‰) as defined by the following expression:

$$\delta Z = [(R_{sample}/R_{standard}) - 1] \times 1{,}000$$

where Z represents ^{15}N or ^{13}C, R_{sample} is the ratio $^{15}N/^{14}N$ or $^{13}C/^{12}C$ for the sample, and $R_{standard}$ is the ratio $^{15}N/^{14}N$ or $^{13}C/^{12}C$ for the corresponding standard. We used USGS 40 and USGS 41 as standards for both N and C to calculate the regression used to convert the mass spectrometer signal to sample $\delta^{15}N$ and $\delta^{13}C$ values. Precision for these standards (standard deviation) was 0.06‰ for $\delta^{15}N$ and 0.03‰ for $\delta^{13}C$. All nitrogen values were ref-

Table 1. Summary of sampling location characteristics and analytical results for Smallmouth Bass in Virginia streams sorted by stream and watershed size, with designated watershed size category indicated. Stream order is approximate Strahler order and TL is total length in mm with sample range in parentheses. The primary consumer baseline $\delta^{13}C$ and δ15N calculated for each stream is provided, as well as mean values for Smallmouth Bass with standard deviation in parentheses.

Watershed size group	Stream	Stream order	Watershed area (km^2)	*n*	Mean TL	Primary consumer baseline		Smallmouth Bass	
						$\delta^{13}C$	$\delta^{15}N$	$\delta^{13}C$	$\delta^{15}N$
Small	Walkers Crk.	2	39	14	170 (120–215)	–32.0	6.16	–23.5 (2.11)	11.9 (0.4)
	Pedlar R.	2	112	24	200 (140–345)	–22.8	3.2	–22.3 (1.9)	9.8 (0.5)
	Piney R.	3	164	16	203 (120–370)	–17.6	4.8	–20.1 (3.8)	11.1 (0.8)
	South R.	3	284	14	231 (185–310)	–31.8	7.1	–25.8 (3.1)	13.2 (0.3)
	Buffalo Crk.	3	293	26	202 (170–245)	–34.7	6.0	–28.5 (2.1)	11.8 (0.4)
	Rockfish R.	3	320	10	164 (110–230)	–27.4	5.9	–23.1 (2.1)	11.1 (0.7)
Medium	Tye R.	4	477	19	195 (115–290)	–21.3	5.7	–17.7 (2.8)	11.7 (0.5)
	Craig Crk.	4	796	10	243.5 (205–310)	–28.5	5.5	–25.3 (1.3)	11.0 (0.4)
	Cowpasture R.	4	949	22	203 (130–255)	–25.7	6.1	–24.5 (1.9)	11.2 (0.3)
Large	Maury R.	5	1277	24	222 (122–300)	–30.48	6.4	–26.6 (2.1)	12.2 (0.6)
	Upper James R.	6	5195	24	233 (125–370)	–27.4	6.7	–26.1 (1.0)	11.9 (0.5)
	Middle James R.	7	15132	20	212 (145–345)	–32.2	6.0	–26.4 (2.5)	12.2 (0.7)

erenced to atmospheric nitrogen ($\delta^{15}N = 0‰$) and carbon values were referenced to Vienna Pee Dee Belemnite ($\delta^{13}C = 7.7‰$).

Data Analysis

Watershed boundaries and upstream catchment area were determined using the ArcGIS Spatial Analysis Hydrology tool (ESRI, Redlands, CA). "Pour point" location for each sample site was defined as the furthest downstream location sampling occurred and was determined in the field using a handheld GPS. We grouped collection locations into categories according to watershed area corresponding roughly with natural 'breaks' in the distribution in our sample. "Small" rivers were those with watersheds <350 km^2 (n = 6), "medium" rivers were characterized by watersheds 350–1,000 km^2 (n = 3), and "large" rivers had watersheds >1,000 km^2 (n = 3). Average total length (TL) of Smallmouth Bass in each sample was compared across watershed size categories by analysis of variance (ANOVA) (α = 0.05). We used Welch's t-tests for posthoc pairwise comparisons between categories to account for unbalanced group sizes, with Bonferroni correction applied when assessing significance.

We used numeric metrics (*sensu* Hyslop 1980) to quantify the frequency of different prey items in the guts of Smallmouth Bass collected. We quantified frequency of occurrence (F_i) of a particular prey item i as the percentage of nonempty stomachs containing that prey item. Recent studies have demonstrated this measure is robust and suitable for the type of comparisons we draw here (Baker et al. 2014). We also quantified the proportion (P_i) each prey item represented among all prey items identified. To simplify comparisons across study sites, we grouped prey into general categories of macroinvertebrates (insects and aquatic larvae), crayfish, and fish prey. Average F_i and P_i for each of these prey groups were calculated for each watershed size category and compared using contingency tables (Everitt 1992). We also calculated the average trophic level of prey consumed weighted by numeric abundance in guts for comparison with stable isotope methods.

We determined trophic position from nitrogen stable isotope data following methods described by Post (2002) which have been subsequently applied in analyses of lotic communities (e.g., Anderson and Cabana 2007; Sabo et al. 2010). Vander Zanden and Rasmussen (2002) provide an excellent summary of the methodology specifically in the context of studying trophic position of Smallmouth Bass. Fractionation processes lead to enriched ^{15}N:^{14}N (abbreviated as $\delta^{15}N$ when standardized to a reference material) in organisms higher in the food chain; increases in $\delta^{15}N$ of 2.5%–3.4% per trophic level are typically observed (Post 2002). Carbon isotopes ($\delta^{13}C$) fractionate less between trophic levels, but can be useful for discriminating different sources of primary production such as autochthonous versus allochthonous carbon sources in lotic food webs. When analyzing $\delta^{15}N$ to calculate trophic position, the selection of appropriate "baseline" or reference organisms from which trophic position is calculated is an important component of the methodology. Typical baseline organisms are primary consumers with known grazing specificity, e.g. feeding primarily upon epilithic algae (Minagawa and Wada 1984; Post 2002; Anderson and Cabana 2007). We chose two common epilithic grazers ("scrapers") as our baseline organisms: pulmonate snails (*Lymnaiidae* and *Physidae*) and aquatic beetles in the family *Psephenidae*. Following recommendations in Anderson and Cabana (2007), we averaged $\delta^{15}N$ from these two organisms and used this as the baseline value for primary consumers. In some locations, only one or the other taxa was collected. We used standard linear regression to develop a predictive model of correspondence between

$\delta^{15}N$ in these taxa across locations ($r^2 = 0.85$), used this model to predict unknown values, and thus calculate a predicted "average" of the two taxa for each sampling location. A trophic fractionation rate for $\delta^{15}N$ of 3.4% enrichment per trophic level was assumed in calculations of trophic position of Smallmouth Bass (Vander Zanden and Rasmussen 2001; Post 2002) applying the standard equation:

$$TP_{consumer} = [(\delta^{15}N_{consumer} - \delta^{15}N_{baseline})/3.4] + 2$$

which assumes a primary consumer baseline ($\delta^{15}N_{baseline}$). Trophic position of each individual was calculated and mean values were compared across watershed size categories by analysis of covariance (ANCOVA) with length as a covariate. Posthoc *t*-tests were used for pairwise comparisons with Bonferroni correction applied to assess significance, and we performed a regression analysis on trophic position as a function of total length to further explore the influence of this variable. During data analysis we noted that individual variation in average trophic position was high among streams and the number of streams in watershed size groups was unbalanced. Therefore, we additionally analyzed the data using a linear mixed model approach to compare average trophic position across watershed size categories. In this analysis, length and watershed size were treated as fixed effects and stream was treated as a random effect. Mixed model analysis was performed using the lme4 library (Bates et al. 2015) in the R statistical platform (R Core Team 2017). Significance of fixed effects was assessed using log-likelihood comparisons of null and alternative models.

We also calculated the isotopic niche of Smallmouth Bass at each sample site by estimating Bayesian standard ellipses of bivariate $\delta^{15}N$ and $\delta^{13}C$ data following Jackson et al. (2011). Ellipse areas were compared across sites to examine for trends in isotopic niche corresponding to watershed size. All calculations were completed using the SIBER package (Jackson et al. 2011) in the R statistical platform (R Core Team 2017).

Using stable isotopes to study trophic ecology requires several assumptions, the most important of which is that the tissue of the consumer of interest (here, Smallmouth Bass) has equilibrated to the isotope composition of its assimilated diet. Nonequilibrium of a consumer's isotope composition with its diet can occur when consumers move to a new habitat or switch to a new forage base (e.g., ontogenetic trophic niche shift from invertivory to piscivory). The time required to approach consumer-diet equilibrium depends on growth and metabolism and in the case of rapidly growing consumers, such as juvenile fish, growth rate dominates equilibration rate (Hesslein et al. 1993; Trueman et al. 2005). Herzka (2005) showed that the relative biomass ratio = W_t/W_0 (ratio of body weight at elapsed time t relative to the initial weight at the time of the diet shift $t = 0$), can provide an indicator of the likelihood of isotopic equilibration to the novel forage base. Relative biomass ratios of 4–6 have been proposed as sufficient for newly accrued biomass to dominate muscle isotope composition of juvenile fish (Herzka 2005). We explored a range of initial body sizes 70–120 mm TL (4.4–21.7 g) at which ontogenetic trophic shifts from invertivorous to piscivorous diets were expected to occur. We estimated isotope equilibration schedules for juvenile Smallmouth Bass following an ontogenetic trophic niche shift by modeling tissue equilibration as a function of somatic growth (Fry and Arnold 1982; Herzka 2005). This approach ignores metabolic effects and focuses on the dominant role of anabolic processes (i.e., dilution). Biomass-specific growth rates were estimated based on empirical length-weight relationships ($W_t = aTL^b$, a = 0.221, b = 2.95; this study) and assumed a daily growth rate of 0.8–0.9 mm/d (Scott and Crossman 1973).

Results

In June and July of 2015, we collected 223 Smallmouth Bass ranging from 110 to 370 mm TL across all twelve locations (Table 1). Average length of fish was 207 mm TL, and it differed across watershed size categories (ANOVA, $F_{2,217}$ = 5.2, P = 0.006). Posthoc comparisons indicated that average length was lower in small watersheds (mean = 197.9 mm, SE = 4.4) compared to large (mean = 223.3 mm, SE = 7.4; t_s = –2.9419, df = 111.12, P = 0.003). Average length in medium-size watersheds (mean = 207.5 mm, SE = 6.2) did not differ from either small or large watersheds. Average sample size across sites was ~18 with a minimum sample size of n = 10 and maximum n = 25 among sampling locations.

We were able to retrieve gut contents from 163 individuals while the remainder (n = 60 or 27%) were found empty. Larval and adult insects were common with the majority of aquatic origin; most common terrestrial insects were in the orders Coleoptera and Hymenoptera. Dominant aquatic insect orders included Megaloptera, Odonata, and Ephemeroptera. Crayfish (Cambaridae) were observed in guts of fish from all sampling streams except the lower James River. Fish remains were difficult to identify but were observed in guts from all streams as well. Among those identified *Notorus* sp. (Ictaluridae) were the most common. Frequency of occurrence (F_i) of and proportional representation (P_i) of macroinvertebrate, crayfish, and fish in guts did not differ across watershed size categories (F_i: df = 4, χ^2 = 3.1, P = 0.55; P_i: df = 4, χ^2 = 4.27, P = 0.37; Figure 2). Average trophic position of prey was 2.7 (SD = 0.45) suggesting trophic position for Smallmouth Bass of 3.7 (following Vander Zanden et al. 1997).

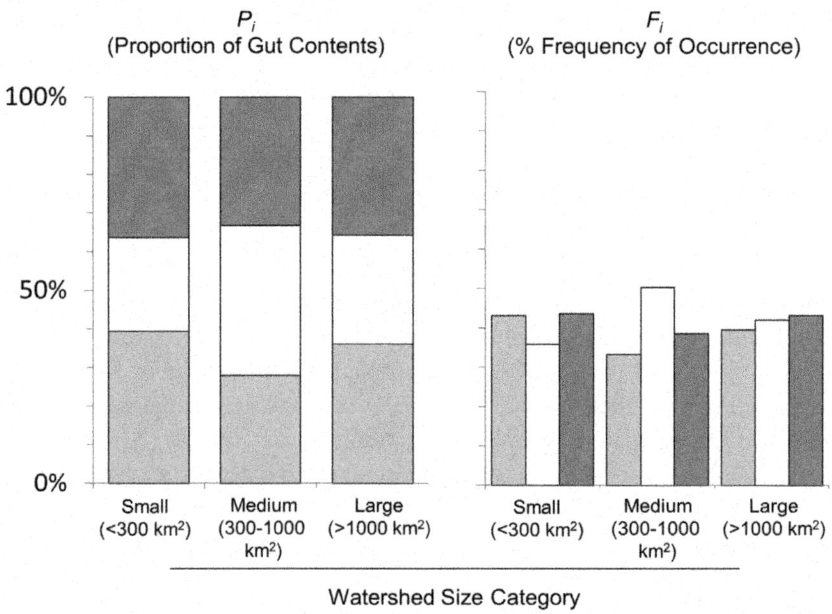

Figure 2. Two quantitative representations of frequency of prey items observed in gut contents of Smallmouth Bass from Virginia streams, grouped into broad taxonomic categories of fish (dark shaded), crayfish (white) and other macroinvertebrates (light shaded). Following Hyslop (1970), proportion of gut contents (P_i) is the percentage of a particular taxon among all items identified in gut contents, and frequency of occurrence (F_i) is the percent of non-empty stomachs containing that taxon. Comparisons across watershed size categories were not statistically significant.

Comparison of metrics calculated from isotopic data were not consistent with our hypothesis of lower trophic position in smaller streams. Isotopic ratios were relatively consistent between replicate tissue samples. The average standard deviation for $\delta^{15}N$ among replicate samples was 0.08 with a coefficient of variation (100•SD/mean) of 0.85%, and for $\delta^{13}C$ average standard deviation was 1.02 and coefficient of variation was 3.88%. Variation across streams was greater for $\delta^{13}C$ than for $\delta^{15}N$ (Table 1; Figure 3). Trophic position estimates from $\delta^{15}N$ data ranged from 3.5 to 3.9 across streams, with the highest value observed in the Pedlar River (Table 2). The average trophic position pooled across all samples was 3.7 (SD = 0.2), identical to that calculated from gut contents. Mean trophic position differed across watershed size categories after controlling for the effect of total length (Table 1; Figure 4; ANCOVA, $F_{2,216}$ = 12.12, $P < 0.001$) and pairwise comparisons indicated trophic position was higher in small watersheds (mean = 3.8, SE = 0.02) than in medium (mean = 3.6, SE = 0.02; $t = 4.7$, df = 114, $P < 0.001$) or large watersheds (mean = 3.7, SE = 0.03; $t = 2.9$, df = 134, $P = 0.004$). Thus, the pattern of variation across stream size was opposite our hypothesis, as we expected trophic position to be higher in larger streams. Total length was significantly and positively related to trophic position ($F_{1,216}$ = 29.1, $P < 0.001$) as expected, though variability among individuals was high (Figure 5). When results are visualized with streams ordered according to watershed size it does not appear that there is any systematic pattern of variation in trophic position corresponding to size of watershed (Figure 4), though there is a great deal of variability among streams within and across categories. The linear mixed model analysis including stream as a random

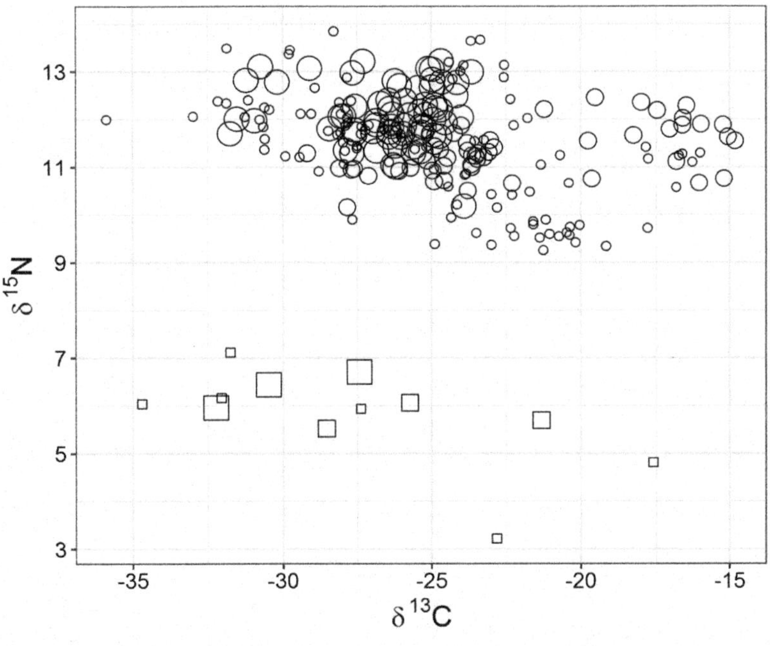

Figure 3. Biplot of tissue $\delta^{13}C$ and $\delta^{15}N$ measured from Smallmouth Bass and a composite primary consumer baseline value calculated from epilithic grazers collected in Virginia streams. Baseline for each stream is indicated by square symbols, while Smallmouth Bass are represented by circles. The size of the symbol corresponds with the watershed size category of the stream from which samples were collected (small, medium, and large).

Table 2. Summary of sampling location characteristics and analytical results for Smallmouth Bass in Virginia streams sorted by stream and watershed size, with designated watershed size category indicated. Mean trophic position calculated from $\delta^{15}N$ data is provided for streams and watershed size groups with standard error in parentheses. Trophic niche width is reported by standard ellipse area of bivariate $\delta^{15}N$ and $\delta^{13}C$ data corrected for small sample size, with 95% Bayesian credible intervals reported in parentheses.

			Mean trophic position	
Watershed size group	Stream	Trophic niche width	Stream	Group
Small	Walkers Crk.	2.2 (1.1–3.7)	3.67 (0.03)	3.77 (0.02)
	Pedlar R.	2.6 (1.6–3.8)	3.95 (0.03)	
	Piney R.	8.2 (4.7–13.3)	3.86 (0.06)	
	South R.	3.0 (1.7–5.2)	3.80 (0.03)	
	Buffalo Crk.	2.9 (1.9–4.2)	3.67 (0.03)	
	Rockfish R.	3.8 (1.8–7.7)	3.52 (0.06)	
Medium	Tye R.	4.6 (2.8–7.3)	3.77 (0.04)	3.62 (0.02)
	Craig Crk.	1.3 (0.6–2.5)	3.60 (0.04)	
	Cowpasture R.	1.9 (1.2–2.9)	3.50 (0.02)	
Large	Maury R.	3.8 (2.4–5.5)	3.69 (0.04)	3.69 (0.03)
	Upper James R.	1.5 (1.0–2.2)	3.53 (0.03)	
	Middle James R.	5.4 (3.3–8.4)	3.86 (0.05)	

effect indicated a positive effect of length on trophic position ($\chi^2 = 49.4$, $P < 0.001$) but did not indicate a significant effect of watershed size ($\chi^2 = 2.7$, $P = 0.36$). Likewise, isotopic niche width estimated from standard ellipse areas showed substantial variation among streams (Table 1), but comparisons across watershed size categories were not significant (ANOVA, $F_{2,9} = 0.33$, $P = 0.7$).

Results of the equilibrium model analysis suggest that Smallmouth Bass in the lower end of the size range in our sample are likely reaching equilibration with their diet following the expected ontogenetic diet shift (Figure 6). The model indicates juvenile Smallmouth Bass of 80–100 mm length require approximately 1.5–3 months for muscle tissues to equilibrate to a new diet in this river system. At the edges of this size range, the time required for juvenile Smallmouth Bass muscle to equilibrate to a novel diet ranged from 55 to 78 d (80 mm individual) to 70–95 d (100 mm individual). Total growth in length over this time period is expected to average 65 mm (range: 44–86 mm), yielding an expected length at equilibrium of approximately 155 mm. Thus, the equilibration model indicates that all but the smallest individuals appear to have experienced sufficient growth to equilibrate to their local diet.

Discussion

The objective of this study was to test the hypothesis that the trophic position of Smallmouth Bass should decrease with stream size, reflecting a decrease in piscivory accompanied by an increase in feeding upon macroinvertebrates in smaller systems. Such a shift would correspond with theoretical differences in prey availability across hierarchies of stream size. Our data do not

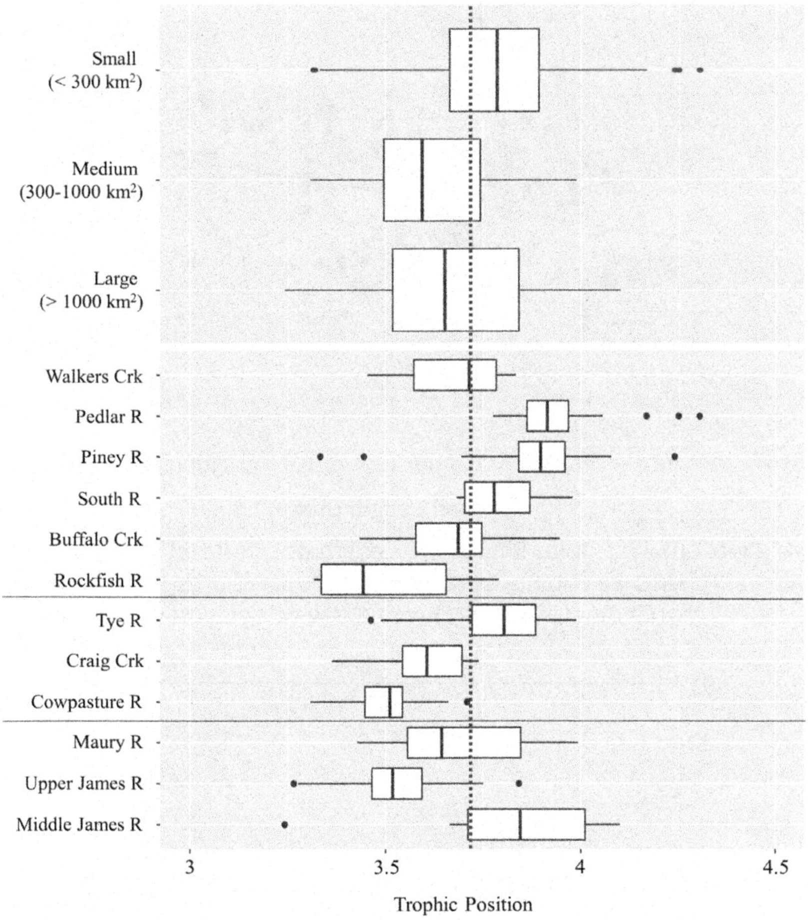

Figure 4. Boxplots summarizing trophic position estimates for samples of Smallmouth Bass in Virginia streams grouped by watershed size (top panel) and collection location (bottom panel). Boxplots indicate range of observations (outliers as points), first and third quartiles (extent of boxes), and median (central line in boxes). The vertical dotted line indicates the approximate position of the global mean trophic position estimated for all samples (3.7). Solid horizontal lines in the bottom panel separate streams into watershed size categories.

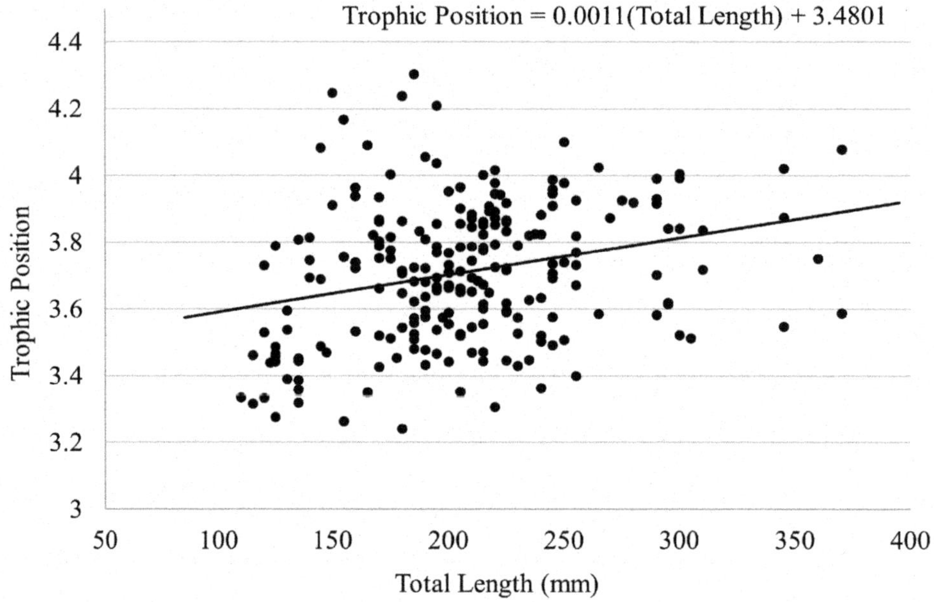

Figure 5. Correlation between total length and estimated trophic position from $\delta^{15}N$ determined for Smallmouth Bass in Virginia streams. The illustrated regression line is positive in slope and statistically significant ($P < 0.001$) with a low correlation coefficient ($r^2 = 0.07$).

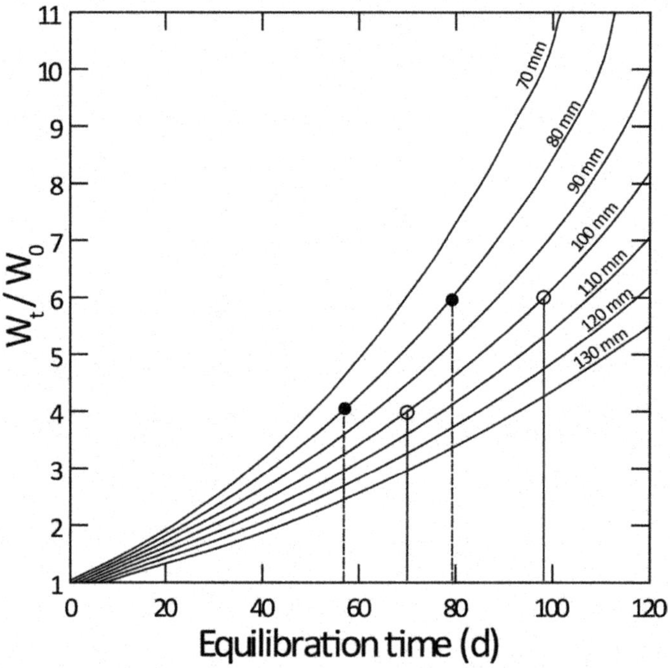

Figure 6. Predicted equilibration isoclines for tissue-diet isotope composition following an ontogenetic diet shift by Smallmouth Bass during the young-of-the-year period at a range of lengths (70–130 mm). Ratios of body weight at time *t* to initial body weight (W_t/W_0) of 4 and 6 and corresponding equilibration time in days are indicated by filled circles and dashed lines (80 mm isocline) and open circles and solid lines (100 mm isocline).

support this hypothesis. Instead, our analysis indicates weak statistical support for the opposite pattern in that average trophic position was slightly higher in the group designated as 'small' watersheds in this study. However the ANCOVA results indicating differences among these artificial watershed size 'categories' were not mirrored in the corresponding GLMM analysis where stream was included as a random variable, suggesting that ANCOVA results were possibly influenced by the combination of high variation among streams and unbalanced groups. In gut content analyses, the representation of prey taxa in did not vary across watershed size groups indicating that prey selection was consistent across watershed groups during the period of collection. Finally, trophic niche width (calculated as standard ellipse area of bivariate $\delta^{15}N$ and $\delta^{13}C$) varied among streams but did not differ among stream size groups. Average trophic position of Smallmouth Bass was consistently between 3.5 and 4.0 in all streams studied (Table 1), and means of trophic position estimates in watershed size groups do not differ meaningfully (i.e., ecologically) from the global mean of 3.7. Taken collectively, these results allow us to reject our initial hypothesis and support a null conclusion that trophic position, and perhaps trophic niche, of Smallmouth Bass was consistent across hierarchical variation in stream size and habitat in our study.

A number of factors can influence the average trophic position measured in these streams via stable isotope methods, and some could have confounded detection of the hypothesized pattern of decreasing trophic position in smaller watersheds. For example, individuals may be moving between systems and immigrating into sampled streams from areas with a different $\delta^{15}N$ baseline. Humston et al. (2010, 2017) documented substantial dispersal of Smallmouth Bass between rivers in this watershed. We selected sampling locations where downstream distance to confluences with larger rivers was fairly long (>8 km) to minimize potential for recent immigration from larger systems. However, confluences with smaller streams may have been relatively close and thus movement among tributaries could influence calculations of trophic position. Alternatively, $\delta^{15}N$ in Smallmouth Bass in smaller streams could be enriched due to the presence of an unsampled primary consumer guild with elevated $\delta^{15}N$ compared to our baseline organisms. Several studies have documented that larger rivers tend to have higher baseline $\delta^{15}N$ and $\delta^{13}C$ compared to smaller, upstream tributaries in the same watershed (Huntsman et al. 2016). Upstream migration of primary consumers (e.g., Uno and Power 2015) could therefore elevate $\delta^{15}N$ in predators that feed upon them, leading to inflated estimates of trophic position. A third possibility involves temperature-dependence of nitrogen trophic fractionation. Barnes et al. (2007) documented an increase in $\delta^{15}N$ at lower water temperatures and note that this could complicate comparisons of trophic position among systems with differing temperature regimes. We generally expect a greater degree of riparian shading in smaller streams, and groundwater or tributary inputs are likely to have a greater proportional effect on local temperatures in smaller systems as well. It is therefore possible that calculations of trophic position in smaller streams are artificially inflated due to temperature dependence. However, the consistency of diet composition across watershed sizes supports the null conclusion of no difference in trophic position across watershed sizes and demonstrates the complementarity of these two methodologies.

We also have to consider the possibility that the observation of slightly higher trophic positions in our smaller watersheds (as detected in ANCOVA analysis) reflects subtle differences in feeding behavior in these systems, perhaps differences that were not detectable in the timescale or sample size integrated in gut content analysis. Sabo et al. (1996) observed that the net rate of energy gain of Smallmouth Bass feeding in a stream was inversely related to water depth. This is likely a function of increased encounter rates with prey in microhabitats characterized by lower water volume. Decreased volume overall in smaller streams could similarly increase encounter rates between Smallmouth Bass and fish prey (or other secondary consumer prey). This could result in elevated trophic position of Smallmouth Bass in small streams, opposite of what we originally hypothesized. The collective evidence in our study is not sufficient to support this conclusion, but perhaps suggest that it warrants further investigation.

The average trophic position of Smallmouth Bass estimated in our study (3.7) suggests a diet dominated by secondary consumer prey and is within the range reported by other studies using similar methodologies. Vander Zanden et al. (1997) reported an average trophic position of 4.02 for adult smallmouth bass collected from 15 lakes in Ontario and Quebec, though lake-specific averages ranged from 3.5 to 4.7. Anderson and Cabana (2007) estimated a trophic position of 3.4 for Smallmouth Bass collected from rivers of the St. Lawrence watershed, with individuals ranging from 2.9 to 4.1. However, their sample size was small (n = 3) and all fish were less than 125 mm. A coarse comparison suggests that average trophic position of Smallmouth Bass may be slightly lower in lotic systems compared to lentic systems. However, such comparisons are equivocal because they cannot account for differences in the size of fish analyzed in these studies.

All statistical analyses indicate a positive relationship between individual size and trophic position, which supports expectations that larger Smallmouth Bass should occupy a higher trophic position than smaller conspecifics. Substantial individual variation exists and the relatively low increase in trophic position with size (i.e., slope) suggests that this difference is likewise of limited ecological relevance. The small group of individuals at the lower extreme of the size range appear to be influential on the regression and could be contributing to the positive slope observed. While these individuals are above the size associated with piscivory (Mittlebach and Persson 1998) our equilibration model results suggest that those <155 mm may still be integrating the isotopic signal of this ontogenetic diet shift.

Our study confirms the importance of crayfish in the diet of Smallmouth Bass, as documented in several previous studies (e.g., Rabeni 1992; Roell and Orth 1993; Brewer and Orth 2015). Crayfish are prey for a variety of fish, turtles, and terrestrial predators in these systems (Englund and Krupa 2000); however, it is not known to what degree predation limits their abundance. The increasing presence of a prolific predator such as Smallmouth Bass could substantially increase predation on crayfish in these systems. As polytrophic omnivores, crayfish have dynamic influence in food webs and exert indirect effects on other taxa through feeding and bioturbation (Whitledge and Rabeni 1997; Usio and Townsend 2004). Crayfish grazing on filamentous macroalgae can facilitate colonization by epilithic diatoms and grazing insects (Creed 1994) or provide resources for "collector" invertebrate guilds by shredding leaf litter (Heard and Richardson 1995). As selective predators, crayfish can likewise influence abundance of lower trophic levels through competitive release and trophic cascades (Usio and Townsend 2004). If the expansion of Smallmouth Bass substantially increases predation on crayfish, the role of crayfish in these aquatic communities could be disrupted with broader consequences for food web dynamics.

Much of the concern surrounding expansion of Smallmouth Bass populations arises from their potential effects on fish assemblages as piscivores, and our gut content data confirm that fish constitute a large proportion of Smallmouth Bass diet in all rivers we sampled. Results of previous studies have documented Smallmouth Bass predation on salmonids (e.g., Rieman et al. 1991; Fritts and Pearsons 2004, 2006) and have demonstrated their effects on fish communities in lacustrine habitats (e.g., Whittier and Kincaid 1999; Vander Zanden et al. 1999; MacRae and Jackson 2001; Vander Zanden et al. 2004). It is difficult to predict what effects might arise from the increasing presence of a piscivorous predator in smaller streams. As thermal regimes shift in coldwater streams, the fish assemblage will likewise shift to one more typical of coolwater habitats (Myers et al. 2018)—suggesting that coldwater communities will retract into upstream reaches unless some degree of adaptation occurs (Stitt et al. 2014). Rubenson and Olden (2017) note the upstream limit of Smallmouth Bass populations is characterized by two distinct fronts: (1) the front edge, where the population is established and persists year-round; and (2) the leading edge, where individuals may appear but the population is not established. Our "small" watershed sampling locations occurred in areas associated with the front edge, characterized by persistent presence of Smallmouth Bass in sizes ranging from young of year to adult. We did not observe salmonids or other characteristic coldwater species during sampling. However, the distance the leading edge extends upstream from the front edge is variable and strongly influenced by climatic conditions within years and dispersal abilities of individuals (Rubenson and Olden 2017). Thus, the leading edge of the Smallmouth Bass population could extend seasonally into coldwater habitats and influence those fish communities through piscivory.

Although our initial hypothesis was not supported, our experimental results shed some light—and raise additional questions—on the potential trophic influence of expanding Smallmouth Bass populations in low-order streams. Our results also demonstrate the high degree of complementarity between gut content and stable isotope data, underscoring their joint utility in studies of trophic ecology (Whitledge and Rabeni 1997; Grey et al. 2002; Davis et al. 2012). Further studies could take a comparative approach to examine how stream communities respond to invasion by Smallmouth Bass. This could be done by comparing assemblages in streams inhabited by Smallmouth Bass with similarly structured reference (uninvaded) streams, ideally integrating pre- and postinvasion observations (sensu Underwood 1992). Simulation studies using ECOPATH or similar approaches could shed light on the potential food web changes a novel predator could effect in stream communities (e.g., Warren et al. 2014). An integrated approach combining these empirical and theoretical methods can generate reasonable expectations for how Smallmouth Bass population expansion can affect stream ecosystems and inform proactive conservation strategies to conserve structure of low-order stream communities.

Acknowledgments

We thank Dawn Kirk (U.S. Forest Service) and Jason Hallacher (Virginia Department of Game and Inland Fisheries) for advice on study design. This manuscript was greatly improved by the comments and advice of two anonymous reviewers. This study was supported by the National Science Foundation under grant no. MRI-1337630 to EWH, Jeffress Memorial Trust grant no. J-1038 to RH, and funding from the Washington and Lee Lenfest Grant program to RH.

References

Anderson, C., and G. Cabana. 2007. Estimating the trophic position of aquatic consumers in river food webs using stable nitrogen isotopes. Journal of the North American Benthological Society 26:273–285.

Baker, R., A. Buckland, and M. Sheaves. 2014. Fish gut content analysis: robust measures of diet composition. Fish and Fisheries 15:170–177.

Barnes, C., C. J. Sweeting, S. Jennings, J. T. Barry, and N. V. C. Polunin. 2007. Effect of temperature and ration size on carbon and nitrogen stable isotope fractionation. Functional Ecology 21:356–362.

Bates, D. M. Maechler, B. Bolker, S. Walker. 2015. Fitting linear mixed-effects models using lme4. Journal of Statistical Software 67:1–48.

Benke, A. C., and C. C. Cushing. 2005. Rivers of North America. Elsevier, Burlington, Massachusetts.

Brewer, S. K., and D. J. Orth. 2015. Smallmouth Bass *Micropterus dolomieu* Lacepède, 1802. Pages 9–26 *in* M. D. Tringali, J. M. Long, T. W. Birdsong, and M. S. Allen, editors. Black bass diversity: multidisciplinary science for conservation. American Fisheries Society, Symposium 82, Bethesda, Maryland.

Brown, T., B. Runciman, S. Pollard, A. D. A. Grant, and M. J. Bradford. 2009. Biological synopsis of Smallmouth Bass (*Micropterus dolomieu*). Canadian Manuscript Report of Fisheries and Aquatic Sciences 2887.

Carey, M. P., B. L. Sanderson, T. A. Friesen, K. A. Barnas, and J. D. Olden. 2011. Smallmouth Bass in the Pacific Northwest: a threat to native species; a benefit for anglers. Reviews in Fisheries Science 19:305–315.

Chu, C., N. E. Mandrak, and C. K. Minns. 2005. Potential impacts of climate change on the distributions of several common and rare freshwater fishes in Canada. Diversity and Distributions 11:299–310.

Creed, R. P., Jr. 1994. Direct and indirect effects of crayfish grazing in a stream community. Ecology 75:2091–2103.

Davis, A. M., M. L. Blanchette, B. J. Pusey, T. D. Jardine, and R. G. Pearson. 2012. Gut content and stable isotope analyses provide complementary understanding of ontogenetic dietary shifts and trophic relationships among fishes in a tropical river. Freshwater Biology 57:2156–2172.

Englund, G., and J. J. Krupa. 2000. Habitat use by crayfish in stream pools: influence of predators, depth and body size. Freshwater Biology 43:75–83.

Everitt, B. S. 1992. The analysis of contingency tables. CRC Press, Boca Raton, Florida.

Farrell, A. P., and C. E. Franklin. 2016. Recognizing thermal plasticity in fish. Science 351:132–133.

Fritts, A. L., and T. N. Pearsons. 2004. Smallmouth Bass predation on hatchery and wild salmonids in the Yakima River, Washington. Transactions of the American Fisheries Society 133:880–895.

Fritts, A. L., and T. N. Pearsons. 2006. Effects of predation by nonnative Smallmouth Bass on native salmonid prey: the role of predator and prey size. Transactions of the American Fisheries Society 135:853–860.

Fry, B., and C. Arnold. 1982. Rapid $^{13}C/^{12}C$ turnover during growth of brown shrimp (Penaeus aztecus). Oecologia 54:200–204.

Gard, M. 2004. Interactions between an introduced piscivore and a native piscivore in a California stream. Environmental Biology of Fishes 71:287–295.

Grey, J., S. J. Thackeray, R. I. Jones, and A. Shine. 2002. Ferox trout (*Salmo trutta*) as 'Russian dolls': complementary gut content and stable isotope analyses of the Loch Ness food web. Freshwater Biology 47:1235–1243.

Heard, S. B., and J. S. Richardson. 1995. Shredder-collector facilitation in stream detrital food webs—is there enough evidence? Oikos 72:359–366.

Herzka, S. Z. 2005. Assessing connectivity of estuarine fishes based on stable isotope ratio analysis. Estuarine Coastal and Shelf Science 64:58–69.

Hesslein, R. H., K. A. Hallard, and P. Ramlal. 1993. Replacement of sulfur, carbon, and nitrogen in tissue of growing Broad Whitefish (*Coregonus nasus*) in response to a change in diet traced by $\delta^{34}S$, $\delta^{13}C$ and $\delta^{15}N$. Canadian

Journal of Fisheries and Aquatic Sciences 50:2071–2076.

Humston, R., S. S. Doss, C. Wass, C. Hollenbeck, S. R. Thorrold, S. Smith, and C. P. Bataille. 2017. Isotope geochemistry reveals ontogeny of dispersal and exchange between main-river and tributary habitats in Smallmouth Bass *Micropterus dolomieu*. Journal of Fish Biology 90:528–548.

Humston, R., B. M. Priest, W. C. Hamilton, and P. E. Bugas, Jr. 2010. Dispersal between tributary and main-stem rivers by juvenile Smallmouth Bass evaluated using otolith microchemistry. Transactions of the American Fisheries Society 139:171–184.

Huntsman, B. M., J. T. Petty, S. Sharma, and E. R. Merriam. 2016. More than a corridor: use of a main stem stream as supplemental foraging habitat by a Brook Trout metapopulation. Oecologia 182:463–473.

Hyslop, E. J. 1980. Stomach content analysis: a review of methods and their application. Journal of Fisheries Biology 17:411–429.

Jackson, A. L., R. Inger, A. C. Parnell, and S. Bearhop. 2011. Comparing isotopic niche widths among and within communities: SIBER– stable isotope Bayesian ellipses in R. Journal of Animal Ecology 80:595–602.

Jackson, D. A., and N. E. Mandrak. 2002. Changing fish biodiversity: Predicting the loss of cyprinid biodiversity due to global climate change. Pages 89–98 *in* N. A. McGinn, editor. Fisheries in a changing climate. American Fisheries Society, Symposium 32, Bethesda, Maryland.

Jenkins, R. E., and N. M. Burkhead. 1994. Freshwater fishes of Virginia. American Fisheries Society, Bethesda, Maryland.

Johnson, B. M., P. J. Martinez, J. A. Hawkins, and K. R. Bestgen. 2008. Ranking predatory threats by nonnative fishes in the Yampa River, Colorado, via bioenergetics modeling. North American Journal of Fisheries Management 28:1941–1953.

Johnson, J. H., and D. S. Dropkin. 1993. Diel variation in diet composition of a riverine fish community. Hydrobiologia 271:149–158.

Johnson, J. H., and D. S. Dropkin. 1995. Diel feeding chronology of six fish species in the Juniata River, Pennsylvania. Journal of Freshwater Ecology 10:11–18.

Lawrence, D. J., J. D. Olden, and C. E. Torgersen. 2012. Spatiotemporal patterns and habitat associations of Smallmouth Bass (*Micropterus dolomieu*) invading salmon-rearing habitat. Freshwater Biology 57:1929–1946.

Lawrence, D. J., B. Stewart-Koster, J. D. Olden, A. S. Ruesch, C. E. Torgersen, J. J. Lawler, D. P. Butcher, and J. K. Crown. 2014. The interactive effects of climate change, riparian management, and a nonnative predator on stream-rearing salmon. Ecological Applications 24:895–912.

Loppnow, G. L., K. Vascotto, and P. A. Venturelli. 2013. Invasive Smallmouth Bass (*Micropterus dolomieu*): history, impacts, and control. Management of Biological Invasions 4:191–206.

MacRae, P. S., and D. A. Jackson. 2001. The influence of Smallmouth Bass (*Micropterus dolomieu*) predation and habitat complexity on the structure of littoral zone fish assemblages. Canadian Journal of Fisheries and Aquatic Sciences 58:342–351.

Minagawa, M., and E. Wada. 1984. Stepwise enrichment of 15N along food chains: further evidence and the relation between $\delta^{15}N$ and animal age. Geochimica et Cosmochimica Acta 48:1135–1140.

Mittlebach, G., and L. Persson. 1998. The ontogeny of piscivory and its ecological consequences. Canadian Journal of Fisheries and Aquatic Sciences 55:1454–1465.

Myers, B. J., C. A. Dolloff, J. R. Webster, K. H. Nislow, B. Fair, and A. L. Rypel. 2018. Fish assemblage production estimates in Appalachian streams across a latitudinal and temperature gradient. Ecology of Freshwater Fish 27:363–377.

Petersen, J. H., and J. F. Kitchell. 2001. Climate regimes and water temperature changes in the Columbia River: bioenergetic implications for predators of juvenile salmon. Canadian Journal of Fisheries and Aquatic Sciences 58:1831–1841.

Post, D. M. 2002. Using stable isotopes to estimate trophic position: models, methods, and assumptions. Ecology 83:703–718.

R Core Team. 2017. R: A language environment for statistical computing. R Foundation for Statistical Computing, Vienna. Available: http://www.R-project.org/. (November 2016).

Rabeni, C. F. 1992. Trophic linkages between stream centrarchids and their crayfish prey. Canadian Journal of Fisheries and Aquatic Sciences 49:1714–1721.

Rieman, B. E., R. C. Beamesderfer, S. Vigg, and T. P. Poe. 1991. Estimated loss of juvenile salmonids to predation by Northern Squawfish, Walleyes, and Smallmouth Bass in John Day Reservoir, Columbia River. Transactions of the American Fisheries Society 120:448–458.

Roell, M. J., and D. J. Orth. 1993. Trophic basis of production of stream-dwelling Smallmouth Bass, Rock Bass, and Flathead Catfish in relation to invertebrate bait harvest. Transactions of the American Fisheries Society 122:46–62.

Rubenson, E. S., and J. D. Olden. 2017. Dynamism in the upstream invasion edge of a freshwater fish exposes range boundary constraints. Oecologia 184:453–467.

Sabo, J. L., J. C. Finlay, T. Kennedy, and D. M. Post. 2010. The role of discharge variation in scaling of drainage area and food chain length in rivers. Science 330:965–967.

Sabo, M. J., D. J. Orth, and E. J. Pert. 1996. Effect of stream microhabitat characteristics on rate of net energy gain by juvenile Smallmouth Bass, *Micropterus dolomieu*. Environmental Biology of Fishes 46:393–403.

Schlosser, I. J. 1987. The role of predation in age- and size-related habitat use by stream fishes. Ecology 68:651–659.

Scott, W. B., and E. J. Crossman. 1973. Freshwater fishes of Canada. Fisheries Research Board of Canada Bulletin 184.

Sharma, S., D. A. Jackson, C. K. Minns, and B. J. Shuter. 2007. Will northern fish populations be in hot water because of climate change? Global Change Biology 13:2052–2064.

Shuter, B. J., and J. R. Post. 1990. Climate, population viability, and the zoogeography of temperate fishes. Transactions of the American Fisheries Society 119:314–336.

Stitt, B. C., G. Burness, K. A. Burgomaster, S. Currie, J. L. McDermid, and C. C. Wilson. 2014. Intraspecific variation in thermal tolerance and acclimation capacity in Brook Trout (*Salvelinus fontinalis*): physiological implications for climate change. Physiological and Biochemical Zoology 87:15–29.

Trueman, C. N., R. A. R. McGill, and P. H. Guyard. 2005. The effect of growth rate on tissue-diet isotopic spacing in rapidly growing animals. An experimental study with Atlantic Salmon (*Salmo salar*). Rapid Communications in Mass Spectrometry 19:3239–3247.

Underwood, A. J. 1992. Beyond BACI: the detection of environmental impacts on populations in the real, but variable, world. Journal of Experimental Marine Biology and Ecology 161:145–178.

Uno, H., and M. E. Power. 2015. Mainstem-tributary linkages by mayfly migration help sustain salmonids in a warming river network. Ecology Letters 18:1012–1020.

Usio, N., and C. R. Townsend. 2004. Roles of crayfish: consequences of predation and bioturbation for stream invertebrates. Ecology 85:807–822.

Vander Zanden, M. J., G. Cabana, and J. Rasmussen. 1997. Comparing trophic position of freshwater fish calculated using stable nitrogen isotope ratios (δ^{15}N) and literature dietary data. Canadian Journal of Fisheries and Aquatic Sciences 54:1142–1158.

Vander Zanden, M. J., J. M. Casselman, and J. B Rasmussen. 1999. Stable isotope evidence for the food web consequences of species invasions in lakes. Nature (London) 401:464–467.

Vander Zanden, M. J., J. D. Olden, J. H. Thorne, and N. E. Mandrak. 2004. Predicting occurrences and impacts of smallmouth bass introductions in north temperate lakes. Ecological Applications 14:132–148.

Vander Zanden, M. J., and J. B. Rasmussen. 2001. Variation in delta N-15 and delta C-13 trophic fractionation: implications for aquatic food web studies. Limnology and Oceanography 46:2061–2066.

Vander Zanden, M. J., and J. B. Rasmussen. 2002. Food web perspectives on studies of bass populations in north-temperate lakes. Pages 173–184 *in* D. P. Philipp and M. S. Ridgway, editors. Black Bass: ecology, conservation, and management. American Fisheries Society, Symposium 31, Bethesda, Maryland.

Vannote, R. L., G. W. Minshall, K. W. Cummins, J. R. Sedell, and C. E. Cushing. 1980. The river continuum concept. Canadian Journal of Fisheries and Aquatic Sciences 37:130–137.

Warren, D. R., C. J. Harvey, M. M. McClure, and B. L. Sanderson. 2014. Use of an ecosystem-based model to evaluate alternative conservation strategies for juvenile Chinook salmon in a headwater stream. North American Journal of Fisheries Management 34:839–852.

Whitledge, G. W., and C. E. Rabeni. 1997. Energy sources and ecological role of crayfishes in an Ozark stream: insights from stable isotopes and gut analysis. Canadian Journal of Fisheries and Aquatic Sciences 54:2555–2563.

Whittier, T. R., and T. M. Kincaid. 1999. Introduced fish in northeastern USA lakes: regional extent, dominance, and effect on native species richness. Transactions of the American Fisheries Society 128:769–783.

American Fisheries Society Symposium 87:75–92, 2019

Role of Recovering River Herring Populations on Smallmouth Bass Diet and Growth

Jonathan M. Watson* **and Stephen M. Coghlan, Jr.**
Department of Wildlife, Fisheries, and Conservation Biology
University of Maine, 5755 Nutting Hall, Room 210
Orono, Maine 04469, USA

Joseph Zydlewski
U.S. Geological Survey, Maine Cooperative Fish and Wildlife Research Unit
and
Department of Wildlife, Fisheries, and Conservation Biology
University of Maine, 5755 Nutting Hall, Room 210
Orono, Maine, 04469, USA

Daniel B. Hayes
Department of Fisheries and Wildlife, Michigan State University
480 Wilson Road, Room 13, East Lansing, Michigan 48824, USA

Daniel S. Stich
Biology Department, SUNY College at Oneonta
112 Science Building 1, Oneonta, New York 13820, USA

Abstract.—Fish assemblages in Atlantic coastal rivers have undergone extensive ecological change in the last two and a half centuries due to human influence, including extirpation of many migratory fish species, such as river herring *Alosa* spp. and introduction of nonnative piscivores, notably Smallmouth Bass *Micropterus dolomieu*. Recently, dam removals and fish passage improvements in the Penobscot River, Maine, have allowed river herring to return to reaches of the river that have been inaccessible since the late 19th century. Alosine populations have increased and this trend is anticipated to continue. This may increase forage in the system which could potentially increase growth for Smallmouth Bass, the

*Corresponding author: Jonathan.Watson@maine.edu

dominant piscivore. We examined the diet and growth of Smallmouth Bass collected from areas of the Penobscot River watershed with and without access to river herring as prey. We collected 765 Smallmouth Bass throughout 2015, examined the stomach contents of 573 individuals, and found notable differences in diet among three river reaches with common seasonal trends. Juvenile river herring composed an average of 19% (SE = ±6%) of stomach contents by mass from Smallmouth Bass collected in the freshwater tidal area but were rarely observed in the diets upstream. We used estimates from von Bertalanffy growth models to examine differences in growth among reaches and found that asymptotic length was the longest (425 mm TL) in the Tidal reach where access to river herring was unrestricted. We then used these data to predict changes to growth associated with increased access to juvenile river herring prey with bioenergetics models. Results indicated that substituting juvenile river herring for less energy-dense prey (e.g., invertebrates) may lead to increases in seasonal growth throughout the watershed as river herring populations continue to rebound in response to dam removal. Our results provide insight into the diet and growth of Smallmouth Bass in a large New England river, and provide a foundation for future work investigating unfolding changes to these characteristics following recent dam removals.

Introduction

Growth and life history patterns of predatory fishes are influenced by prey availability (Adams et al. 1982a; Boisclair and Leggett 1989; Dunlop et al. 2005), prey size (Michaletz 1997; Pazzia et al. 2002), and prey energy density (Rand et al. 1994). Diet composition can influence growth (Boisclair and Leggett 1989) and size at maturity (Shuter et al. 2016) of predators. Differences in diet can result in growth differences across spatial (Yako et al. 2000; Glover and DeVries 2013) and temporal (Martin 1970; Shuter et al. 2016) scales. Therefore, the introduction or restoration of forage fish may result in changes to predator diet and growth. The degree to which seasonal influxes of sea-run forage fishes influence these characteristics in freshwater resident predators is, however, poorly documented.

One such piscivorous fish, the Smallmouth Bass *Micropterus dolomieu*, native to the St. Lawrence and Mississippi drainages (Werner 2004), has been introduced widely throughout the world (Jackson 2002) and has been implicated in declines of abundance and diversity of native minnows, salmonines, and other sensitive taxa (Whittier and Kincaid 1999; Weidel et al. 2007). They also prey upon anadromous fishes in areas outside of their native range such as the Pacific Northwest (e.g., Fritts and Pearsons 2004; Tabor et al. 2007). In Maine, Smallmouth Bass were introduced in the late 19th century and have spread throughout most watersheds since then (Warner 2005). During this period, construction of numerous large dams fragmented habitat and contributed to the decline or extirpation of anadromous fish populations (Saunders et al. 2006). Recently, the Penobscot River in Maine has been the focus of a restoration project, including two large dam removals and upgraded fish passage facilities at several remaining dams. This project is collectively known as the Penobscot River Restoration Project—PRRP (see Opperman et al. 2011 for description).

Anadromous river herring (Alewife *Alosa pseudoharengus* and Blueback Herring *A. aestivalis*), once abundant in the Penobscot River watershed, were at historically low abundances (Saunders et al. 2006) prior to recovery efforts that included dam removal and stocking. These efforts increased connectivity for sea-run fishes, resulting in substantial population growth for anadromous river herring *Alosa* spp. in this system, with returns increasing by more than one million spawning adults in just a few years following dam removal (M. Simpson, Maine Department of Marine Resources, personal communication). This change in prey dynamics has the potential to influence diet and growth of Smallmouth Bass populations in the catchment.

Anadromous river herring can influence resident fishes and freshwater ecosystems via several pathways. Migration of adults and freshwater rearing of juveniles can influence nutrient budgets (Twining et al. 2013) and shift zooplankton communities (Post et al. 2008; Howeth et al. 2013). Both adult and juvenile life stages can provide energy-dense forage for piscivorous fish (Hall et al. 2012). In freshwater they are preyed upon by Largemouth Bass *Micropterus salmoides* (Yako et al. 2000), Blue Catfish *Ictalurus furcatus* (MacAvoy et al. 2000), and other species (Mattocks et al. 2017). Landlocked river herring populations are important prey for Salmonines in the Great Lakes (Stewart and Ibarra 1991; Rand et al. 1994; Savitz 2009), Striped Bass *Morone saxatilis* (Cyterski et al. 2002), and Chain Pickerel *Esox niger* (Brodersen et al. 2015). Other studies (e.g., Yako et al. 2000; Mattocks et al. 2017) have described predation by piscivorous fishes on juvenile anadromous river herring in their natal lentic environments. Notably, Yako et al. (2000) found that juvenile anadromous alewives constituted a large portion of Largemouth Bass diets in Massachusetts lakes with fish growing larger when anadromous river herring were present.

Interactions between anadromous river herring and Smallmouth Bass likely extend beyond that of a predator–prey relationship, but previous studies have produced conflicting results. Hanson and Curry (2005) hypothesized that competition between young-of year (YOY) river herring and Smallmouth Bass might be significant based on diet overlap. In contrast, Willis (2009) noted little diet overlap between these species and observed no difference in Smallmouth Bass condition in lakes with or without river herring. The uncertainty surrounding competition during early life history has influenced policy decisions about fish passage in North America. For example, in the St. Croix River, the eastern border between Maine and New Brunswick, concerns over a perceived decline in Smallmouth Bass abundance prompted the state to close fish passage structures and exclude river herring from natal spawning grounds, although restoring river herring access to historic habitat is a top priority of Native American tribes and New Brunswick provincial fisheries agencies (Willis 2009).

Rebounding populations of river herring are expected to provide an important prey source for Smallmouth Bass, the dominant piscivore the lower Penobscot River watershed (Kiraly et al. 2014; Watson et al. 2018), because juvenile river herring are available as prey throughout the growing season due to extended emigration timing (Iafrate and Oliveira 2008). Rearing and out-migration of anadromous river herring is protracted throughout the summer and fall in the Northeastern United States (Yako et al. 2002), when water temperatures are optimal for Smallmouth Bass consumption (~22°C; Whitledge et al. 2003). Furthermore, Smallmouth Bass living in freshwater tidal areas may benefit from extended access to pulses of juvenile river herring exported from lakes and ponds upstream. Though juveniles are typically thought to migrate directly to the ocean, several studies of river herring have revealed prolonged presence in freshwater tidal waters (e.g., Limburg 1998; Gahagan et al. 2012; Turner and Limburg 2016)

Smallmouth Bass predation on juvenile anadromous river herring in rivers has not been well described, though these interactions have the potential to greatly influence population dynamics of both species. The purpose of this study was to assess diet and growth of Smallmouth Bass, in the context of watershed-scale efforts to recover river herring populations. We used the Penobscot River watershed as a study system to test the extent to which recent increases in river herring populations may influence Smallmouth Bass growth and diet. To do so we (i) assessed the diet of Smallmouth Bass in three river reaches with different access to river herring as prey, (ii) compared growth of Smallmouth Bass among those reaches, and (iii) used a bioenergetics model to determine potential Smallmouth Bass growth under scenarios of increasing consumption of anadromous river herring following projected population increases.

Methods

Collection and Processing

We used three study reaches in the Penobscot River catchment based on river herring presence (both current and historic) and locations relative to several main-stem dams (Kiraly et al. 2014). We collected Smallmouth Bass from two reaches of the Penobscot River (Figure 1), corresponding to strata delineated by Kiraly et al. (2014): (i) Orono, which is currently below the lowermost dam and, as of 2014, was accessible to river herring, and (ii) the Tidal reach, which was freshwater, downstream from the head of tide, and has been historically accessible to "remnant" populations of migratory fishes including river herring (Grote et al. 2014). We also sampled the Piscataquis River (Piscataquis reach), a major tributary to the Penobscot River, which was inaccessible to anadromous river herring during our collections, but supports a regionally valuable Smallmouth Bass fishery. We collected individuals from an additional reach (Argyle) and used those individuals only in growth analyses.

We captured a total of 764 Smallmouth Bass with electrofishing and angling from May to October 2015. We used electrofishing to capture 372 individuals in conjunction with fish community surveys described by Watson et al. (2018). We used angling with curly-tailed grub lures (white, 7.6 cm length) to collect an additional 393 individuals from July until September. We used 572 individuals ranging from 145- to 464-mm TL for stomach analysis, with 72% of those fish captured during summer angling collections. Angling began at sunrise and continued until minimum sample sizes were met or four hours had elapsed. This time was chosen because Smallmouth Bass are crepuscular predators (Reynolds and Casterlin 1976) and water temperatures are lowest at night, which slows digestion. We attempted to collect ten individuals in each 50-mm size-class from each site, starting at 150 mm, in order to obtain a representative range of sizes and diets. For common size classes (e.g., 200–300 mm TL), we frequently released fish once we met our goal. However, for uncommon sizes (e.g., TL > 350 mm) or during periods of low catch due to low water temperatures (e.g., October sampling), we did not always meet our catch quotas.

We measured total length (TL) of each fish and, if they met our size criteria, we euthanatized them by lethal overdose with buffered MS-222. The fish were then placed in a cooler with ice water and brought back

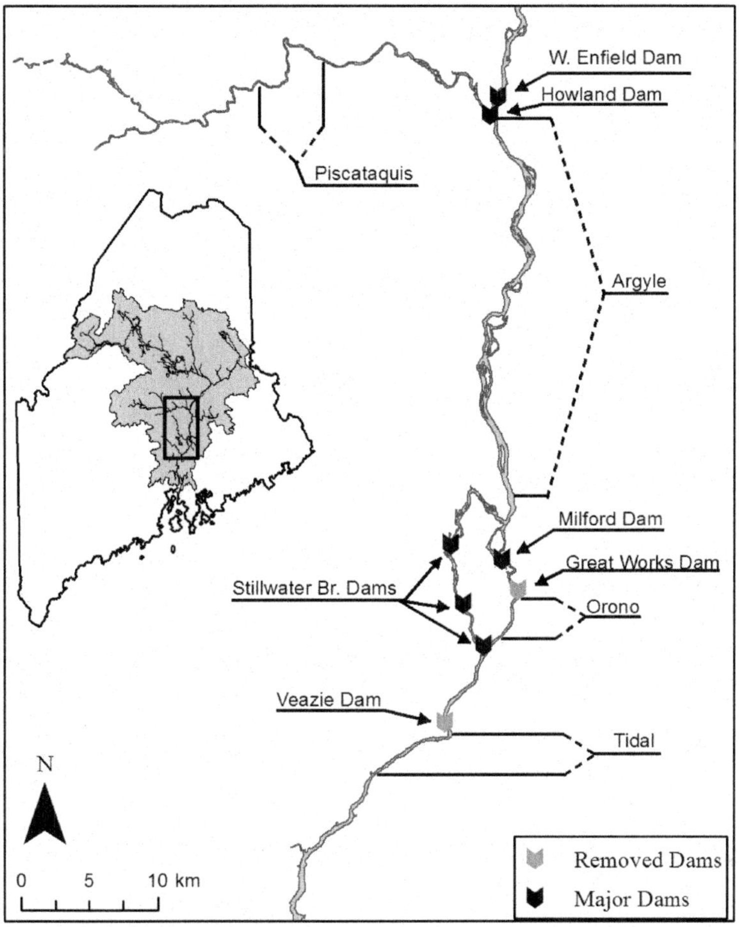

Figure 1. The Penobscot River watershed, Maine (inset) and lower mainstem river including the Piscataquis River. Also included are the locations of dams both present and former and the delineations of the sample reaches where Smallmouth Bass were collected.

to the laboratory for dissection. Upon arrival, fish were dissected immediately or frozen for later dissection. We measured a subset of fish (n = 75) before and after freezing and confirmed total length measurements were not detectably different (two sample t-test, P = 0.43). During dissection, we removed sagittal otoliths and removed stomachs by cutting the esophageal and pyloric sphincters. Stomachs were wrapped in muslin cloth, fixed in a buffered 10% formaldehyde solution for 72 h, rinsed with water, and stored in 70% ethanol before further dissection and sorting.

Diet Analysis

We removed all preserved contents from each stomach and sorted into one of five categories: river herring, other fish, insect, crayfish, and other. Items were classified as "other" if the prey was either unidentifiable or did not fit into one of the other four prey categories. The "insect" prey category included both terrestrial (e.g., caterpillar) and aquatic (e.g., stoneflies) insect species. The "other fish" category included all species other than river herring and unidentifiable fish tissue. The distinctive body form of river herring (i.e., deep bodied, laterally compressed, deeply forked tail) facilitated identification in gut contents. Each prey type from each stomach was then placed in a tared aluminum container and dried at 60°C for 24 h. Dry mass was recorded to the nearest 0.001 g.

We calculated average frequency occurrence (O_i) of each prey type i in each reach j, defined as:

$$O_i = \frac{J_{ij}}{P_j}$$

where J_{ij} was the number of fish containing prey i in reach j and P_j was the number of fish with food in their stomachs in reach j. This measure describes how often each prey type was consumed and compliment measures of prey mass. We also calculated the ratio of the total mass of each prey type i in group j following the ratio estimation procedure detailed by Hansen et al. (2007). Fish were grouped according to collection month and size-class (TL < 224, 225–274, 275–324, and <325 mm TL) in each reach and ratios were calculated as:

$$\hat{R} = \frac{\Sigma_{i=1}^{n} y_{ij}}{\Sigma_{i=1}^{n} x_j}$$

Where y_{ij} was the mass of prey type i in group j and x_j was the total mass of all prey types in group j. The standard error for this ratio (Hansen et al. 2007) was approximated as:

$$\mathrm{SE}\,(\hat{R}) = \frac{1}{\sqrt{n_j}\;\bar{x}_j}\sqrt{\frac{\Sigma_{i=1}^{n}\,(y_{ij} - \hat{R}x_j)^2}{n_j - 1}}$$

Where n_j is the number of stomachs in group j. Unless otherwise noted, all results of diet study are reported as mean ± SE.

Otolith Preparation

We chose to use sagittal otoliths for aging as opposed to nonlethal aging structures (e.g., scales) to reduce aging error and increase precision for Smallmouth Bass older than age five (Long and Fisher 2011). Otoliths were sectioned along the dorsal-ventral axis and photographed along with an external length standard using a Spot 3.1 camera (SPOT Imaging, Sterling Heights, Michigan) mounted on the trinocular port of a MEIJI Techno EMZ-13TR stereomicroscope. We measured to the distal edge of the most recent annulus along the medial axis and measured total medial radius of each otolith with ImageJ (Schneider et al. 2012) and the ObjectJ plugin (Vischer and Nastase 2015). During digital measuring, we simultaneously viewed the photograph of the sectioned otolith and the corresponding slide of sections from the same otolith using a dissecting microscope to ensure that each annuli and margin were marked appropriately. Each otolith was aged by two independent readers and discrepancies in age determination were reconciled prior to measurement. The measurement of each otolith was assigned a qualitative confidence level (1 through 10, low to high) and only measurements with a confidence level of eight or higher were considered in this analysis.

Growth Modeling

We used the von Bertalanffy growth model (VBGM; von Bertalanffy 1938) to estimate growth curves for fish from each reach. We used the Modified Fry method detailed by Vigliola and Meekan (2009) to estimate length for the most recent annulus to account for the variability in capture dates and corresponding seasonal growth differences. We used size at age data from individuals captured in the three reaches from the diet study: Tidal (n = 166), Orono (n = 257), and Piscataquis (n = 132). We also included size at age data from the Argyle reach (n = 167) to improve parameter estimates, but do not present those results here for brevity.

We fit the VBGM as a Bayesian hierarchical model with a Markov chain Monte Carlo (MCMC) approach in JAGS (Plummer 2003) using the R2jags package (Su and Yajima 2015) in R (R Core Team 2016). We used a hierarchical model specification (see He and Bence 2007) to facilitate information sharing among reaches while allowing separate estimation of VBGM parameters for fish within each reach.

The length of each fish (L_i) at age t_i was estimated using the VBGM as:

$$L_i = L_{\infty j}\,(1 - e^{-K_j(t_i - t_{0j})})$$

where $L_{\infty j}$ was the asymptotic length of fish in each reach j, K_j was the Brody growth coefficient in each reach, and t_{0j} was the age at length zero in each reach.

We used uninformative hyperpriors for all model parameters. We confirmed model convergence using the Gelman-Rubin statistic ($\hat{r} < 1.10$), and by graphical inspection of mixing among chains (Kruschke 2011). Effective sample size was sufficient to construct posterior estimations of all parameters. We assessed differences in growth parameters by calculating the difference in parameter estimates between all paired reaches and assessed statistical significance based on lack of overlap of 95% CRI with zero (Kéry 2010).

Bioenergetics Modeling

We used the Wisconsin mass-balance bioenergetics model (Hanson et al. 1997) and metabolic values reported by Whitledge et al. (2003) to model annual Smallmouth Bass growth under different diet scenarios. We used daily average temperature values from two USGS gauging stations, one on the Piscataquis River and the other on the mainstem Penobscot River (USGS 2016a, 2016b). All models were run for the period in which average water temperatures were greater than 8°C (May 1–October 31). Because Smallmouth Bass are largely inactive, and presumably not feeding below 10°C (Roell and Orth 1993), we assumed that all annual growth would occur during the modeled period, though other studies (e.g., Adams et al. 1982b) have shown that weight loss does occur during winter. We also substituted daily average temperature measurements from the two gauges in the Piscataquis and Tidal reaches to determine whether differences affected estimates.

Modeled diet proportions were taken from individuals with a total length of 225–325 mm at time of capture to avoid the influence of different feeding strategies employed by the largest and smallest fish in our samples. Because 93% of "other" diet items by mass were found in only 4% of stomachs, we excluded these prey types from diet composition in bioenergetics analyses. We interpolated diet proportions for each day between sampling events. Because we did not collect diet samples in May and October in the Piscataquis reach, we estimated diet composition in those months based on seasonal trends observed in the other two reaches.

We ran all models using average prey energy content values reported by Yako et al. (2000) for four prey types: river herring (5.6 kJ/g), other fish (4.1 kJ/g), crayfish (3.2 kJ/g), and insect (3.2 kJ/g). Prey energy content was assumed to be constant throughout the modeling period. Predator energy content was held constant at 4.2 kJ/g (wet mass) through all models, a common assumption for models of centrarchid bioenergetics (Whitledge et al. 2003). We chose to model annual growth for individuals ranging from age 2 to age 5 due to the dominance of those age classes in our samples. We used results from the VBGM and a length-mass relationship ($R^2 = 0.99$) developed for Smallmouth Bass in the Penobscot River watershed (Watson et al. 2018) to estimate seasonal growth in grams.

We first ran models using observed diet data for each reach to produce an estimate of the percent of maximum consumption (%

C_{max}) for each age-class required to achieve average annual growth. Next, we ran a model with proportions of juvenile river herring observed in Orono diets substituted in the Piscataquis reach diets to simulate a relatively modest increase in river herring consumption. Finally, we ran two models with proportions of juvenile river herring observed in diets of Smallmouth Bass captured in the Tidal reach substituted for the diet in both Orono and Piscataquis reaches to simulate large increases in consumption of this prey. Given that estimates calculated by Opperman et al. (2011) for carrying capacity of river herring in this system following restoration efforts are an order of magnitude larger than was observed at the most downstream fishway in 2015, we felt estimates of consumption of river herring in the Tidal reach were reasonable for possible future consumption in upstream reaches. In all instances where diet was manipulated, % C_{max} was held at the rates previously modeled and the proportion of remaining diet items consisted of the observed diets rescaled to represent the proportion of the diet not composed by juvenile river herring.

Results

Diet

Insects were the most frequently consumed prey type in every reach (Table 1), though many fish contained minimal insect matter, which resulted in lower empty stomach occurrence. On average, we found insects in 76% of stomachs that contained prey. Fish were, on average, the next most commonly occurring prey type, and were found in 23% of stomachs containing prey. The proportion of empty stomachs from each sampling occasion ranged from 4% to 33%, with an average of 19% across all sampling occasions. We removed data from fish with empty stomachs from further diet analyses.

Average stomach content mass was highest in the Tidal reach, relative to other reaches and we also found novel prey types. Average dried stomach content mass ranged from 0.39 ± 0.13 g in the Piscataquis reach to 0.57 ± 0.16 g in the Tidal reach and was highly variable among individuals. Items in the "other" prey category composed a large portion of the diets (>20% by mass) in several months due to the occurrence of several large prey items in relatively few individuals. Several novel food items dominated the mass of this category, including three rodents, three frogs, and two juvenile snapping turtles with an average dried mass of 5.5 ± 3.8 g, 2.0 ± 1.2 g, and 2.1 ± 1.0 g, respectively.

Smallmouth Bass consumed juvenile anadromous river herring to a variable extent, depending on availability and location in the watershed. Smallmouth Bass consumed the most river herring in the Tidal reach across all months and size classes. In the Orono reach, river herring were only found in diets during July samples and only in 3% of stomachs. We found river herring in 14% of stomachs con-

Table 1. Average frequency of occurrence (%) of each prey type found in all Smallmouth Bass stomachs containing prey in three sampled reaches of the Penobscot River watershed, Maine.

Reach	River Herring	Other Fish	Crayfish	Insect	Other
Tidal	14	30	8	68	20
Orono	3	18	4	76	19
Piscataquis	0	15	15	83	15

taining prey from the Tidal reach. During the months in which they were available (July–October), juvenile river herring averaged 19 ± 6% of prey by mass in the Tidal reach, 4 ± 1.9% in the Orono reach, and were not found in the stomachs collected in the Piscataquis reach (Figure 2). Consumption of adult river herring was limited to one occurrence in the Tidal reach in May.

River herring were more prevalent in the diets of Smallmouth Bass in the Tidal reach than the upstream reaches. In the Tidal reach, river herring composed 12% to 55% of the diet by mass in each size-class (Figure 2B). Juvenile river herring composed 44 ± 28% of the diet by mass for the smallest size-class (TL < 225 mm) sampled in the Tidal reach. In contrast, the diet proportion by mass of the

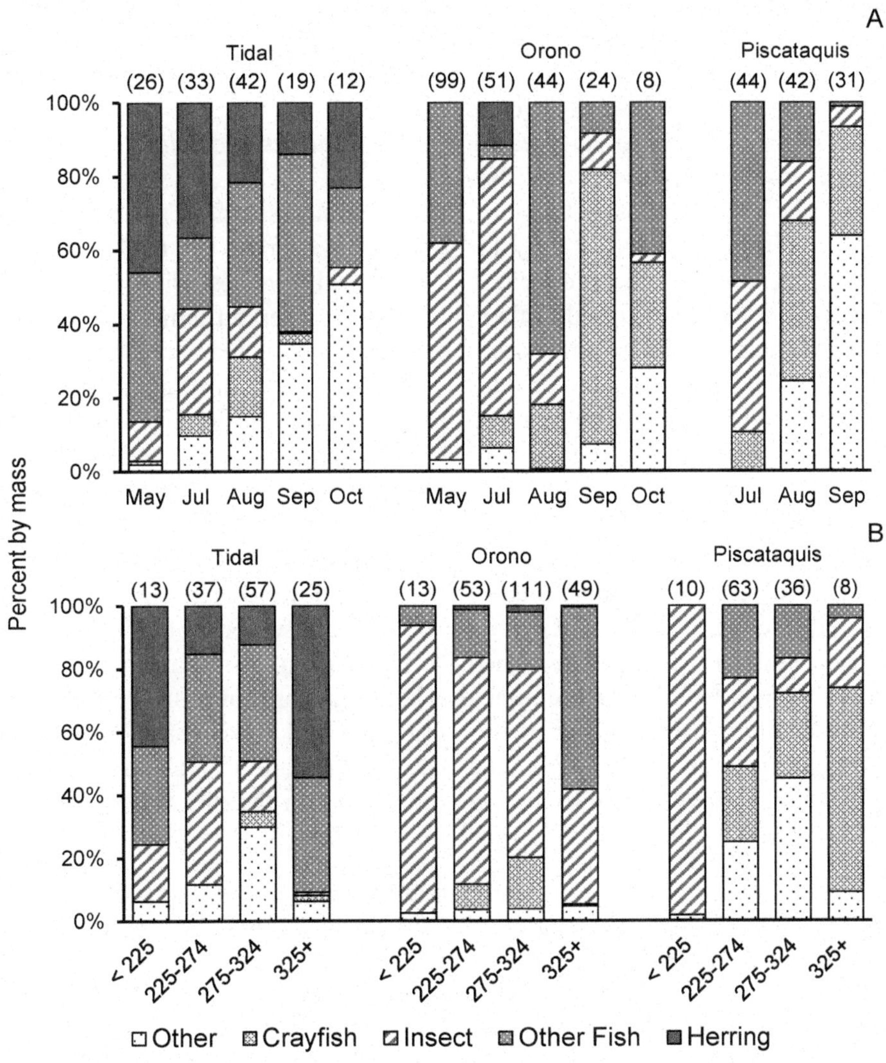

Figure 2. Percent diet composition by mass (g) of Smallmouth Bass reported by month (A) and 50-mm size-class (B) in three different reaches (Tidal, Orono, Piscataquis) of the Penobscot River watershed, Maine. The number of stomachs containing prey items in each sample is shown in parentheses above each bar.

smallest size-class in the other two reaches was dominated by insects, with their contribution decreasing in larger size classes. The contribution of other fish to the diets of different size classes of Smallmouth Bass in the Tidal reach remained relatively consistent, ranging from 31 ± 20% to 37 ± 13% by mass. In the Orono reach, other fish composed the largest portion of the diet by mass (R = 58 ± 13%) for the largest size-class (TL ≥ 325 mm). In the Piscataquis reach, crayfish composed the largest portion of the diet by mass in the largest size-class (R = 65 ± 21%).

We also observed more insects in diets across reaches earlier in the growing season followed by increasing proportions of different prey types. Insects composed the largest observed proportion by mass in July for each reach and decreased in the proceeding months (Figure 2A). We also found "other" prey items increasing in relative proportion throughout the sampling period, primarily due to large individual prey items during September and October sampling periods.

Otolith Analysis and Growth Modeling

We used back calculations from a total of 722 otoliths to fit the hierarchical VBGM. Of a total of 765 otoliths processed, we assigned low confidence ratings to 24 measurements, and they were omitted from further analysis. An additional 19 otoliths were collected from age-0 fish and were only used to inform back-calculations. Smallmouth Bass grew to the largest asymptotic length in the Tidal reach (Table 2) and length-at-age varied accordingly (Figure 3). As confirmation of this trend, we observed only a modest difference in average asymptotic size among reaches, with the largest difference (3.6 cm) occurring between the Tidal and Orono reaches. Other VBGM parameter estimates (K, t_0) were not significantly different between reaches. Interestingly, we observed fish up to age 20 in the Orono reach, but only age 13 and 10 in the Piscataquis and Tidal reaches, respectively.

Bioenergetics Modeling

Estimates of % C_{max} decreased with age and ranged from 0.4 to 0.9 across modeled ages in all reaches using observed diet data. Average observed temperature was lower in the Piscataquis reach (17.4°C) than the Tidal reach (18.6°C) during the modeled period, though these differences did not result in large changes to estimated growth (≤5% ending mass difference) when temperature data were substituted among reaches with estimated consumption rates held constant.

When we substituted the river herring-rich diet observed in the Tidal reach into diets of fish in Orono and Piscataquis reaches (holding consumption rates constant) we simulated that seasonal growth would increase substantially (Figure 4). Increased river herring in the diet for the Piscataquis and

Table 2. Pairwise comparisons of the difference in posterior distributions of maximum length (L_∞) and associated 95% credible intervals (CRI) among three reaches (Tidal, Orono, Piscataquis) of the Penobscot River catchment, Maine USA.

Comparison	Mean difference	Lower 95% CRI	Upper 95% CRI
Orono - Piscataquis	–4.02	–32.80	16.16
Orono - Tidal *	–36.59	–71.88	–14.19
Piscataquis - Tidal *	–32.57	–60.65	–7.09

*Indicates statistical significance ($\alpha < 0.05$)

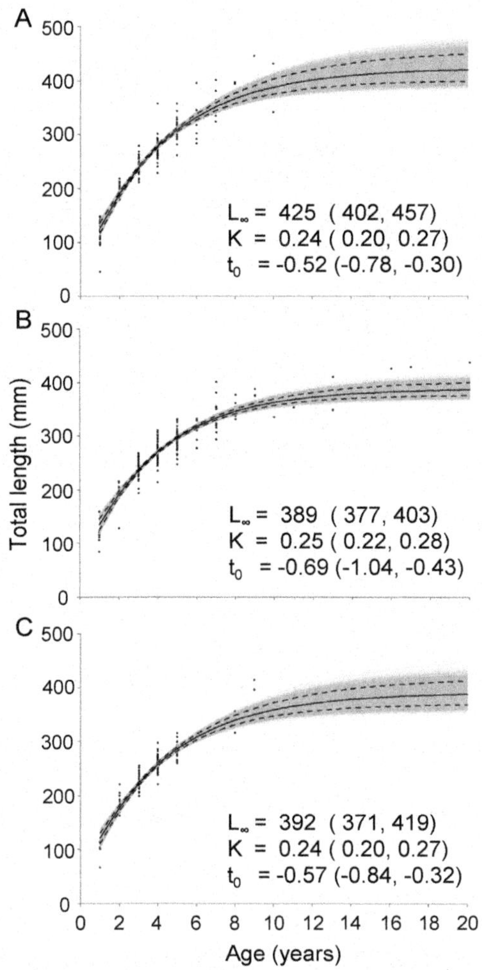

Figure 3. Posterior predictions from von Bertalanffy growth models fit to back-calculated size-at-age data for Smallmouth Bass in the Tidal (A), Orono (B), and Piscataquis (C) reaches of the Penobscot River, Maine. Points represent raw data, gray lines represent posterior predictive von Bertalanffy growth model (VBGM) curves, black solid lines represent average posterior predictive VBGM curves, and the dashed lines represent the 95% credible interval. Upper and lower 95% credible intervals are provided in parentheses next to each estimated growth parameter.

Orono reaches resulted in a modeled average seasonal growth increase of 35% and 31% by mass, respectively. In contrast however, substituting the lower observed prevalence of river herring found in the Orono diet into the Piscataquis diet resulted in a moderate increase in growth, with an average increase of 7% by mass across age classes. It is important to note that assumptions made about consumption are a substantial source for errors in growth estimates (Bajer et al. 2004), so results from this model represent only an estimate of possible future growth.

Estimates of % C_{max} were influenced by the diet energy density. We estimated that fish in the Piscataquis reach fed at the highest rate, followed by the Orono and Tidal reaches, respectively. This is likely because Smallmouth Bass in the Piscataquis reach ate primarily insects and crayfish, which have

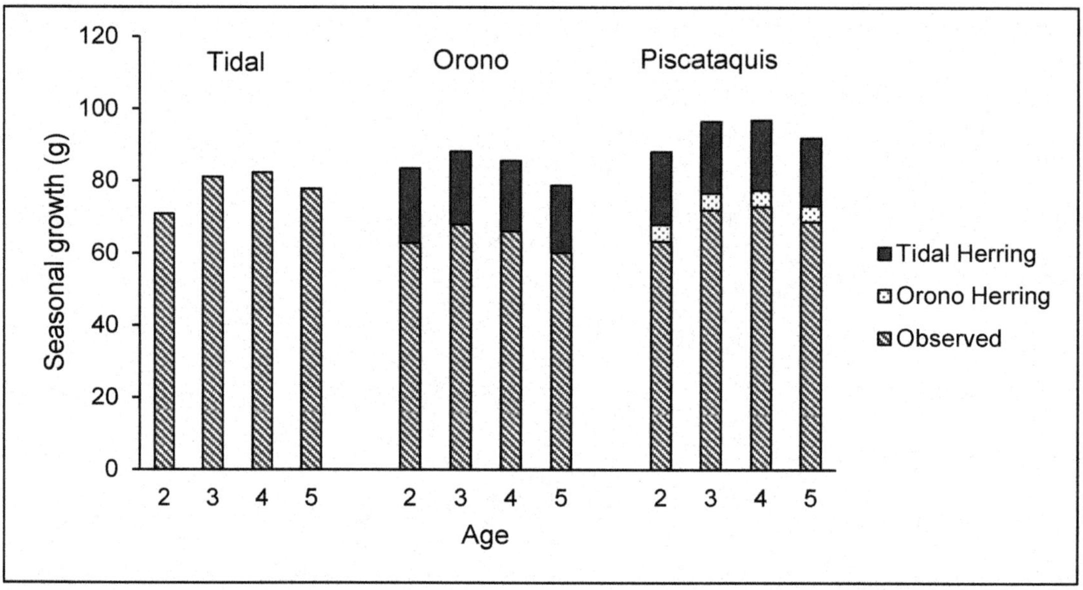

Figure 4. Observed and predicted seasonal growth (g) of Smallmouth Bass ages 2–5 in three reaches (Tidal, Orono, Piscataquis) of the Penobscot River watershed, Maine. Growth was predicted for two separate bioenergetics models (Tidal Herring, Orono Herring) under conditions of increased availability of river herring prey for restored reaches. The Tidal Herring model substituted proportions of river herring in the diet from the Tidal reach into the diets of the two upstream reaches. The Orono Herring model substituted the modest consumption of river herring in the Orono reach into the diet of the Piscataquis reach. Consumption, prey energy density, and metabolic rates were held constant for model predictions.

a relatively low energy density. In contrast, diets in the Tidal reach are composed primarily of fish (including river herring) which are more energy-dense. Thus, Smallmouth Bass in this reach are estimated to feed at a lower rate, because they consume more energy-dense prey and have similar seasonal growth relative to the other reaches.

Discussion

Our study demonstrates that Smallmouth Bass feed upon energy-dense juvenile anadromous river herring in the Tidal reach, which may contribute to modestly larger asymptotic size. Smallmouth Bass appear to feed upon these prey when they are available and abundant, though we found river herring in only 3% of Smallmouth Bass stomachs in the Orono reach during the early stages of restoration. Bioenergetics models indicated that substituting river herring in place of less energy-dense prey results in increased seasonal growth potential, but empirical growth estimates (K) did not vary significantly among reaches with varying levels of access to river herring, which may have potentially been associated with differences in consumption and habitat. These data were collected two years following dam removals and fish passage improvement when adult anadromous river herring returns at the lowermost dam were approximately 15% of the potential abundance (Opperman et al. 2011). Thus, this study provides a valuable baseline for Smallmouth Bass growth and utilization of anadromous river herring as abundance of alosine species is projected to increase further.

Several studies have described diet patterns of stream-dwelling Smallmouth Bass similar to those detailed here, including decreasing importance of insects in the diet of older fish (Roell and Orth 1993) and later in the growing season (Probst et al. 1984). These studies also indicate that crayfish comprise the majority of the diet among older fish. This was the case in our most upriver site (Piscataquis) but fish (including river herring) comprised the largest portion of the diet in the Tidal reach (across all sizes). These differences in diet may forecast upstream changes in Smallmouth Bass growth patterns as river herring populations continue to increase.

Our study contributes to a growing body of evidence that river herring may provide an important source of forage for resident freshwater piscivores which may impact their growth and abundance. Fish that consume larger or more energy-dense prey grow faster (Martin 1970; Boisclair and Leggett 1989) and have higher winter survival (Shuter and Post 1990). River herring are an energy-dense forage fish relative to other prey types (Adams et al. 1982b; Cyterski et al. 2002), and similar studies have shown that piscivores prey on seasonally available, energy-dense clupeids in various freshwater systems (Yako et al. 2000; Cyterski et al. 2002; Trippel et al. 2015). Yako et al. (2000) attributed larger asymptotic size of Largemouth Bass in lakes to the presence of juvenile anadromous Alewife. In this study, we estimated the largest asymptotic length for Smallmouth Bass in the Tidal reach, which may be associated with a persistent, albeit remnant population of river herring in that area. Brodersen et al. (2015) observed that Chain Pickerel also preyed extensively on landlocked Alewife and, as a result, exhibited greater lipid content than their counterparts in lakes without Alewife.

Juvenile anadromous river herring exhibit multiple, distinct emigration events from natal habitats (Iafrate and Oliveira 2008; Turner and Limburg 2016) throughout the summer, coinciding with the growing season of Smallmouth Bass. These migration timings may have limited our ability to detect juveniles in Smallmouth Bass diets in upstream reaches where swift stream flows expedite out-migration. In this study, freshwater tidal areas seemed to present the largest opportunity for predator–prey overlap. Other studies of river herring out-migration patterns (Limburg 1998; Turner and Limburg 2016) have described the protracted presence of river herring in tidal areas throughout the late summer and early fall, which may present piscivores in these habitats with additional forage.

In this study, most VBGM parameter estimates were similar between reaches and were comparable to the growth standard reported by Jackson et al. (2008). Relative to this standard, our parameterization of the VBGM produced estimates that were above the mean value (75 percentile) for ages 1 and 2 but below the mean value (25 percentile) for ages older than 3 in all modeled reaches despite similar growth coefficient estimates ($K \approx 0.2$). Similarly, Jackson et al. (2008) reported a standard L_∞ estimate of 499 mm TL for Smallmouth Bass, which is greater than the upper bounds of the 95% CRI estimated for any reaches modeled here. Slower growth at older ages estimated in this study may be the result of Smallmouth Bass growth in a colder climate, where winter mortality is higher for smaller individuals and growth is slower (Beamesderfer and North 1995).

Diet is likely a contributing factor to the differences in estimated maximum length among sites, though other ecological mechanisms are also likely to be causal. Rennie et al. (2009) demonstrated that a suite of biological (e.g., prey availability, predator density) and environmental (e.g., temperature, habitat) factors must be considered to determine the impact of food web changes on resident fish. For example, population density constrains growth rates of individuals (Lorenzen and Enberg 2002; Dunlop et al. 2005). While we made no direct measure of Smallmouth Bass population densities in this study, both Kiraly et al. (2014) and Watson et al. (2018) observed the highest biomass caught per unit effort in the Orono reach with lower densities in the remaining two reaches. Furthermore, the Tidal reach may present conditions more favorable for larger fish (e.g., slower current velocities).

River herring populations in the Penobscot River have only recently rebounded following a major effort to restore river connectivity. It is important to note that we described the growth of Smallmouth Bass in the upper reaches (i.e., Orono, Piscataquis) when the adult river herring returns were beginning to rebound due to stocking efforts and increasing river connectivity. The year we sampled Smallmouth Bass diets (2015) was only the second year after dam removal. Adult river herring returns counted at the lowermost dam were above 100,000 individuals in 2015 (M. Simpson, Maine Department of Marine Resources, personal communication). However, that number more than doubled in 2016, and will likely continue to increase. By 2017 the river herring run exceeded 1.2 million adult fish upstream of Milford Dam. The carrying capacity of river herring for the Penobscot River is estimated to be more than 4 million returning adults (Opperman et al. 2011). Therefore, the changes in growth we anticipate to be associated with changing diets upriver would likely take several years to be detectable through length at age analysis in Smallmouth Bass.

Our bioenergetics models exhibit a positive relationship between river herring consumption and seasonal growth, and it is important to note that these models are limited in their ability to forecast growth under different diet scenarios, especially when accounting for seasonal prey availability (Chipps and Wahl 2008). Thus, results of our bioenergetics model should be interpreted within the context of our assumptions (constant consumption, prey energy density, metabolism). Because we estimated that consumption varied between reaches and metabolism likely varies with habitat and flow, these assumptions were not strictly met and our ability to accurately estimate future growth is limited. Predator–prey interactions are also inherently complex and there are several factors in this study that limit our ability to accurately predict Smallmouth Bass growth in the presence of increasing anadromous river herring availability. Though restoring connectivity for anadromous river herring can be a source of forage for resident fish (Mattocks et al. 2017), other studies have indicated that introduction may create interspecific competition with juvenile sportfish for planktonic prey (Hanson and Curry 2005; Creque and Czesny 2012). Although there are uncertainties about predator–prey interactions, we anticipate that recovering river herring populations will increase growth potential of Smallmouth Bass.

Acknowledgments

We would like to thank the editor and several reviewers for comments that improved this manuscript. We also thank L. Katz, G. Innes, D. Whitaker, and numerous students from the University of Maine for field and laboratory assistance. We thank R. Peabody for land access. This work is based upon research supported in part by the Penobscot River Restoration Trust and the University of Maine Department of Wildlife, Fisheries, and Conservation Biology. The U.S. Geological Survey Maine Cooperative Fish and Wildlife Research Unit provided logistical support. The views expressed herein are those of the author(s) and do not necessarily reflect the views of the Penobscot River Restoration Trust, or any of their members. This research was performed under subpermit from the

U.S. Fish and Wildlife Service Section 10(a)(1)(A) Permit #TE-697823, Maine Department of Inland Fish and Wildlife Scientific Collections Permit, and University of Maine approved Institutional Animal Care and Use Committee Protocol Number A2014–08–04. This is the Maine Agriculture and Forest Experiment Station Publication Number 3574, the University of Maine, Orono, Maine, USA. Any use of trade, firm, or product names is for descriptive purposes only and does not imply endorsement by the U.S. Government.

References

Adams, S. M., R. B. McLean, and M. M. Huffman. 1982a. Structuring of a predator population through temperature-mediated effects on prey availability. Canadian Journal of Fisheries and Aquatic Sciences 39:1175–1184.

Adams, S. M., R. B. Mclean, and J. A. Parrotta. 1982b. Energy partitioning in Largemouth Bass under conditions of seasonally fluctuating prey availability. Transactions of the American Fisheries Society 111:549–558.

Bajer, P. G., G. W. Whitledge, R. S. Hayward. 2004. Widespread consumption-dependent systematic error in fish bioenergetics models and its implications. Canadian Journal of Fisheries and Aquatic Sciences 61:2158–2167.

Beamesderfer, R., and J. North. 1995. Growth, natural mortality, and predicted response to fishing for Largemouth Bass and Smallmouth Bass populations in North America. North American Journal of Fisheries Management 15:688–704.

Boisclair, D., and W. C. Leggett. 1989. Among-population variability of fish growth: II. influence of prey type. Canadian Journal of Fisheries and Aquatic Sciences 46:468–482.

Brodersen, J., J. G. Howeth, and D. M. Post. 2015. Emergence of a novel prey life history promotes contemporary sympatric diversification in a top predator. Nature Communications 6:1–9.

Chipps, S. R., and D. H. Wahl. 2008. Bioenergetics modeling in the 21st century: reviewing new insights and revisiting old constraints. Transactions of the American Fisheries Society 137:298–313.

Creque, S. M., and S. J. Czesny. 2012. Diet overlap of non-native Alewife with native Yellow Perch and Spottail Shiner in nearshore waters of southwestern Lake Michigan, 2000–2007. Ecology of Freshwater Fish 21:207–221.

Cyterski, M., J. Ney, and M. Duval. 2002. Predator demand for clupeid prey in Smith Mountain Lake, Virginia. Fisheries Research 59:1–16.

Dunlop, E. S., J. A. Orendorff, B. J. Shuter, F. H. Rodd, and M. S. Ridgway. 2005. Diet and divergence of introduced Smallmouth Bass (*Micropterus dolomieu*) populations. Canadian Journal of Fisheries and Aquatic Sciences 62:1720–1732.

Fritts, A. L., and T. N. Pearsons. 2004. Smallmouth Bass predation on hatchery and wild salmonids in the Yakima River, Washington. Transactions of the American Fisheries Society 133:880–895.

Gahagan, B. I., J. C. Vokoun, G. W. Whitledge, and E. T. Schultz. 2012. Evaluation of otolith microchemistry for identifying natal origin of anadromous river herring in Connecticut. Marine and Coastal Fisheries: Dynamics, Management, and Ecosystem Science 4:358–372.

Glover, D. C., and D. R. DeVries. 2013. Growth of Largemouth Bass in a dynamic estuarine environment: an evaluation of the relative effects of salinity, diet, and temperature. Canadian Journal of Fisheries and Aquatic Sciences 501:485–501.

Grote, A. B., M. M. Bailey, J. D. Zydlewski, and J. E. Hightower. 2014. Multibeam sonar (DIDSON) assessment of American Shad (*Alosa sapidissima*) approaching a hydroelectric dam. Canadian Journal of Fisheries and Aquatic Sciences 71:545–558.

Hall, C. J., A. Jordaan, and M. G. Frisk. 2012. Centuries of anadromous forage fish loss: consequences for ecosystem connectivity and productivity. BioScience 62:723–731.

Hansen, M. J., T. D. Beard Jr., and D. B. Hayes. 2007. Sampling and experimental design. Pages 51–120 *in* C. S. Guy and M. L. Brown, editors. Analysis and interpretation of freshwater fisheries data. American Fisheries Society, Bethesda, Maryland.

Hanson, P. C., T. B. Johnson, D. E. Schindler, and J. Kitchell. 1997. Fish Bioenergetics 3.0. Technical report WISCU-T- 97–001. Madison, Wisconsin.

Hanson, S. D., and R. A. Curry. 2005. Effects of size structure on trophic interactions between age-0 Smallmouth Bass and juvenile anadromous Alewives. Transactions of the American Fisheries Society 134:356–368.

He, J., and J. Bence. 2007. Modeling annual growth variation using hierarchical Bayesian and the von Bertalanffy growth function, with application to Lake Trout in southern Lake Huron. Transactions of the American Fisheries Society 136:318–330.

Howeth, J. G., J. J. Weis, J. Brodersen, E. C. Hatton, and D. M. Post. 2013. Intraspecific phenotypic variation in a fish predator affects multitrophic lake metacommunity structure. Ecology and Evolution 3:5031–5044.

Iafrate, J., and K. Oliveira. 2008. Factors affecting migration patterns of juvenile river herring in a coastal Massachusetts stream. Environmental Biology of Fishes 81:101–110.

Jackson, D. A. 2002. Ecological effects of *Micropterus* introductions: the dark side of black bass. Pages 221–232 *in* D. P. Phillip and M. S. Ridgway, editors. Black bass: ecology, conservation, and management. American Fisheries Society, Symposium 31, Bethesda, Maryland.

Jackson, Z. J., M. C. Quist, and J. G. Larscheid. 2008. Growth standards for nine North American fish species. Fisheries Management and Ecology 15:107–118.

Kiraly, I. A., S. M. Coghlan Jr., J. Zydlewski, and D. Hayes. 2014. An assessment of fish assemblage structure in a large river. River Research and Applications 31(3). DOI: 10.1002/rra2738.

Kéry, M. 2010. Introduction to WinBUGS for ecologists: A Bayesian approach to regression, ANOVA, mixed models and related analyses. Academic Press, Upper Saddle River, New Jersey.

Kruschke, J. K. 2011. Doing Bayesian data analysis: a tutorial with R and BUGS. Academic Press, San Diego, California.

Limburg, K. E. 1998. Anomalous migrations of anadromous herrings revealed with natural chemical tracers. Canadian Journal of Fisheries and Aquatic Sciences 55:431–437.

Long, J. M., and W. L. Fisher. 2011. Precision and bias of Largemouth, Smallmouth, and Spotted Bass ages estimated from scales, whole otoliths, and sectioned otoliths. North American Journal of Fisheries Management 21:636–645.

Lorenzen, K., and K. Enberg. 2002. Density-dependent growth as a key mechanism in the regulation of fish populations: evidence from among-population comparisons. Proceedings of the The Royal Society B 269:49–54.

MacAvoy, S. E., S. A. Macko, S. P. McIninch, and G. C. Garman. 2000. Marine nutrient contributions to freshwater apex predators. Oecologia 122:568–573.

Martin, N. V. 1970. Long-term effects of diet on the biology of the Lake Trout and the fishery in Lake Opeongo, Ontario. Journal of the Fisheries Research Board of Canada 27(69):125–146.

Mattocks, S., C. J. Hall, and A. Jordaan. 2017. Damming, lost connectivity, and the historical role of anadromous fish in freshwater ecosystem dynamics. BioScience 67:713–728.

Michaletz, P. H. 1997. Influence of abundance and size of age-0 Gizzard Shad on predator diets, diet overlap, and growth. Transactions of the American Fisheries Society 126:101–111.

Opperman, J. J., J. Royte, J. Banks, L. R. Day, and C. Apse. 2011. The Penobscot River, Maine, USA: a basin-scale approach to balancing power generation and ecosystem restoration. Ecology and Society 16(3):7.

Pazzia, I., M. Trudel, M. Ridgway, and J. B. Rasmussen. 2002. Influence of food web structure on the growth and bioenergetics of Lake Trout (*Salvelinus namaycush*). Canadian Journal of Fisheries and Aquatic Sciences 59:1593–1605.

Plummer, M. 2003. JAGS: A program for analysis of Bayesian graphical models using Gibbs sampling. *In* K. Hornik, F. Leisch, and A. Zeileis, editors. Proceedings of the 3rd International Workshop on Distributed Statistical Computing (DSC 2003), March 20–22, Vienna, Austria.

Post, D. M., E. P. Palkovacs, E. G. Schielke, and S. I. Dodson. 2008. Intraspecific variation in a predator affects community structure and cascading trophic interactions. Ecology 89:2019–2032.

Probst, W. E., C. F. Rabeni, W. G. Covington, and R. E. Marteney. 1984. Resource use by stream-dwelling Rock Bass and Smallmouth Bass. Transactions of the American Fisheries Society 113:283–294.

R Core Team. 2016. R: A language and environment for statistical computing. R Foundation for Statistical Computing, Vienna, Austria. URL: https://www.R-project.org/. (September 2018).

Rand, P. S., B. F. Lantry, R. O'Gorman, R. W. Owens, and D. J. Stewart. 1994. Energy density and size of pelagic prey fishes in Lake Ontario, 1978–1990: implications for Salmonine energetics. Transactions of the American Fisheries Society 123:519–534.

Rennie, M. D., W. G. Sprules, and T. B. Johnson. 2009. Factors affecting the growth and condition of Lake Whitefish (*Coregonus clupeaformis*). Canadian Journal of Aquatic Sciences 66:2096–2108.

Reynolds, W. W., and M. E. Casterlin. 1976. Activity rhythms and light intensity preferences of *Micropterus salmoides* and *M. dolomieui*. Transactions of the American Fisheries Society 105:400–403.

Roell, M. J., and D. J. Orth. 1993. Trophic basis of production of stream-dwelling Smallmouth Bass, Rock Bass, and Flathead Catfish in relation to invertebrate bait harvest. Transactions of the American Fisheries Society 122:46–62.

Saunders, R., M. A. Hachey, and C. W. Fay. 2006. Maine's diadromous fish community: past, present, and implications for Atlantic Salmon recovery. Fisheries 31:537–547.

Savitz, J. 2009. Diets of Lake Michigan salmon and maximum size of Alewife prey. Journal of Freshwater Ecology 24:563–566.

Schneider, C. A., W. S. Rasband, and K. W. Eliceiri. 2012. NIH Image to ImageJ: 25 years of image analysis. Nature Methods 9:671–675.

Shuter, B. J., H. C. Giacomini, D. De Kerckhove, and K. Vascotto. 2016. Fish life history dynamics: shifts in prey size structure evoke shifts in predator maturation traits. Canadian Journal of Fisheries and Aquatic Sciences 73:693–708.

Shuter, B. J., and J. R. Post. 1990. Climate, population viability, and the zoogeography of temperate fishes. Transactions of the American Fisheries Society 119:314–336.

Stewart, D. J., and M. Ibarra. 1991. Predation and production by Salmonine fishes in Lake Michigan, 1978–88. Canadian Journal of Fisheries and Aquatic Sciences 48:909–922.

Su, Y., and M. Yajima. 2015. R2jags: Using R to run 'JAGS'. R package version 0.5–7. Available: https://CRAN.R-project.org/package=R2jags. (September 2018).

Tabor, R. A., B. A. Footen, K. L. Fresh, M. T. Celedonia, F. Mejia, D. L. Low, and L. Park. 2007. Smallmouth Bass and Largemouth Bass predation on juvenile Chinook Salmon and other Salmonids in the Lake Washington basin. North American Journal of Fisheries Management 27:1174–1188.

Trippel, N. A., M. S. Allen, and R. S. McBride. 2015. Importance of resident and seasonally transient prey to Largemouth Bass in the St. Johns River, Florida. Transactions of the American Fisheries Society 144:140–149.

Turner, S. M., and K. E. Limburg. 2016. Juvenile river herring habitat use and marine emigration trends: comparing populations. Oecologia 180:77–89.

Twining, C. W., D. C. West, and D. M. Post. 2013. Historical changes in nutrient inputs from humans and anadromous fishes in New England's coastal watersheds. Limnology and Oceanography 58:1286–1300.

USGS (U.S. Geological Survey). 2016a. National Water Information System data available on the World Wide Web (USGS Water Data for the Nation). Available: http://waterdata.usgs.gov/nwis/uv?01031500. (September 2018).

USGS (U.S. Geological Survey). 2016b. National Water Information System data available on the World Wide Web (USGS Water Data for the Nation). Available: http://waterdata.usgs.gov/nwis/uv?01036390. (September 2018).

Vigliola, L., and M. G. Meekan. 2009. The back-calculation of fish growth from otoliths. Pages

174–211 *in* B. S. Green, B. D. Mapstone, G. Carlos, and G. A. Begg, editors. Tropical fish otoliths: information for assessment, management and ecology. Springer, New York.

Vischer, N. and S. Nastase. 2015. ObjectJ: Non-destructive marking and linked results. Version 1.03s. Available: http://sils.fnwi.uva.nl/bcb/objectj/. (September 2018).

von Bertalanffy, L. 1938. A quantitative theory of organic growth (inquiries on growth laws II). Human Biology 10(2):181–213.

Warner, K. 2005. Smallmouth Bass introductions in Maine. Fisheries 30(11):20–26.

Watson, J. M., S. M. Coghlan, J. Zydlewski, D. B. Hayes, I. A. Kiraly. 2018. Dam removal and fish passage improvement influence fish assemblages in the Penobscot River, Maine. Transaction of the American Fisheries Society 147:525–540.

Weidel, B. C., D. C. Josephson, and C. E. Kraft. 2007. Littoral fish community response to Smallmouth Bass removal from an Adirondack lake. Transactions of the American Fisheries Society 136:778–789.

Werner, R. G. 2004. Freshwater fishes of the northeastern United States. Syracuse University Press, Syracuse, New York.

Whitledge, G. W., R. S. Hayward, R. D. Zweifel, and C. F. Rabeni. 2003. Development and laboratory evaluation of a bioenergetics model for subadult and adult Smallmouth Bass. Transactions of the American Fisheries Society 132:316–325.

Whittier, T. R., and T. M. Kincaid. 1999. Introduced fish in northeastern USA lakes: regional extent, dominance, and effect on native species richness. Transactions of the American Fisheries Society 128:769–783.

Willis, T. V. 2009. How policy, politics, and science shaped a 25-year conflict over Alewife in the St. Croix River, New Brunswick-Maine. Pages 793–811 *in* A. J. Haro, K. L. Smith, R. A. Rulifson, C. M. Moffitt, R. J. Klauda, M. J. Dadswell, R. A. Cunjack, J. E. Cooper, K. L. Beal, and T. S. Avery, editors. Challenges for diadromous fishes in a dynamic global environment. American Fisheries Society, Symposium 69. Bethesda, Maryland

Yako, L. A., M. E. Mather, and F. Juanes. 2000. Assessing the contribution of anadromous herring to Largemouth Bass growth. Transactions of the American Fisheries Society 129:77–88.

Yako, L. A., M. E. Mather, and F. Juanes. 2002. Mechanisms for migration of anadromous herring: an ecological basis for effective conservation. Ecological Applications 12:521–534.

American Fisheries Society Symposium 87:93–110, 2019

Changes in Body Condition and Diet of Lotic Smallmouth Bass across Two Flow Regimes during Summer Months at the Southern Extent of Their Native Range

Christopher R. Middaugh*
Arkansas Cooperative Fish & Wildlife Research Unit
Department of Biological Sciences, Fayetteville, Arkansas 72701, USA

Daniel D. Magoulick
U.S. Geological Survey, Arkansas Cooperative Fish & Wildlife Research Unit
Department of Biological Sciences, University of Arkansas
Fayetteville, Arkansas 72701, USA

Abstract.—The Ozark Plateau is located at the southern extent of native Smallmouth Bass *Micropterus dolomieu* range and water temperature and drought conditions during summer months may potentially affect growth of Smallmouth Bass in this region. Groundwater streams in the region do not warm to the same extent as runoff streams during summer months and could provide a thermal refuge habitat for Smallmouth Bass from high summer temperatures and drought conditions. Our study objective was to examine differences in body condition and diet of Smallmouth Bass through summer months between groundwater and runoff streams. We sampled Smallmouth Bass from eight streams across two flow regimes monthly from June–September from 2014 to 2016 in the Ozark Plateau of Arkansas and Missouri. Relative weights were calculated and diet contents were examined for each fish. Linear mixed model analyses indicated that relative weights declined in both stream types in 2014 and 2015, but not in 2016. Surprisingly, there was no significant difference in change in relative weights between runoff and groundwater streams in any year. No diet shifts over the course of the summer were noted in any year, and no differences were seen between

*Corresponding author: Christopher.middaugh@agfc.ar.gov
Current address: Research, Evaluation, and Compliance Division, Arkansas Game and Fish Commission, Little Rock, Arkansas 72205, USA

stream types. Our results suggest that further work should investigate the refuge qualities of groundwater streams for Smallmouth Bass in this region as Smallmouth Bass from both stream types may currently respond similarly to summer conditions.

Introduction

Flow regime, a measure of hydrologic classification, is a critical determinant of ecological processes within a stream (Olden et al. 2012). It can structure the fish assemblages present (Marchetti and Moyle 2001) and species abundances (Jowett et al. 2005). Flow regime could also affect the response of fish populations to climate change (Wenger et al. 2011). Because of the ecological importance of the flow regime, many streams across the United States have been classified according to their flow regimes. For example, in the Ozark-Ouachita Interior Highlands streams have been classified into three broad groups, groundwater, runoff, and intermittent (Leasure et al. 2016). Groundwater streams typically are more thermally regulated than runoff and intermittent streams with warmer temperatures in the winter and cooler temperatures during summer months (Whitledge et al. 2006; Middaugh et al. 2018). In the Ozark Plateau (hereafter Ozarks), many runoff and intermittent streams dry to a series of isolated pools during summer months (Homan et al. 2005; Dauwalter and Fisher 2008b; Hafs et al. 2010), whereas groundwater streams typically maintain a more stable base flow. These differences in physical habitat conditions could lead to seasonal effects on fishes within the streams.

Water temperature and flow variability differ among flow regimes and both structure many aspects of lotic systems. Water temperature affects growth and survival of fishes (Magnuson et al. 1979; Christie and Regier 1988) as well as structuring distributions (e.g., Smallmouth Bass *Micropterus dolomieu*, Shuter et al. 1980). Similarly, flow related disturbance events, such as drought, can affect assemblage composition (Resh et al. 1988). Pool isolation caused by drought could potentially lead to increased competition (Zaret and Rand 1971) and increased susceptibility to predation (Harvey and Stewart 1991) for fishes as biotic and abiotic conditions can be severe in these pools (Magoulick and Kobza 2003). In a review, Walters (2016) found that 65% of studies on lotic fish body condition response to low flow events showed a decline in body condition (Walters 2016).

Smallmouth Bass are an economically and ecologically important species and are locally abundant throughout the Ozarks (Robison and Buchanan 1988). However, this region is located at the southern extent of Smallmouth Bass native range and growth may be reduced in some systems during summer months as water temperatures exceed optimal levels (Zweifel et al. 1999; Middaugh et al. 2018). Climate change could further affect growth as air temperatures warm through the coming century, and many lotic systems in the United States are already warming above historical temperatures (Kaushal et al. 2010). In addition to increased water temperatures, climate change could affect drought conditions by prolonging the duration and increasing the severity of seasonal droughts (Girvetz et al. 2009). Changing temperatures and discharge patterns could favor competitor species such as Largemouth Bass *Micropterus salmoides* over Smallmouth Bass in this region (Sowa and Rabeni 1995), but groundwater streams could provide a thermal refuge to Smallmouth Bass if temperatures in runoff streams warm past habitable conditions.

We utilize a measure of body condition, relative weight, to evaluate differences in Smallmouth Bass over the course of the summer within and among streams. Relative weight is commonly used by fisheries biologists as it is relatively easy to measure and useful in describing population condition (Blackwell et al. 2000). Relative weights have been used to evaluate prey abundance, to make management decisions, and to compare condition across populations (Blackwell et al. 2000). Relative weights and other condition factors have been used to effectively measure seasonal and inter-annual change in body condition for a variety of fishes (e.g., Atlantic Cod *Gadus morhua*, Lambert and Dutil 1997; Barbel *Barbus sclateri*, Encina and Granado-Lorencio 1997; Striped Bass *Morone saxatilis*, Smith 2012; Yellow Perch *Perca flavescens*, Staton et al. 2014). Lotic Smallmouth Bass relative weight can vary by season (Orth et al. 1983; Austen and Orth 1988; Dauwalter and Fisher 2008b). Two studies from Oklahoma found that relative weights were highest in the summer compared to other seasons (Austen and Orth 1988; Dauwalter and Fisher 2008b). In contrast, another Oklahoma based study found low relative weights during late summer and early fall (Orth et al. 1983). None of these studies incorporated multiple streams across flow regimes into investigations of Smallmouth Bass body condition.

In a review of North American inland fish climate change literature, Lynch et al. (2016) recommend investigating sources of resilience for fishes. Our study compares Smallmouth Bass body condition and diets across streams from two flow regimes in order to examine whether groundwater streams show greater refuge potential than runoff streams. We hypothesized 1) that Smallmouth Bass body condition would decline in runoff streams but not groundwater streams over the course of summer and 2) that Smallmouth Bass diet would shift in runoff streams but not groundwater streams over the course of summer. We expected to find differences between stream types because of differing hydrologic patterns over the course of summer and differing thermal conditions within both stream types. Groundwater streams have a more stable flow and typically a higher base flow than runoff streams (Leasure et al. 2016). In this region, many runoff streams have no flow during late summer and fall each year (Leasure et al. 2016). Low flow conditions could lead to sub-optimal conditions and higher competition in pool habitats affecting available prey density and type (Zaret and Rand 1971; Magoulick and Kobza 2003). Runoff streams typically warm to higher temperatures than groundwater streams and are expected to warm more in response to climate change, potentially affecting thermally sensitive fishes more than in groundwater streams (Carlson et al. 2017a; Carlson et al. 2017b; Middaugh et al. 2018). In the Ozarks of Arkansas and Missouri, runoff streams often exceed optimal thermal conditions for Smallmouth Bass whereas groundwater streams stay much closer to optimal temperatures for growth (22°C, Zweifel et al. 1999; Middaugh et al. 2018). We expected these flow and temperature differences between stream types to lead to differences in body condition over the course of the summer.

Methods

Study Site

During June–September 2014, 2015, and 2016 we collected Smallmouth Bass from streams in the Ozarks of northern Arkansas and southern Missouri (Figure 1) that had previously been classified into runoff and groundwater flow regimes (Leasure et al. 2016). These flow regime classifications were developed from 10 hydrologic variables and provide a classification of streams into three broad categories, groundwater, runoff,

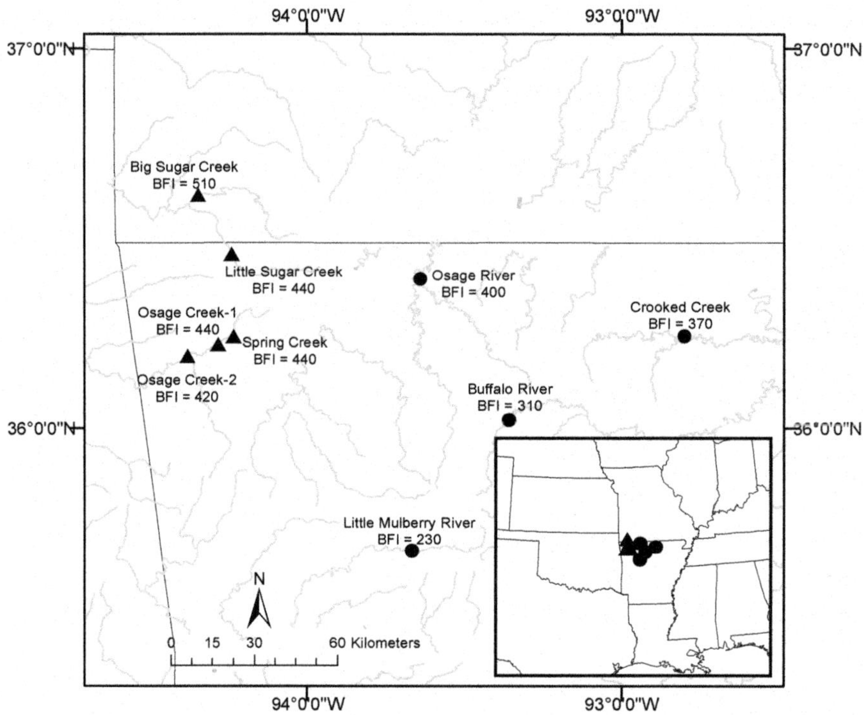

Figure 1. Map of study sites across the Ozarks region of Arkansas and Missouri. Triangles represent groundwater streams and circles represent runoff streams. Base flow index (BFI) values are show by stream names.

and intermittent with subcategories within each (Leasure et al. 2016).

We selected stream sites from each stream type based on preliminary sampling where we were able to effectively capture adequate numbers of adult Smallmouth Bass. Though Crooked Creek is classified as groundwater flashy by Leasure et al. (2016), many of the characteristics (summer drying, relatively large daily fluctuations in water temperature) are more similar to runoff streams and thus we grouped it into the runoff category. Two sites were on the same stream (Osage Creek-1 and Osage Creek-2) but were located ~13 km apart. Another site (Spring Creek) was located on a tributary of Osage Creek-1, and was located ~5.6 km away. All sites were either third or fourth Strahler stream order (Strahler 1957; third order: Buffalo River, Little Mulberry River, Osage River, and Spring Creek; fourth order: Crooked Creek, Big Sugar Creek, Osage Creek-1, Osage Creek-2, and Little Sugar Creek). Groundwater sites were located within two distinct hydrologic unit code (HUC) 8-digit watersheds and runoff sites were spread across four HUC-8 watersheds. We attempted to spread sample sites apart as much as possible, but were limited by the availability of regional groundwater sites where we were able to effectively capture Smallmouth Bass.

Fish Collection and Sample Processing

Our stream sites were too deep to effectively backpack electroshock or seine and too shallow for a motorized boat which restricted our sampling options. We determined that the most effective way to capture fish was through hook and line sampling. At each collection event, we used hook and line sampling to capture Smallmouth Bass using soft-plastic baits fished with spinning rods. We captured a small minority of fish on minnow imitation artificial baits or crayfish imitation patterns fished with a fly-rod. We collected Smallmouth Bass twice in June 2014, and monthly thereafter, typically collecting fish in the first two weeks of each month. Upon arriving at a site, we typically sampled upstream, fishing in all habitat units (i.e., pool, riffle, and run). We attempted to capture approximately ten fish on each sampling occasion and the stream distance sampled depended on the rate of capture of fish, but ranged from approximately 0.4 km to 2.4 km. Sites where samples were not taken in either August or September were excluded from all analyses. We excluded several samples from relative weight analyses due to few fish collected (<6) and several samples were missed due to high water or other logistical constraints (Table 1). One site was sampled in 2014–2015 (Osage Creek-2) but not in 2016 because of access limitation, and so we added an additional site (Big Sugar Creek) in 2016 (Table 1).

After each fish was captured, we recorded total length (to the nearest mm) and wet weight (to the nearest 0.1 g). We used these measurements to calculate relative weight (W_r) for each fish based on the equation proposed by Kolander et al. (1993). The relationship is valid for fish >150 mm total length and we excluded any fish shorter than this length. After measuring weight and length, we collected the diet contents of each fish using gastric lavage (Kamler and Pope 2001). We performed gastric lavage by inserting a flexible plastic tube of varying diameter based on fish size through the mouth and into the stomach of the fish. We used a 1 L laboratory wash bottle with a long nozzle to flush water into the stomach. We caught diet contents in a fine mesh (1 mm) net as they exited the stomach. After washing the stomach, we checked the fish's stomach by inserting a finger through the mouth into the stomach to feel for additional contents which would subsequently be washed out with the wash bottle. In the event of difficulty washing a diet item out of the stomach (e.g., large crayfish), we used long forceps to remove the diet item. We chose to use this simplistic form of gastric lavage rather than a more complicated washing mechanism (e.g., Light et al. 1983; Van Den Avyle 1980) due to our need for light-weight and nonbulky sampling equipment. After collecting the diet, fish were returned alive to the stream.

Diet contents were stored in 70% ethanol in the field and then examined later in the laboratory. We classified diet contents of each fish into the broad categories of crayfish, fish, invertebrate, and an "other" category because the heavily digested state of many samples prevented species level identification and our limited sample size prevented fine-scale analysis of the data. We measured wet weight to the nearest 0.001 g of each category of diet item for each fish and noted the number of individual diet items (e.g., number of crayfish). Diet contents were then dried in a 70°C oven for 48 h and dry weight to the nearest 0.001 g was measured for each category of diet item for each fish. For each site, we calculated the mean proportional dry mass for each broad prey category for subsequent analyses.

Discharge and Temperature Data Collection

We collected data on historic discharge levels (1950–present) and air temperatures (1950–present) within our study region in order to compare our study time period to average conditions. We plotted discharge in the King's river (USGS gauge number 07050500) as a representative runoff stream for the region because none of our runoff sites had historic gauge data and this stream is in close proximity to our sample sites and is similar in terms of characteristics

Table 1. Numbers of Smallmouth Bass >150 mm captured at each sampling event. Numbers in parenthesis are number of fish with diet contents. Values noted with an * indicate samples not included in relative weight analyses because of <6 fish collected or failure to sample that site in late summer (August or September). Sampling events noted with n/a indicate missed samples due to high water, site access restriction, or other logistical failure to sample the site during that period.

	2014					2015				2016			
	June	Mid-June	July	Aug.	Sept.	June	July	Aug.	Sept.	June	July	Aug.	Sept.
Buffalo River	9 (8)	10 (6)	14 (9)	9 (9)	14 (2)	11 (7)	12 (5)	11 (8)	10 (9)	11 (7)	10 (7)	11 (4)	12 (8)
Crooked Creek	10 (8)	8 (6)	10 (9)	12 (9)	11 (2)	11 (7)	12 (11)	10 (9)	10 (2)	12 (7)	8 (5)	n/a	12 (8)
Little Mulberry	n/a	11 (9)	11 (6)	n/a	13 (9)	10 (9)	12 (9)	11 (8)	10 (5)	13 (10)	13 (3)	12 (5)	12 (6)
Osage River	13 (10)*	n/a	11(6)*	n/a	n/a	n/a	12 (8)	10 (5)	12 (9)	10 (4)	8 (4)	11 (10)	10 (6)
Big Sugar Creek	n/a	n/a	n/a	n/a	n/a	n/a	n/a	n/a	n/a	12 (8)	12 (7)	11 (8)	10 (5)
Osage Creek-1	n/a	n/a	13 (8)*	n/a	4 (3)*	12 (6)	12 (6)	11 (8)	11 (8)	10 (9)	12 (8)	10 (8)	13 (9)
Osage Creek-2	8 (4)*	8 (6)*	7 (5)*	n/a	n/a	15 (13)	n/a	n/a	10 (6)	n/a	n/a	n/a	n/a
Spring Creek	10 (6)	6 (6)	14 (10)	7 (4)	n/a	9 (6)	12 (7)	10 (6)	n/a	11 (9)	n/a	11 (8)	10 (7)
Little Sugar Creek	8 (5)	9 (4)	13 (7)	9 (6)	n/a	n/a	10 (4)	12 (7)	n/a	11 (6)	10 (7)	10 (7)	10 (7)

and magnitude. We used site Osage Creek-1 (USGS gauge number 07195000) as a representative groundwater stream for the region in terms of discharge. We collected air temperature data from current and historic NOAA NCDC climate data for Drake Air Field (GHCND:USW00093993) in Fayetteville, Arkansas as this was relatively centrally located to all sample sites. We visually compared mean monthly historic high air temperatures and number of days with air temperature hotter than an upper thermal climate threshold previously identified as limiting the presence of Smallmouth Bass (>31°C; Bouska et al. 2015) with air temperatures collected by the same weather station during the study period.

Data Analyses

We assessed change in relative weight over time using linear mixed effects models with the program lme4 (Bates et al. 2015) in program R (R Core Team 2015). This type of analysis has been previously used to examine change in relative weight over time (e.g., Dauwalter and Fisher 2008b). We first created three mixed models (one for each year) to test for a change in relative weight by month. The response variable was relative weight and we used the variable month and the interaction between month and stream type as fixed effects. We included the month effect of site as a random effect with random slopes and intercepts for each site. The random slopes and intercepts for each site accounted for variation among sites and allowed each site to vary independent of other sites. We calculated *P*-values to test for a significant effect of month using likelihood ratio tests of the full model against reduced models. We next used the same mixed models as described above to test for an effect of stream type by calculating *P*-values of the full model against reduced models.

We created one additional mixed model to test for a significant difference in change in relative weight among years. Using all three years of combined data, the response variable was relative weight and we used month as a fixed effect. We included the month effect of site and the month effect of year as random effects with random slopes and intercepts used for each random effect to account for variation among sites and years. We calculated a *P*-value to test for a significant effect of year using likelihood ratio tests of the full model against a reduced model. Residual plots were examined for all models and no deviations from normality or homoscedasticity were noted.

Smallmouth Bass diets were analyzed in two ways. First, we tested for differences in diets among months using a permutational multivariate analysis of variance (PERMANOVA) for each year. The response variables were the proportion of each prey group by dry weight, and the predictor variables were month, stream type, and the interaction of month and stream type. We included site as a stratifying variable to account for nestedness of sites within stream type. This causes randomizations in the PERMANOVA calculations to occur within each site and not across all sites (Oksanen et al. 2017). We used Bray-Curtis distance measures and 999 permutations for each test. We used the VEGAN package in R to conduct these analyses (Oksanen et al. 2017). We next examined whether the proportion of fish with empty stomach contents was related to month using logistic regression. Six logistic regression models were created, one for each stream type within each year. Chi-square ANOVA tables were used to determine overall significance of final logistic regression models. All analyses were conducted in program R.

Results

We collected a total of 828 Smallmouth Bass over the course of three years across eight sites. We collected 218 Smallmouth Bass in

2014, 288 in 2015, and 322 in 2016 (Table 1). Mean total length of fish did not vary greatly among months or among sites (mean = 264.9 mm, range = 154–448 mm runoff streams; mean = 268.6, range = 152–403 mm groundwater streams). We found significant declines in Smallmouth Bass relative weight over the course of summer in 2014 and 2015, but not in 2016 (2014 X^2 = 13.23, P < 0.001; 2015 X^2 = 27.15, P < 0.001; 2016 X^2 = 1.25, P = 0.26). These analyses examine changes in all streams combined within each year. When differences between stream types were analyzed, no significant differences were found between runoff and groundwater streams in any year (2014 X^2 = 0.29, P = 0.59; 2015 X^2 = 3.56, P = 0.06; 2016 X^2 = 0.57, P = 0.45). This indicates that in 2014 and 2015 both groundwater and runoff streams declined in relative weight similarly, but in 2016 neither stream type declined significantly in relative weight (Figure 2). Two runoff streams, the Buffalo River and the Little Mulberry River, were consistent in showing declining relative weights over summer in all three years, though other sites did not decline in relative weight in 2016. Though we found a significant decline in relative weights from smallmouth bass collected in 2014 and 2015 but not in 2016, when data from all years were combined there were no significant differences in change in relative weight over the course of summer among years (X^2 = 6.78, P = 0.08).

We collected diets from 137 fish in 2014, 188 fish in 2015, and 207 fish in 2016 (Table 1). Logistic regression model results indicated no significant effect of month on proportion of empty stomachs for all but one year and stream type (chi-square: 2014 groundwater P = 0.63; 2015 groundwater P = 0.37; 2015 runoff P = 0.62; 2016 groundwater P = 0.41; 2016 runoff P = 0.40). The only significant model was 2014 runoff (chi-square P= 0.008). The only significant month coefficient within that model was September (z = –2.85, P = 0.004) and the proportion of empty stomachs was higher in this month than in previous months (Figure 3).

No significant effects of month on diet composition were found in any year (PERMANOVA: 2014 F_4 = 0.58, P = 0.82, 2015 F_3 = 0.85, P = 0.49; 2016 F_3 = 1.04, P = 0.40). Similarly, no significant effects of stream type were found in any year (PERMANOVA: 2014 F_1 = 0.47, P = 0.80; 2015 F_1 = 2.90, P = 0.30; 2016 F_1 = 1.35, P = 0.38). Smallmouth Bass consumed mostly crayfish and fish prey during the majority of months across both stream types (Figure 4). Crayfish prey was often the highest proportion of the diet by weight. Invertebrate and other prey were typically a small part of the diet, but in a few months invertebrate prey substantially contributed to the total proportion of diets (Figure 4).

About half of the months in the three years that we sampled fell within typical ranges for mean monthly discharge levels in both groundwater and runoff streams in this region (Figure 5). After comparing our sample period to historical averages, we found that the King's River had above average discharge levels in June and July 2015, but in August and September discharge fell to near normal levels. Similarly, 2014 had high discharge in June, but more typical levels in later months. In contrast, 2016 had low- to average discharge levels in early summer, but then above average levels in August and September. Discharge at Osage Creek-1 was typically above average, but much less variable among years than in the runoff stream. Air temperatures were typically within expected values based on historical data (Figure 6). In general, 2016 was slightly warmer and had lower discharge levels in early summer compared to 2014 and 2015, but both 2014 and 2015 had lower discharge levels later in the summer.

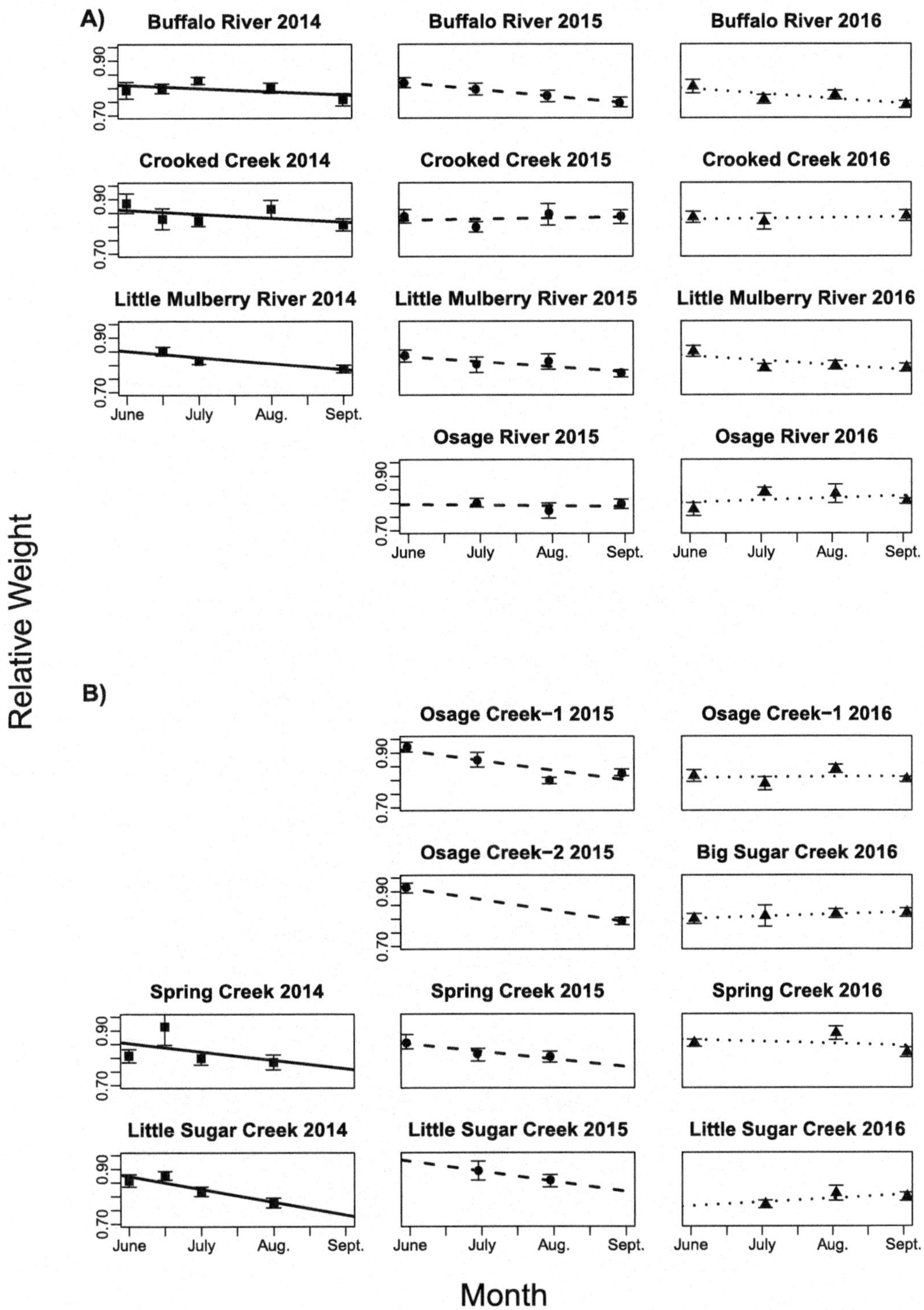

Figure 2. Change in mean monthly Smallmouth Bass relative weight over summer months for each study site in each year. Runoff streams are shown in A) and groundwater in B). Error bars show standard error.

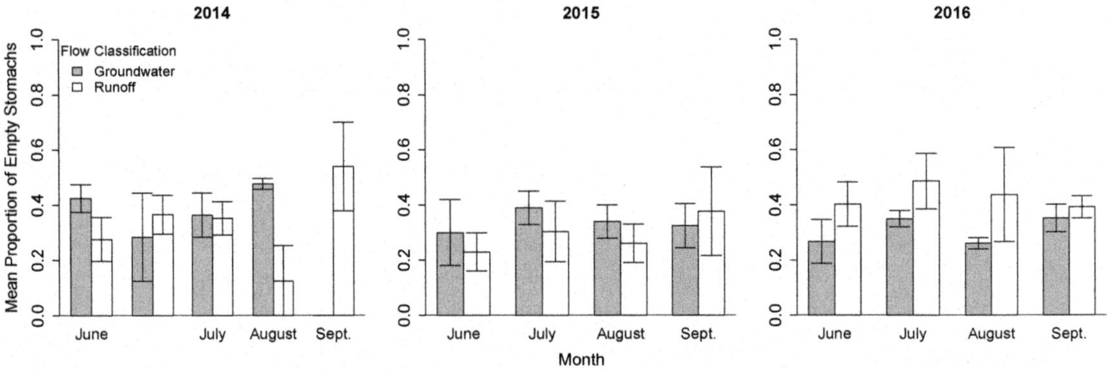

Figure 3. Proportion of collected Smallmouth Bass with no diet contents in both groundwater (dark bars) and runoff (white bars) sites in each study year. Error bars show standard error.

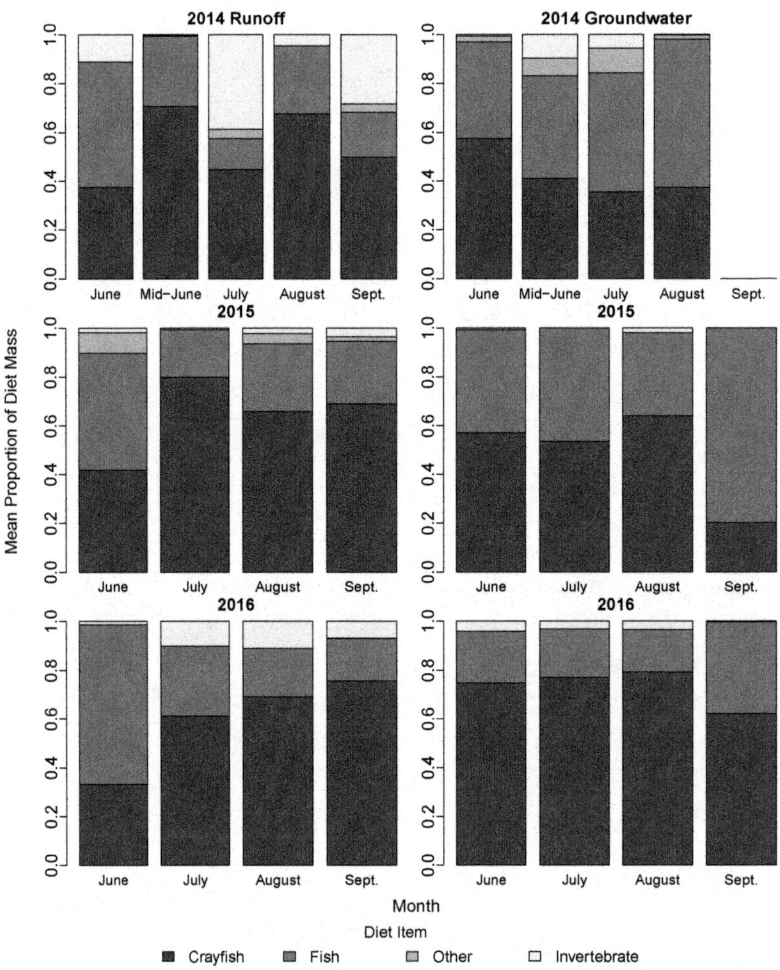

Figure 4. Monthly diet composition of Smallmouth Bass in runoff and groundwater streams each year. Diets are shown by prey type and are based on dry mass of prey. No significant trends were found for changes over the course of the summer or between stream types.

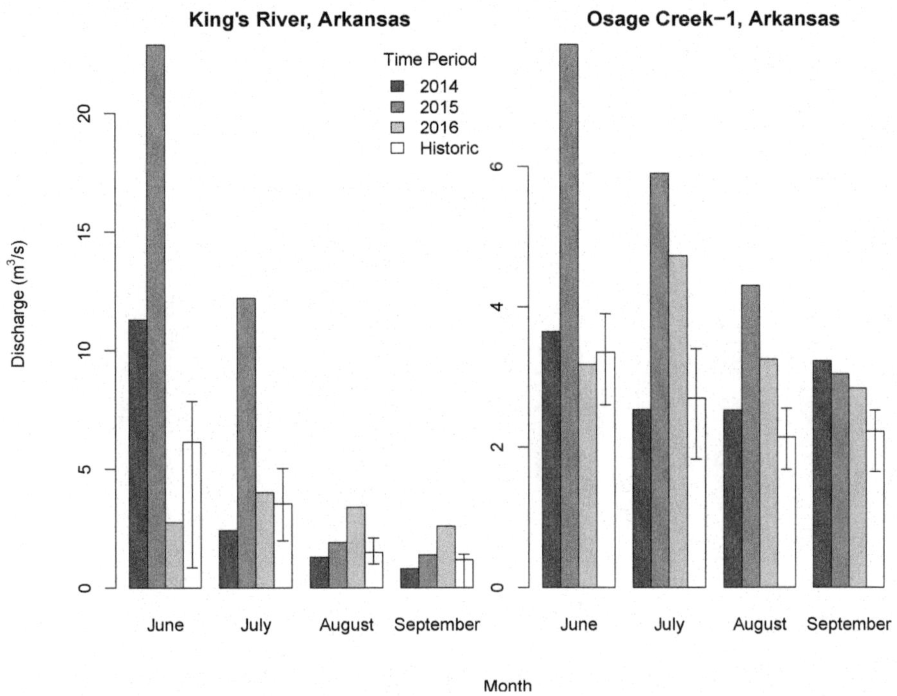

Figure 5. Observed and historical median monthly discharge values for a runoff stream (King's River, Arkansas; USGS gage number 07050500) and a groundwater stream (Osage Creek-1, Arkansas; USGS gage number 07195000) in the sample region. Historical values span from 1950–2016. Error bars show 95% confidence intervals of the median values calculated using a bootstrap procedure with 1,000 replicates.

Discussion

We found a significant decline in relative weight of Smallmouth Bass over the course of the summer for two out of three years and no differences in relative weight change over the course of the summer between groundwater and runoff streams in any year. Our results were surprising as we hypothesized that we would observe a decline in relative weight within runoff streams, but not in groundwater streams. Dauwalter and Fisher (2008b) examined relative weights in a runoff and groundwater stream in Oklahoma across a range of orders (order 3–5; upstream to downstream) and found differences in relative weights at the third order portion of the stream (with the groundwater stream having higher mean relative weights), but not in higher order (order 4–5) locations. Our streams were all either third or fourth order. Though we did not monitor water temperatures at each site, we assume that there were differences in water temperatures between groundwater and runoff streams as is commonly the case in this region (Whitledge et al. 2006; Middaugh et al. 2018). However, it is possible that high water levels during some months could prevent differences in temperatures between stream types from occurring (e.g., June and July 2015). Although within long-term average temperature and discharge levels, these three study years were all relatively mild compared to other recent years. We anticipate if a year of more severe drought or warmer temperatures were experienced, it could lead to differences in relative weight between flow regimes. Further research will be required to address this issue.

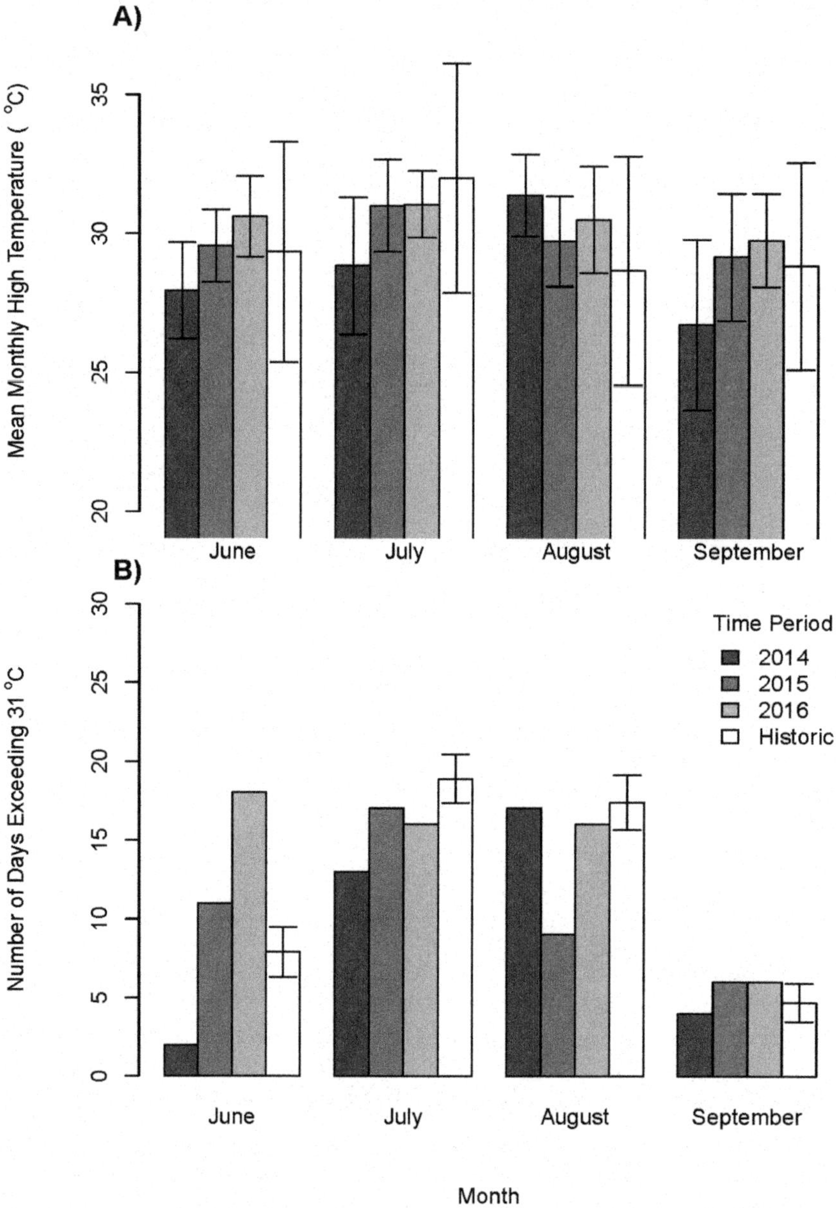

Figure 6. Observed and historical mean monthly high temperatures during summer months at Fayetteville, Arkansas at Drake Air Field (June–September; A) and observed and historical mean monthly number of days exceeding 31°C (B). Historical values span from 1950–2016. Error bars show 95% confidence intervals.

Our relative weight values were similar to those seen for lotic Smallmouth Bass from a range of locations (Oklahoma, Orth et al. 1983; Dauwalter and Fisher 2008b; Arkansas, Johnson et al. 2009; Iowa, Jansen et al. 2008). Orth et al. (1983) sampled Glover Creek, a runoff flashy creek in Oklahoma and found a decline in relative weight over the course of the summer during one year. Temperatures during that study were typically within optimal levels for Smallmouth Bass growth, so the authors postulated that the change may be due to food limitations. Dauwalter and Fisher (2008b) sampled a groundwater stream and a runoff stream in Oklahoma. They found that survival in both was lowest from summer to fall and they postulated that this could either be due to high temperatures or low water conditions, among other factors such as high angling pressure. They found lowest relative weights shortly before the spring spawn. Relative weight was higher during the summer sample and stayed at nearly the same level for the fall sample. Their results could be due to the coarse resolution that they used (seasonal rather than monthly sampling).

Our relative weight results have a number of management implications. A decline in relative weight of Smallmouth Bass during summer months could reduce size-at-age and limit the number of trophy fish produced. These population changes could affect angler satisfaction, an important consideration in the region as many Ozark streams are popular fisheries for Smallmouth Bass. Type of water body can affect the relative weights of fish (Neumann et al. 2012) and in Ozark stream systems Smallmouth Bass populations typically have lower W_r values than lentic and more northern populations, consistent with our results (Orth et al. 1983; Dauwalter and Fisher 2008b; Johnson et al. 2009). Two of our streams (Buffalo River and Crooked Creek) are designated "Blue Ribbon Smallmouth Bass Streams" in Arkansas and are considered to be some of the best Smallmouth Bass fisheries in the state (Quinn et al. 2004). We were surprised to find that relative weights of Smallmouth Bass in these two sites were about the same as the other populations we sampled, and even lower than some groundwater streams in early summer samples (Figure 2).

Diet composition did not change over the course of the summer in any year of our study. Similarly, proportion of fish with empty stomachs did not significantly change over the course of summer, except in runoff streams in 2014. We hypothesized that Smallmouth Bass from runoff streams would shift diets in response to low flow conditions as we expected high competition in refuge pool habitats (Magoulick and Kobza 2003) that would affect prey availability (Zaret and Rand 1971). However, we did not observe pool isolation lasting throughout the summer in any of our study years, likely preventing high levels of competition from occurring. Overall, diets were dominated by crayfish and fish prey, with a few months having higher contributions of invertebrate prey. Invertebrates seemed to be site dependent with several sites (Crooked Creek and Little Mulberry) having invertebrate prey items more commonly found in diets of fish. It is possible that our coarse resolution of data prevented us from observing fine scale shifts of prey use within diet categories. Our diet results were similar to other Smallmouth Bass diets collected in Arkansas. Johnson et al. (2009) found 39% of Smallmouth Bass had empty stomachs, and crayfish and fish were consumed in approximately equal numbers, followed distantly by insect prey. Similar to relative weights, diet patterns of Smallmouth Bass can vary seasonally. In an Oklahoma Ozark stream, Dauwalter and Fisher (2008a) found adult and sub-adult Smallmouth Bass diets varied seasonally, though they primarily consumed a mixture of fish and crayfish.

Capturing fish by angling could have introduced bias into our results. Vulnerability to angling is a heritable trait in some fishes (e.g., Largemouth Bass, Cooke et al. 2007; Philipp et al. 2009). Various other factors can affect fish vulnerability to angling including fish size (Lennox et al. 2017), activity level (Alós et al. 2012), and metabolism (Cooke et al. 2007). However, factors affecting vulnerability can be species specific (Lennox et al. 2017) and no research has been conducted on this topic for Smallmouth Bass. In order to minimize bias in the fish captured, we typically used the same type of bait at each site and among time periods. Smallmouth Bass are an aggressive and opportunistic feeder and we often captured fish that had very full stomachs. Our diet results are similar to other studies of lotic Smallmouth Bass diets (Dauwalter and Fisher 2008a; Johnson et al. 2009), indicating that our choice of bait and method of sampling likely did not influence our diet results. Though a subset of the populations that we sampled may not have been vulnerable to capture by angling, we believe our sampling efforts provide a reasonably diverse sample over time.

Classifying our streams in a categorical way ignores the continuous nature of the differences in various flow characteristics in streams. One reason that our results may have been similar between stream types is that the streams selected may not have been as different as the categorical classification suggests. The base flow index, a measure of long-term base flow relative to stream flow (Leasure et al. 2016; Figure 1), was lower in all of our runoff sites compared to groundwater sites, but there is a relatively wide range of values in base flow index among sites, indicating that our sites were varied in their relative classifications. Interestingly, the two sites that were most consistent in declining relative weight all three summers (Buffalo River and Little Mulberry River) had the lowest base flow index values.

There is a chance that our repeated sampling could have affected our results, either because we recaptured individual fish multiple times each summer, or by causing increased mortality or other behavioral effects within a population. We did not mark individuals in any way and therefore do not know whether any fish were resampled. However, we were likely capturing a very small number of fish from the total population at each given sampling event. We typically sampled a large section of stream, spreading the relatively few captures over a large area, reducing the likelihood of recaptures. It is also unlikely that we had an effect on the population as a whole as we observed very little capture mortality (e.g., from "gut hooking"; < 5%) and previous work with Largemouth Bass indicates that catch-and-release angling likely does not significantly affect growth (Pope and Wilde 2004).

Our results were surprising in that groundwater and runoff streams appeared to show very similar trends in body condition during summer months. Water temperatures in many runoff streams in the Ozarks currently exceed optimal levels for growth of Smallmouth Bass which could result in declines in growth and seasonal decreases in body weight (Middaugh et al. 2018). Warming water temperatures due to climate change could positively affect Smallmouth Bass annual growth in some parts of their range, potentially even in streams near their southern range extent (Pease and Paukert 2014). However, under future climate conditions in the Ozarks of Arkansas, Oklahoma, and Missouri Smallmouth Bass growth during summer months is predicted to strongly decline in runoff streams as temperatures increase, but decline moderately in groundwater streams, potentially providing a thermal refuge for Smallmouth Bass (Middaugh et al. 2018). Our results indicate that groundwater streams may not provide as much of a growth refuge during summer months as predicted by Middaugh et al. (2018). Smallmouth Bass have been observed using large groundwater springs as a refuge from cold winter water temperatures (Westhoff et al. 2014) and could potentially use these refuge habitats during summer months. Smaller groundwater seepages could also provide smaller scale refuges within streams (Dauwalter and Fisher 2008b).

Our lack of differences between stream types could be due to the specific streams selected or other factors we do not take into account. For example, our groundwater streams were only located within two HUC-8 watersheds whereas runoff streams were spread throughout four HUC-8 watersheds, potentially increasing the variability among streams. The streams we selected all had robust populations of Smallmouth Bass which could have affected our results. We did not attempt to quantify the abundance of Smallmouth Bass in our streams and our results may not apply directly to streams with marginal habitat and limited Smallmouth Bass populations. We also did not take into account differences among years in seasonal conditions before summer months. This could affect relative weights of fish in initial June sampling each year. These differences could have led to the lack of a decline in relative weight in Smallmouth Bass from groundwater streams in 2016, however, initial relative weights of Smallmouth Bass in runoff stream samples were very consistent among years.

In conclusion, we found that Smallmouth Bass relative weights declined in both groundwater and runoff streams in two years of our study, and in all three years for select runoff sites. We did not observe any trends in diet composition over the course of summer months or any differences in diets of Smallmouth Bass between stream types. It is possible that years with higher temperatures and stronger seasonal drying, similar to conditions expected under climate change, could lead to differences in seasonal change in relative weight and diet between stream types. Future work could attempt to examine body condition and diet of Smallmouth Bass during strong drought years and focus on how isolated pool habitats in runoff streams could affect diet and body condition. Further work should also be conducted to determine refuge potential of groundwater streams in the region by comparing habitat conditions such as water temperature between runoff and groundwater streams during summer months.

Acknowledgments

We thank the numerous people who assisted with fieldwork, including T. Douglas, B. Kessinger, S. Koenigbauer, T. Pittman, R. Middaugh, J. Middaugh, and M. Middaugh. We would also like to thank M. Siepker, A. Hafs, and two anonymous reviewers for providing comments that improved this manuscript. This work was conducted under IACUC protocol # 14046. Any use of trade, firm, or product names is for descriptive purposes only and does not imply endorsement by the U.S. Government.

References

Alós, J., M. Palmer, and R. Arlinghaus. 2012. Consistent selection towards low activity phenotypes when catchability depends on encounters among human predators and fish. PLoS ONE 7:e48030.

Austen, D. J., and D. J. Orth. 1988. Evaluation of a 305-mm minimum-length limit for Smallmouth Bass in the New River, Virginia and West Virginia. North American Journal of Fisheries Management 8:231–239.

Bates, D., M. Maechler, B. Bolker, and S. Walker. 2015. Fitting linear mixed-effects models using lme4. Journal of Statistical Software 67:1–48.

Blackwell, B. G., M. L. Brown, and D. W. Willis. 2000. Relative weight (*Wr*) status and current use in fisheries assessment and management. Reviews in Fisheries Science 8:1–44.

Bouska, K. L., G. W. Whitledge, and C. Lant. 2015. Development and evaluation of species distribution models for fourteen native central U.S. fish species. Hydrobiologia 747:159–176.

Carlson, A. K., W. W. Taylor, K. M. Hartikainen, D. M. Infante, T. D. Beard, Jr., A. J. Lynch. 2017a. Comparing stream-specific to generalized temperature models to guide salmonid management in a changing climate. Reviews in Fish Biology and Fisheries 2:443–462.

Carlson, A. K., W. W. Taylor, K. M. Schlee, T. G. Zorn, and D. M. Infante. 2017b. Projected impacts of climate change on stream salmonids with implications for resilience-based management. Ecology of Freshwater Fish 26:190–204.

Christie, G. C., and H. A. Regier. 1988. Measures of optimal thermal habitat and their relationship to yields for 4 commercial fish species. Canadian Journal of Fisheries and Aquatic Sciences 45:301–314.

Cooke, S. J., C. D. Suski, K. G. Ostrand, D. H. Wahl, and D. P. Phillip. 2007. Physiological and behavioral consequences of long-term artificial selection for vulnerability to recreational angling in a teleost fish. Physiological and Biochemical Zoology 80:480–490.

Dauwalter, D. C., and W. L. Fisher. 2008a. Ontogenetic and seasonal diet shifts of Smallmouth Bass in an Ozark stream. Journal of Freshwater Ecology 23:113–121.

Dauwalter, D. C., and W. L. Fisher. 2008b. Spatial and temporal patterns in stream habitat and Smallmouth Bass populations in Eastern Oklahoma. Transactions of the American Fisheries Society 137:1072–1088.

Encina, L., and C. Granado-Lorencio. 1997. Seasonal changes in condition, nutrition, gonad maturation and energy content in Barbell, *Barbus sclateri*, inhabiting a fluctuating river Environmental Biology of Fishes 50:75–84.

Girvetz, E. H., C. Zganjar, G. T. Raber, E. P. Maurer, P. Kareiva, and J. J. Lawler. 2009. Applied climate-change analysis: the climate wizard tool. PLoS ONE 4:e8320.

Hafs, A. W., C. J. Gagen, and J. K. Whalen. 2010. Smallmouth Bass summer habitat use, movement, and survival in response to low flow in the Illinois Bayou, Arkansas. North American Journal of Fisheries Management 30:604–612.

Harvey, B. C., and A. J. Stewart. 1991. Fish size and habitat depth relationships in headwater streams. Oecologia 87:336–342.

Homan, J. M., N. M. Girondo, and C. J. Gagen. 2005. Quantification and prediction of stream dryness in the interior highlands. Journal of the Arkansas Academy of Sciences 59:95–100.

Jansen, A. C., M. C. Quist, and J. Kopaska. 2008. Assessment of Smallmouth Bass populations in Iowa interior rivers. Journal of the Iowa Academy of Science 11:17–23.

Johnson, R. L., A. D. Christian, S. D. Henry, and S. W. Barkley. 2009. Distribution, population characteristics, and physical habitat associations of black bass (*Micropterus*) in the lower Eleven Point River, Arkansas. Southeastern Naturalist 8:653–670.

Jowett, I. G., J. Richardson, and M. L. Bonnett. 2005. Relationship between flow regime and fish abundances in a gravel-bed river, New Zealand. Journal of Fish Biology 66:1419–1436.

Kamler, J. F., and K. L. Pope. 2001. Nonlethal methods of examining fish stomach contents. Reviews in Fisheries Science 9:1–11.

Kaushal, S. S., G. E. Likens, N. A. Jaworski, M. L. Pace, A. M. Sides, D. Seekell, K. T. Belt, D. H. Secor, and R. L. Wingate. 2010. Rising stream and river temperatures in the United States. Frontiers in Ecology and the Environment 8:461–466.

Kolander, T. D., D. W. Willis, and B. R. Murphy. 1993. Proposed revision of the standard weight (*Ws*) equation for smallmouth bass. North American Journal of Fisheries Management 13:398–400.

Lambert, Y., and J. D. Dutil. 1997. Condition and energy reserves of Atlantic Cod (*Gadus morhua*) during the collapse of the northern Gulf of St. Lawrence stock. Canadian Journal of Fisheries and Aquatic Sciences 54:2388–2400.

Leasure, D. R., D. D. Magoulick, and S. D. Longing. 2016. Natural flow regimes of the Ozark-Ouachita Interior Highlands region. River Research and Applications 32:18–35.

Lennox, R. J., J. Alós, R. Arlinghaus, A. Horodysky, T. Klefoth, C. T. Monk, and S. J.

Cooke. 2017. What makes fish vulnerable to capture by hooks? A conceptual framework and a review of key determinants. Fish and Fisheries DOI10.1111/faf.12219.

Light, R. W., P. H. Adler, and D. E. Arnold. 1983. Evaluation of gastric lavage for stomach analyses. North American Journal of Fisheries Management 3:81–85.

Lynch, A. J., B. J. E. Myers, C. Chu, L. A. Eby, J. A. Falke, R. P. Kovach, T. J. Krabbenhoft, T. J. Kwak, J. Lyons, C. P. Paukert, and J. E. Whitney. 2016. Climate change effects on North American inland fish populations and assemblages. Fisheries 41:346–361.

Magnuson, J. J., L. B. Crowder, and P. A. Medvick. 1979. Temperature as an ecological resource. American Zoologist 19:331–343.

Magoulick, D. D., and R. M. Kobza. 2003. The role of refugia for fishes during drought: a review and synthesis. Freshwater Biology 48:1186–1198.

Marchetti, M. P., and P. B. Moyle. 2001. Effects of flow regime on fish assemblages in a regulated California stream. Ecological Applications 11:530–539.

Middaugh, C. R., B. Kessinger, and D. D. Magoulick. 2018. Climate-induced seasonal changes in Smallmouth Bass growth rate potential at the southern range extent. Ecology of Freshwater Fish 27:19–29.

Neumann, R. M., C. S. Guy, and D. W. Willis. 2012. Length, weight, and associated indices. Pages 637–676 *in* A. V. Zale, D. L. Parrish, and T. M. Sutton, editors. Fisheries techniques, 3rd edition. American Fisheries Society, Bethesda, Maryland.

Olden, J. D., M. J. Kennard, and B. J. Pusey. 2012. A framework for hydrologic classification with a review of methodologies and applications in ecohydrology. Ecohydrology 5:503–518.

Orth, D. J., D. D. Oakey, and O. E. Maughan. 1983. Population characteristics of Smallmouth Bass in Glover Creek, Southeastern Oklahoma. Proceedings of the Oklahoma Academy of Science 63:37–41.

Oksanen, J., F. G. Blanchet, M. Friendly, R. Kindt, P. Legendre, D. McGlinn, P. R. Minchin, R. B. O'Hara, G. L. Simpson, P. Solymos, M. Henry, H. Stevens, E. Szoecs, and H. Wagner. 2017. Vegan: Community ecology package. R package version 2.4–3.

Pease, A. A., and C. P. Paukert. 2014. Potential impacts of climate change on growth and prey consumption of stream-dwelling Smallmouth Bass in the central United States. Ecology of Freshwater Fish 23:336–346.

Philipp, D. P., S. J. Cooke, J. E. Claussen, J. B. Koppelman, C. D. Suski, and D. P. Burkett. 2009. Selection for vulnerability to angling in Largemouth Bass. Transactions of the American Fisheries Society 138:189–199.

Pope, K. L., and G. R. Wilde. 2004. Effect of catch-and-release angling on growth of Largemouth Bass, *Micropterus salmoides* Fisheries Management and Ecology 11:39–44.

Quinn, J., B. Wagner, and S. Filipek. 2004. Characteristics of fish populations in the rivers and streams of Arkansas. Stream Fisheries Program Report STP2004–01.

R Development Core Team. 2015. R: a language and environment for statistical computing. R Foundation for Statistical Computing, Vienna, Austria. ISBN 3–900051-07–0. http://www.R-project.org.

Resh, V. H., A. V. Brown, A. P. Covich, M. E. Gurtz, H. W. Li, G. W. Minshall, S. R. Reice, A. L. Sheldon, J. B. Wallace, and R. C. Wissmar. 1988. The role of disturbance in stream ecology. Journal of the North American Benthological Society 7:433–455.

Robison, H. W., and T. M. Buchanan. 1988. Fishes of Arkansas. The University of Arkansas Press, Fayetteville.

Shuter, B. J., J. A. MacLean, F. E. J. Fry, and H. A. Regier. 1980. Stochastic simulation of temperature effects on first year survival of Smallmouth Bass. Transactions of the American Fisheries Society 109:1–34.

Smith, N. G. 2012. Seasonal condition of adult striped bass relative to thermal habitat and forage availability. Proceedings of the Southeastern Association of Fish and Wildlife Agencies 66:69–74.

Sowa, S. P., and C. F. Rabeni. 1995. Regional evaluation of the relation of habitat to distribution and abundance of Smallmouth Bass and Largemouth Bass in Missouri streams. Transactions of the American Fisheries Society 124:240–251.

Staton, J. M., C. R. Roswell, D. G. Fielder, M. V. Thomas, S. A. Pothoven, and T. O. Höök. 2014. Condition and diet of Yellow Perch in Saginaw Bay, Lake Huron (1970–2011). Journal of Great Lakes Research 40:139–147.

Strahler, A. N. 1957. Quantitative analysis of watershed geomorphology. Transactions of the American Geophysical Union 38:913–920.

Van Den Avyle, M. J. 1980. Evaluation of a simple method for removing food items from live black bass. The Progressive Fish-Culturist 42:222–223.

Walters, A. W. 2016. The importance of context dependence for understanding the effects of low-flow events on fish. Freshwater Science 35:216–228.

Wenger, S. J., D. J. Isaak, C. H. Luce, H. M. Nevill, K. D. Fausch, J. B. Dunham, D. C. Dauwalter, M. K. Young, M. M. Elsner, B. E. Rieman, A. F. Hamlet, and J. E. Williams. 2011. Flow regime, temperature, and biotic interactions drive differential declines of trout species under climate change. Proceedings of the National Academy of Sciences 108:14175–14180.

Westhoff, J. T., C. Paukert, S. Ettinger-Dietzel, H. Dodd, and M. Siepker. 2014. Behavioural thermoregulation and bioenergetics of riverine Smallmouth Bass associated with ambient cold-period thermal refuge. Ecology of Freshwater Fish 25:72–85.

Whitledge, G. W., C. F. Rabeni, G. Annis, and S. P. Sowa. 2006. Riparian shading and groundwater enhance growth potential for Smallmouth Bass in Ozark streams. Ecological Applications 16:1461–1473.

Zaret, T. M., and A. S. Rand. 1971. Competition in tropical stream fishes: support for the competitive exclusion principle. Ecology 52:336–342.

Zweifel, R. D., R. S. Hayward, and C. F. Rabeni. 1999. Bioenergetics insight into black bass distribution shifts in Ozark border region streams. North American Journal of Fisheries Management 19:192–197.

American Fisheries Society Symposium 87:111–128, 2019

Exploitation of Smallmouth Bass from Selected Ozark Streams and Simulated Effects of Higher Minimum Length Limits

Chris Williamson*
Missouri Department of Conservation
8965 Highway 36, Hannibal, Missouri 63401, USA

Nick Girondo
Missouri Department of Conservation
12655 State Route Y, Rolla, Missouri 65401, USA

Paul Cieslewicz, Dave Knuth, and Mike Reed
Missouri Department of Conservation
2302 County Park Drive, Cape Girardeau, Missouri 63701, USA

John Ackerson, and A.J. Pratt
Missouri Department of Conservation
551 Joe Jones Boulevard, West Plains, Missouri 65775, USA

Sarah Peper
Missouri Department of Conservation
2360 Highway D, St. Charles, Missouri 63304, USA

Jen Girondo
Missouri Department of Conservation
3220 S Highway 185, Sullivan, Missouri 63080, USA

*Corresponding author: Chris.Williamson@mdc.mo.gov

Abstract.—Angler harvest and population characteristics of Smallmouth Bass *Micropterus dolomieu* were assessed through electrofishing surveys and tagging 3,027 fish with reward tags at six sites on five Ozark streams. Growth, exploitation, and mortality were estimated for each site. Predicted population responses to higher length limits were simulated using Fishery Analysis and Modeling Simulator software. Tag return rates ranged between 37% and 64%, angler release rates ranged between 63% and 94%, and annual exploitation ranged between 5% and 26%. The median time at-large for tags returned within one year of tagging ranged from 22 to 47 d of the tagging date. Growth rates were relatively slow, as mean time to reach 305 mm was 4.9 years and mean time to reach 381 mm was 7.8 years. Total annual mortality estimates ranged from 37% to 55%. Annual natural mortality estimates ranged from 13% to 33%. Predicted responses to higher length limits varied considerably by site because of differences in estimated rate functions. Although simulations predicted small increases (0.54–2.73 fish/100 recruits >381 mm) in the number of larger fish with the 381-mm length limit at five of six sites, predicted increases were substantial (17 fish/100 recruits >381 mm) and yield increased 6% at the Current River-Powder Mill site. Individuals in the Current River-Powder Mill site were not reaching their full growth potential due to growth overfishing, while simulations of the remaining five populations indicated no growth overfishing under current conditions and regulations. The combined effects of natural mortality and slow growth limited the effectiveness of higher length limits. Under most conditions, the statewide length limit of 305 mm was adequate to balance the desire of quality fishing and harvest opportunities on most Ozark streams. Our study indicates that fisheries at select stream reaches may be improved by higher length limits where exploitation is high, growth is adequate, and natural mortality is low.

Introduction

Black bass *Micropterus* spp. fisheries are the most popular recreational fisheries in the United States (U.S. Fish and Wildlife Service 2011). Managing these species often involves length limits and creel limits to protect or enhance the fisheries (Wilde 1997). Early efforts to manage these fisheries were intended to protect them from overharvest (Paukert et al. 2007). However, angler attitudes have changed, and catch and release has become more popular. Allen et al. (2008) found that fishing mortality rates of Largemouth Bass *Micropterus salmoides* declined by half since 1990 as more anglers were practicing catch and release. Many current management strategies are designed to enhance the number and size of larger fish in the population (Carlson and Isermann 2010). These strategies may increase the population density, but their effectiveness at increasing the number of larger fish in a population may be limited by lower angler harvest (Allen et al. 2008). Thus, understanding dynamic rate functions like recruitment, growth, and mortality that effect a population are very important in applying appropriate harvest regulations.

Kruse and DeiSanti (2002) showed that restrictive harvest regulations are only effective in increasing the number of large Smallmouth Bass *Micropterus dolomieu* if exploitation and growth rates are high and natural mortality rates are low. When the natural mortality rate is higher than the exploitation rate, restrictive harvest regulations are much less effective at increasing the number of larger fish because angler harvest is not the primary factor affecting mortality (Beamesderfer and North 1995; Isermann and Paukert 2010). It takes time for fish to grow to larger sizes, while at the same time they are susceptible to natural mortality. When fish growth is slow and natural mortality is high, most fish will not live long enough to reach the larger desired size. With these conditions, there will not be an appreciable increase in the number of larger Smallmouth Bass. Moreover, imposing unnecessarily more restrictive harvest regulations creates an unrealized expectation of better fishing from anglers and adds unnecessary regulatory complexity. Modification of regulations can be justified if the harvest regulations do indeed improve fishing.

Although many studies have reported growth and mortality rates of Smallmouth Bass from streams in the United States, very few studies have reported exploitation rates. Exploitation estimates can be expensive and difficult to obtain. Some researchers suggest the exploitation question should be circumvented altogether (Miranda et al. 2002). However, exploitation is the one variable fishery managers can manipulate through fishing regulations. Thus, it is imperative that exploitation of population be understood before regulations are implemented to improve size structure or prevent overfishing. Early efforts to quantify harvest of Smallmouth Bass showed that harvest estimates exceeded the estimated population size on the Courtois Creek (Fajen 1975a). Fishing mortality ranged from 35 to 56% in the Tennessee River (Weathers and Bain 1992). Exploitation estimates averaged 19% in the Maquoketa River in Iowa (Paragamian 1984). More recently, researchers in Idaho reported exploitation rates between 12 and 34% from five reservoirs between 2006 and 2009 (Meyer and Schill 2014). The two most recent investigations from Missouri report low exploitation estimates of Smallmouth Bass. Smith (1998) reported exploitation rates of 7% and 8% for 1995 and 1996, respectively, from the Bourbeuse River in Missouri. Only 2% of Smallmouth Bass were harvested annually from the lower Salt River in 2009 (Williamson 2012). Kruse and DeiSanti (2002) did not report exploitation rates, but they did report that tag return rates varied between 8 and 45% on the Jacks Fork and Gasconade rivers during 1990–2001.

Smallmouth Bass in Missouri Ozark streams have been managed under a variety of harvest regulations over time. Starting in 1961, a daily creel limit of six black bass was implemented, followed by a closed harvest season in 1965, and in 1974, a 305-mm minimum length limit was implemented. In 1992, 12 special management areas were established with either 381-mm or 457-mm minimum length limits. However, a segment of Missouri anglers continued to voice concern that Smallmouth Bass populations are overharvested and not reaching their full potential under current statewide regulations. Anglers requested that higher length limits and reduced creel limits be implemented to protect or improve these fisheries, but information on the potential impacts of more restrictive regulations was lacking. Accurate information was needed to simulate the potential effects of more restrictive harvest regulations.

The goals of this effort were to estimate exploitation, growth, and annual mortality, and then simulate the effects of increased minimum length limits on Smallmouth Bass in each stream to determine if higher length limits are reasonable. This approach has been used successfully for a variety of sport fish species (Maceina et al. 1998a; Maceina et al. 1998b; Slipke et al. 1998).

Methods

The Ozark Highlands region is characterized by rugged hills and covers most of southern Missouri, northern Arkansas, and parts of Illinois, Oklahoma, and Kansas (Nigh and Schroder 2002). Annual precipitation ranges between 107 and 122 cm with an average temperature of 13°C (Nigh and Schroder 2002). Karst topography is common throughout the Ozark Highlands and the streams are meandering within steeply-sided valleys (Funk and Fleener 1974). Most Ozark streams are spring fed and are characterized by clear water, gravel and boulder substrate, and distinct pool, riffle, and run habitats.

We estimated exploitation rates of Smallmouth Bass at six sites on five streams during 2011 and 2013 (Figure 1). Six sites were selected with drainage areas between 352 and 2,850 km^2 to ensure a diversity of streams and angler preferences (Table 1). Because recent studies suggested exploitation rates were generally low (<10%), biologists selected sites where heavy fishing pressure occurred and exploitation rates should be highest. Selected sites were not adjacent to existing special management areas that had higher length limits and reduced creel limits. We excluded these areas because harvest pressure could have changed in these reaches when the special management areas were created. Statewide regulations for Black Bass (305-mm minimum length limit, 6 fish per day creel limit, and a closed harvest season from March 1 to the Friday before the fourth Saturday in May) were in effect at all study sites.

Exploitation rates were estimated using tag returns from anglers. We captured fish by targeted electrofishing, and only tagged fish at least 305 mm TL. The intention was to tag fish in May before the harvest season began on Memorial Day weekend. However, budget-timing limitations resulted in most of the fish being tagged within two weeks after Memorial Day weekend during 2011. In 2013, all fish were tagged prior to opening weekend. Fish were tagged with Carlin dangler tags, placed under the anterior dorsal fin. Each numbered tag listed a contact phone number and the reward amount. We measured all fish captured for total length during tagging runs to develop an age-length key used in later modeling. To avoid sacrificing fish to take otoliths, scales were removed from up to five fish per 13 mm group during 2013 to obtain estimates of age structure and growth rates.

We relied on self-reporting of catch by anglers to estimate fishing mortality. Our exploitation estimates included adjustments for tag loss and angler reporting rates. Tag loss was estimated by monitoring tag retention of 24 tagged Smallmouth Bass placed into a hatchery raceway at Mammoth Spring National Fish Hatchery. A large rootwad, a rootwad with a log attached, a treetop, and two boulder clusters were added to simulate natural habitat. Fish were checked for tag retention at three months, six months, nine months, and one year after tagging. Respective tag loss rates (0–3 months, 4–6 months, etc.) were applied to all sites throughout the study.

A high reward tagging program, developed from Pollock et al. (2001), was used to estimate angler reporting rate in 2011. A standard reward of U.S.$25 (75% of total tags) and a high reward of $75 (25% of total tags) were used at each site. As suggested by Taylor et al. (2006), we assumed that $75 was adequate to elicit a 100% return rate from anglers. Meyer et al. (2012) also suggested reward rates between $50 and $100 were needed to elicit maximum reporting rates. The reporting rate for the low reward tags was calculated by dividing the return rate for the

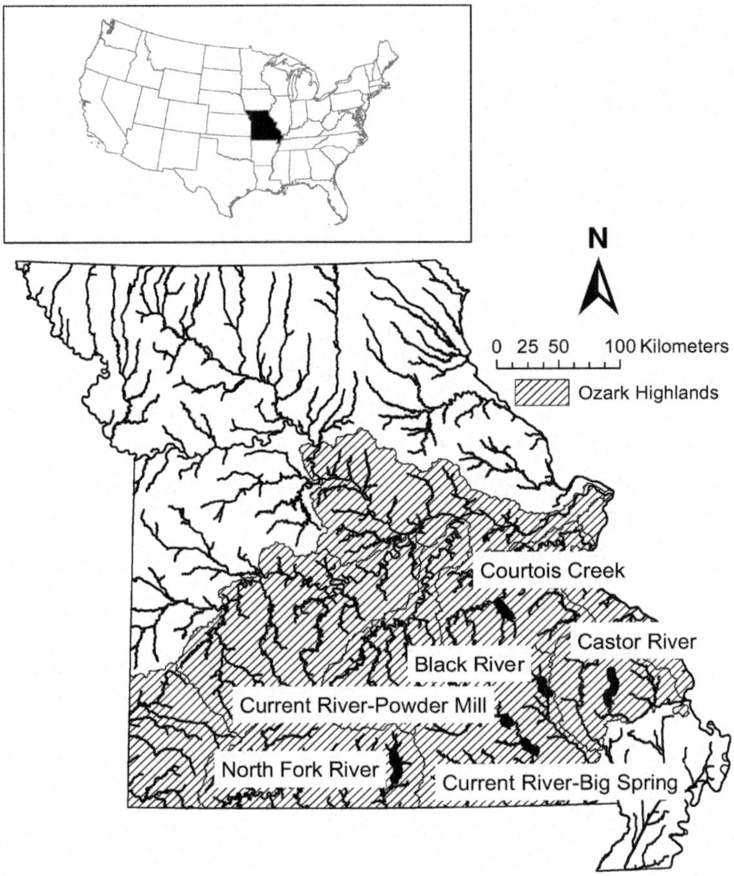

Figure 1. Sites on selected Ozark streams for harvest evaluation of Smallmouth Bass.

Table 1. Study sites for Smallmouth Bass harvest evaluation from selected Ozark streams, 2011 and 2013.

Site	Drainage area (km^2)	Site boundaries	Reach length (km)
Black River	800	Confluence of Middle Fork and West Fork of the Black River to State Highway K	24
Castor River	370	US Highway 72 to US Highway 34	40
Courtois Creek	352	Missouri Highway 8/Berryman Access to Huzzah Conservation Area	16
Current River-Big Spring	2,850	Van Buren Riverfront Park to Cataract Landing	21
Current River-Powder Mill	2,266	Two Rivers to Roberts Field	16
North Fork River	689	Osborn Crossing to Rainbow Spring	34

low reward tags by the return rate for the high reward tags. This methodology saved money and decreased the likelihood of increased angler effort due to high rewards. Only $25 tags were used in 2013 and the 2011 reporting rate from each site was used to adjust return rates of the 2013 tagging cohort.

Annual exploitation estimates were calculated by summing monthly exploitation estimates, which were calculated by dividing the number of tagged fish harvested, corrected for reporting rate, by the number of tagged fish at-large for each month. The number of tagged fish at-large each month was reduced for tag loss and natural mortality (Sammons 2016). The following equation was used to calculate the number of tagged fish at-large each month (N_t):

$$N_t = [(N_{t-1} * R) - \mathrm{H}][1 - (\tfrac{M}{12})]$$

Where R equals the respective tag retention rate, H equals the corrected harvest from the previous month and M equals natural mortality rate (0.39) from Beamesderfer and North (1995). How natural mortality rates changed throughout the year was not known, so the correction was applied evenly throughout the year. Tag returns from fish captured during the closed harvest season or after one year at-large were not used in annual exploitation estimates, but were used to estimate reporting rates.

We estimated age, age frequencies, and growth rates from scales and length measurements of fish collected in 2013 for each site. Ages were estimated by two readers independently without knowledge of fish length. Differences in age estimates were resolved by conference. Age frequency distributions were calculated by applying an age-length key to the entire electrofishing sample for each site. Growth was modeled from mean lengths at age using the von Bertalanffy equation ($l_t = L_\infty[1-e^{-K(t-to)}]$, where l_t is length at time t, K is a growth parameter, L_∞ is the maximum theoretical length, and t_0 is the time at which l_t is 0) using the Fishery Analysis and Modeling Simulator 1.0 (FAMS) Excel add-in (von Bertalanffy 1938; Slipke and Maceina 2010).

Total annual mortality and maximum age for each site was estimated using a weighted linear regression of age frequencies. We pooled 2011 and 2013 samples to help meet the assumptions of the catch curve method. Annual natural mortality was estimated by subtracting the annual exploitation estimate from the total mortality estimate. Conditional fishing mortality (*cf*) and conditional natural mortality (*cm*) were calculated as described in Slipke and Maceina (2010).

The Jones (1957) modification of the Beverton–Holt (1957) yield-per-recruit models in FAMS were used to simulate the response of the six Smallmouth Bass populations to 381 mm and 457 mm minimum length limits compared to the current 305 mm minimum length limit (Allen and Hightower 2010; Slipke and Maceina 2010). Recruitment was arbitrarily set at a constant 100 fish/year for all simulations. Weight-length relationship parameters from Kolander et al. (1993) were used for all populations (a = –5.329, b = 3.200). These simulations allowed us to identify fishing and natural mortality rates at which growth overfishing is expected to occur. These simulations also allowed us to predict changes in the number of fish at a given size under different minimum length limits.

Results

We tagged 3,027 Smallmouth Bass ranging between 305 mm and 521 mm TL at the six sites (Table 2). Total tag return rates ranged from 37% to 64%, and reporting rates ranged from 78 to >100%. The high reward tag return rates averaged 58% (range: 33–69%). The estimated reporting rate for the Castor River site was >100%. This estimate was adjusted to the next high-

Table 2. Tag returns, reporting rate, percent released, and annual exploitation rates for Smallmouth Bass populations from selected Ozark streams, 2011 and 2013.

Site	Year	Fish tagged	Percent tags returned	Reporting rate	Percent reported released	Annual exploitation rate
Black River	2011	169	54%	80%	68%	23%
	2013	200	45%	-	63%	24%
Castor River	2011	178	46%	>100[a]	84%	10%
	2013	200	64%	-	74%	26%
Courtois Creek	2011	199	52%	87%	88%	10%
	2013	200	47%	-	94%	5%
Current River-Big Spring	2011	286	56%	78%	80%	13%
	2013	407	57%	-	79%	15%
Current River-Powder Mill	2011	388	59%	81%	73%	26%
	2013	400	56%	-	71%	26%
North Fork River	2011	200	51%	80%	72%	15%
	2013	200	37%	-	82%	10%

[a]Calculated rate was greater than 100%, the next highest rate (87%) was used for exploitation calculations.

est reported rate (87%) because it could not be >100%. The adjusted rate was used to calculate exploitation estimates at this site. Of the tags that were returned, anglers reported releasing 63–94% of the fish. Tag retention was estimated to be 100% at 3 and 6 months, 94.7% at 9 months, and 88.8% at one year. Because no fish died during the first week of the tag loss study, a tagging mortality adjustment was not applied to exploitation estimates.

Annual exploitation was variable and ranged from 5% to 26% (Table 2). There was little difference in the exploitation estimates between the two years at each site. However, the 2011 estimate for Castor River was likely biased low because the majority of tags were not at large when most angling effort took place. High water delayed the majority of tagging until after the harvest season opened. Thus, only the 2013 exploitation estimate was used in the remaining analyses for the Castor River.

The majority of tag returns and reported harvest occurred within the first 3 months of the harvest season. The median time at-large for tags returned within one year of tagging ranged from 22 to 47 d of the tagging date (Table 3). In some streams, a large percentage of tag returns occurred within the first month of the harvest season. For example, 53% of the tag returns from the Black River 2011 tagging cohort occurred within the first 16 d of the harvest season, which encompasses the first three weekends.

Ages of 615 Smallmouth Bass were estimated. Age ranged from 2 to 11 years (Table 4). Growth coefficients (K) ranged from

Table 3. Median days at-large, percent of returns reported within the first 16 days of the harvest season, and average total lengths of fish tagged from selected Ozark streams. Only fish returned within one year of tagging date were reported.

Site	Tag cohort	Median days at-large	Percent of returns reported within first 16 days of harvest season	Average total length tagged (mm)
Black River	2011	28.5	53%	348
	2013	26	39%	366
Castor River	2011	22	8%[a]	345
	2013	26	38%	351
Courtois River	2011	47	27%	371
	2013	25	23%	353
Current River-Big Spring	2011	31.5	3%[a]	371
	2013	25	38%	366
Current River-Powder Mill	2011	28	16%[a]	361
	2013	25	34%	358
North Fork	2011	35	43%	351
	2013	36.5	37%	356

[a]Tagging effort not complete during this 16-day period.

0.078 to 0.120 (Table 5). Maximum theoretical lengths (L_∞) ranged between 601 and 640 mm. The calculated L_∞ values from the Current River sites were unrealistic (greater than 726 mm), so the average of the other four sites was used to calculate the growth equations for these sites. Growth rates were relatively slow; fish reached a mean length of 305 mm in 4.9 years and 381 mm in 7.8 years. Total annual mortality estimates ranged from 37 to 55%, and annual natural mortality estimates ranged from 13 to 33% (Table 6).

The predicted effect of the 457 mm length limit resulted in substantial percentage increases (83–2,250%) in the number of 457 mm fish, but also resulted in decreased yields between 70 and 94% at five of the six sites (Table 7). A 457-mm minimum length limit would result in a mostly catch-and-release fishery and further analyses concentrated on the 381 mm length limit.

Simulations, using parameters for each population (Table 5 and Table 6), predicted an increase in the number of 381 mm Smallmouth Bass at the 381-mm length limit compared to the current 305-mm minimum length limit at all sites, however the magnitude of these differences varied greatly. Relatively small increases (<65%) in the number of fish greater than 381 mm were predicted at the Courtois Creek, Current River-Big Spring, and North Fork River sites. The number of fish greater than 381 mm at Black River, Castor River, and Current River-Powder Mill were predicted to more than double when compared to the current length limit.

Table 4. Mean length-at-age, standard deviation (SD), minimum length (Min), maximum length (Max), and number of Smallmouth Bass per age class (N) from selected Ozark streams from 2013. All measurements are in millimeters.

Site		Age 1	Age 2	Age 3	Age 4	Age 5	Age 6	Age 7	Age 8	Age 9	Age 10	Age 11
Black River	Mean Length	-	226	262	293	298	312	311	393	436	452	-
	SD	-	36	50	58	67	62	65	3	25	4	-
	Min	-	180	178	198	216	206	211	391	422	450	-
	Max	-	274	401	455	478	483	500	396	488	455	-
	N	-	108	244	270	202	148	97	10	6	2	-
Castor River	Mean Length	199	219	242	275	307	327	353	367	399	-	-
	SD	3	39	44	47	54	36	50	4	5	-	-
	Min	191	178	178	208	236	292	310	358	394	-	-
	Max	201	290	401	391	417	424	424	376	404	-	-
	N	15	141	415	253	229	78	23	14	3	-	-
Courtois River	Mean Length	-	198	219	253	286	315	314	395	377	428	412
	SD	-	11	27	42	49	43	47	28	38	18	23
	Min	-	178	178	178	216	249	254	356	325	419	384
	Max	-	213	274	376	437	429	432	439	467	465	460
	N	-	109	259	268	246	156	97	40	54	6	27
Current River-Big Spring	Mean Length	-	206	260	282	324	354	397	430	433	465	498
	SD	-	18	28	24	36	39	41	34	36	37	-
	Min	-	178	191	241	272	310	330	371	381	414	498
	Max	-	239	328	325	401	455	480	503	493	518	498
	N	-	166	488	244	173	103	70	50	28	17	1

Table 4. Continued.

Site		Age 1	Age 2	Age 3	Age 4	Age 5	Age 6	Age 7	Age 8	Age 9	Age 10	Age 11
Current River-Powder Mill	Mean Length	-	195	244	273	305	328	350	420	450	433	-
	SD	-	8	25	25	20	27	37	33	37	39	-
	Min	-	178	183	244	282	297	310	356	401	368	-
	Max	-	211	290	404	340	411	470	467	480	475	-
	N	-	130	571	268	171	195	177	77	11	17	-
North Fork	Mean Length	-	237	221	279	297	332	358	405	397	-	432
	SD	-	3	30	28	32	20	49	21	24	-	0
	Min	-	231	178	224	254	307	203	373	381	-	432
	Max	-	239	302	328	363	427	442	452	460	-	432
	N	-	10	320	149	182	123	154	31	18	0	2

Table 5. von Bertalanffy growth equation and mortality parameters for six populations of Smallmouth Bass from selected Ozark stream sites. L_∞ = maximum theoretical length, K = growth parameter, and t_0 = the time at which l_t is zero.

Site	L_∞	K	t_0
Black River	601	0.099	–2.42
Castor River	620	0.078	–3.62
Courtois Creek	648	0.081	–2.24
Current River- Big Spring	627[a]	0.120	–1.23
Current River- Powder Mill	627[a]	0.108	–1.36
North Fork River	640	0.084	–2.77

[a]Held L_∞ constant because computed value was unrealistic. Average of other four values.

Table 6. Estimated mortality and exploitation rates from selected Ozark stream sites. Data from 2011 and 2013. A = total annual mortality, u = mean annual exploitation, M = annual natural mortality, cf = conditional fishing mortality, and cm = conditional natural mortality.

Site	A	u	M	cf	cm	Maximum age
Black River	0.55	0.24	0.32	0.29	0.36	11.6
Castor River	0.54	0.26[a]	0.30	0.31	0.33	11.3
Courtois Creek	0.37	0.07	0.30	0.08	0.31	16.5
Current River- Big Spring	0.43	0.14	0.29	0.17	0.32	14.1
Current River- Powder Mill	0.39	0.26	0.13	0.28	0.15	15.9
North Fork River	0.45	0.12	0.33	0.15	0.35	13.4

[a]Only 2013 tag returns were used due to timing of 2011 tagging effort.

Table 7. Fishery Analysis and Modeling Simulator (FAMS) results for selected Ozark stream sites. Recruitment is number of recruits (*N*) per 100 recruits/year for all simulations. MLL = minimum length limit.

Site	MLL (mm)	Yield (kg)	*N* at 305 mm	*N* at 381 mm	*N* at 457 mm
Black River	305	3.1	12.1	1.1	<0.1
	381	1.4	12.1	3.2	0.1
	457	0.2	12.1	3.2	0.5
Castor River	305	3.7	13.5	0.9	<0.1
	381	1.5	13.5	3.3	0.1
	457	0.3	13.5	3.3	0.5
Courtois Creek	305	1.6	12.4	3.0	0.5
	381	0.8	12.4	3.9	0.6
	457	0.2	12.4	3.9	0.8
Current River-Big Spring	305	4.7	19.0	5.3	0.9
	381	3.1	19.0	8.0	1.4
	457	1.4	19.0	8.0	2.5
Current River-Powder Mill	305	23.6	45.7	13.4	2.5
	381	24.9	45.7	30.5	5.7
	457	20.6	45.7	30.5	17.5
North Fork River	305	2.1	12.1	2.0	0.2
	381	1.0	12.1	3.2	0.3
	457	0.2	12.1	3.2	0.6

Although the percentage increase was large, the magnitude of the increases at the Black and Castor river sites were extremely small. At these sites, the number of fish greater than 381 mm increased between 2.04 and 2.41 per 100 recruits. Therefore, an angler who catches 100 fish per year would only catch two additional fish greater than 381 mm under a 381 mm length limit.

The Current River-Powder Mill site exhibited an increase in the number of fish greater than 381 mm (17 fish/100 recruits) and was also the only site with a predicted increase in yield under a 381 mm length limit. This suggests that growth overfishing is occurring under the current length limit, and that a higher length limit at the Current River-Powder Mill site could improve both yield and size structure. Simulations of the remaining five populations indicated that growth overfishing does not occur under a 305-mm length limit (Figure 2).

Discussion

There is a long history of using restrictive Smallmouth Bass harvest regulations to manage and improve fisheries for the enjoyment of all anglers in Missouri Ozark streams (Fajen 1975b, 1981; Fleener 1975; Meneau 1998, 2009; Kruse and DeiSanti 2002). These restrictive harvest regulations

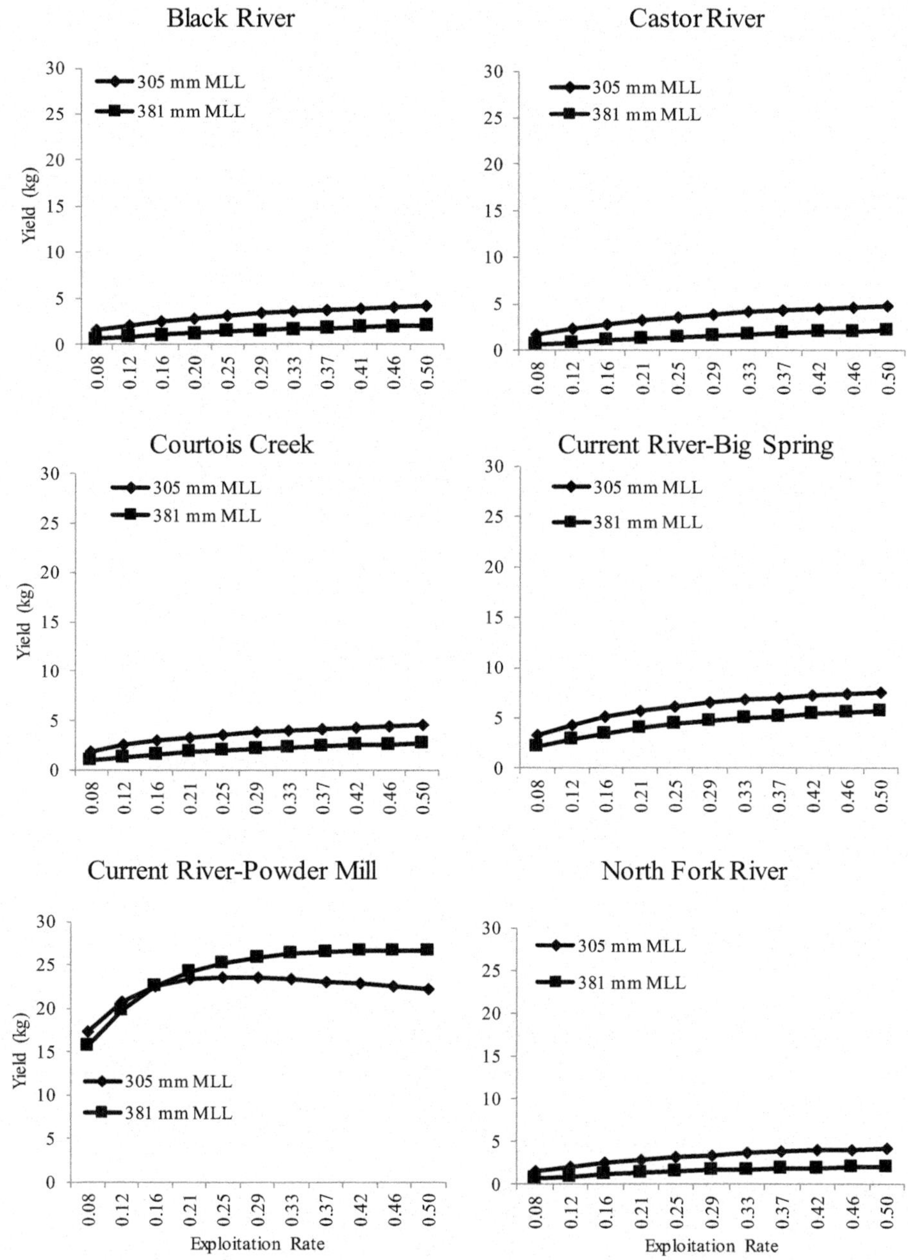

Figure 2. Smallmouth Bass population yield plots from selected Ozark streams. MLL = minimum length limit.

are predicated on population assessments that suggest they will improve fishing. The management challenge is to balance quality population size structure with an opportunity to harvest Smallmouth Bass. Recent survey data indicate that 72% of respondents harvested Smallmouth Bass, whereas 47% of respondents supported increasing the minimum length limit (Ackerson et al. 2012). This survey data, along with our study, indicate that harvest is still an important component of some Smallmouth Bass fisheries.

Tag return rates and reporting rates were much higher than recent studies. Kruse and DeiSanti (2002) reported an average return rate of 27% using reward tags valued between \$5 and \$100 on the Jacks Fork and Gasconade rivers, while Williamson (2012) reported a 20% return rate using \$25 reward tags. On the Zumbro River in Minnesota, anglers returned 19% of Smallmouth Bass tags (Hayes et al. 1997). Our higher return rates indicate either higher reporting rates, angler effort, or catch rate. Printing the reward amount and a telephone number on the tag likely increased reporting rate. Very few studies have measured reporting rates from tagged Smallmouth Bass. Meyer and Schill (2014) had an average reporting rate of 52% from Smallmouth bass tags in Idaho, with estimates varying between 23% and 80%. We assumed the estimated reporting rate on the Castor River >100% occurred by chance due to the small number of high reward tags in the population. As described in Meyer and Schill (2014), anglers could have encountered fewer high reward tags than they should have simply by chance.

Although tag return and reporting rates were similar between sites, we saw a wide range of exploitation rates. This was expected, as different groups of anglers likely fish different streams. Current study results, as well as those from the recently completed angler survey (Ackerson et al. 2012), indicate that Smallmouth Bass harvest may be more prevalent than many predicted. However, this harvest is not spread evenly across all sites or streams. The exploitation rate at the Powder Mill site was almost double the rate at the Big Spring site. This was not expected given the close proximity of the sites. Anecdotal evidence suggests there are more tournaments held at the Big Spring site, which could explain some of the differences as tournament anglers usually release their fish. Still, the observed differences between two sites on the same river indicate that reach-scale management may be appropriate. Determination of reach size and boundaries of special regulation zones should also consider the influence of springs, because they can have profound impacts on movement of Smallmouth Bass in Ozark streams (Westoff et al. 2016).

The authors knew using scales to estimate age of Smallmouth Bass was not ideal, however we had significant concern about negative public reaction to sacrificing fish for otolith removal for this high-profile study. Smallmouth Bass scales are known to be less accurate (Heidinger and Clodfelter 1987) and less precise (Long and Fisher 2001; Maceina and Sammons 2006) than otoliths at older ages. As we tried to balance science and public support, we were concerned about the accuracy of age estimates from scales, and some of the estimates highlight potential error. For example, the range of lengths within specific ages from the Castor and Black rivers was very wide. The standard deviations of mean length at ages from age 3–7 were very high (50–67 mm) at the Black River site. L_∞ values were unrealistically large for two of the populations so we used the mean of the other four populations, suggesting potential error. While some researchers have held L_∞ values at the TL of the largest fish captured (Fiss et al. 2001), we chose to use the calculated values. Previous research has shown that scales typically underage older Smallmouth Bass (Heidinger and Clodfelter 1987), which would cause our growth estimates to be overestimated and total mortality estimates to be higher than actual values. Faster growth rates could have increased the potential benefits of a higher length limit, while higher total mortality estimates, and subsequent higher natural mortality rates could have decreased the potential benefits. Although our concerns over potential errors in age estimates are valid, we do not feel the potential errors were likely large enough to alter our recommendations. Furthermore, scales are the most commonly used structure to estimate age of black bass by state and provincial fisheries agencies (Maceina et al. 2007) and have been validated for younger ages of Smallmouth Bass (Boucher 1998).

Estimated growth rates of Smallmouth Bass in our six study sites were slower than those reported in the literature for many North America populations, but consistent with previous studies in Ozark streams. Mean lengths at age 7 ranged between the 10 and 25 percentiles of North American Smallmouth Bass populations (Jackson et al. 2008). Growth coefficients (*K*) from the von Bertalanffy growth equation were all less than the median of 0.16 from Beamesderfer and North (1995). However, growth coefficients were similar to the Jacks Fork (0.12) and Gasconade (0.14) from Kruse and DeiSanti (2002). The slower growth of Ozark stream Smallmouth Bass is likely a reflection of the nutrient-poor soils and lower primary productivity of these systems (Purkett 1958; Herlihy and Sifneos 2008). Improving growth rates will be difficult in these systems.

Natural mortality rates at five sites were near 30% and were higher than their respective exploitation rate. These natural mortality rates might seem high, but they are less than the median value (39%) from 409 Smallmouth Bass populations from North America (Beamesderfer and North 1995). The combined effects of natural mortality and slow growth limit the effectiveness of higher length limits. The Castor River population is an example of this interaction between growth and natural mortality. If a 381 mm minimum length limit were implemented at this site, 305 mm fish would be protected from harvest for another 3.5 years. If a population started with 100, 305 mm fish, only 29 fish would remain 3.5 years later. Natural mortality would have claimed a majority of the fish during the 3.5-year period, which significantly decreases the effectiveness of the higher length limit. Simulation model results indicate that increasing the length limit at this site would only add 2.4 fish per 100 recruits while decreasing yield by 60%. It is unlikely anglers would notice the difference in the number of larger fish, but anglers would definitely notice the decrease in yield. Since a majority of surveyed anglers continue to support regulated harvest of Smallmouth Bass (Ackerson et al. 2012), no regulation change was recommended at this site or the Black River site, which had similar population rates and simulation results.

The low exploitation rates at the Courtois, Current River-Big Spring, and North Fork sites were between one-half and one-quarter their respective total annual mortality rate at each site. When exploitation is a small portion of total annual mortality and growth rates are slow, higher length limits are less effective at increasing the number of larger fish within a population. Simulation results indicated increasing the length limit to 381 mm at these sites would only increase the number of 381 mm fish by 29–64% while decreasing yield by 33–56%. Achieving widespread angler support for a higher length limit would be difficult at these sites. Therefore, we did not recommend changing harvest regulations at these three sites.

Simulation modeling results indicated only the Current River-Powder Mill site would substantially benefit from a higher length limit. This site had one of the highest rates of exploitation while also having the lowest natural mortality rate of the six sites. Simulation results indicated that growth overfishing was occurring under current conditions at this site. Maximizing yield is not an appropriate goal of this fishery, but it is important to realize that the fishery could be producing more and larger fish. Therefore, we recommended engaging stakeholders to determine angler interest in a 381 mm minimum length limit at this site.

This study provided a scientific-based assessment of Smallmouth Bass harvest at six Ozark stream sites. The dynamic rate function estimates generated from this work and the subsequent simulation modeling provide information needed to manage these Smallmouth Bass fisheries more effectively. This study not only provides information on these six sites, but also gives some insight into the variability of Smallmouth Bass population dynamics among Ozark streams. Under most conditions, the statewide length limit of 305 mm is adequate to balance the desire of quality fishing and harvest opportunities on most Ozark streams. When high exploitation rates, adequate growth rates, and low natural mortality rates exist, a higher length limit can be used to improve a fishery. Finding this balance with anglers will continue to be a challenge as conditions and attitudes change over time.

Acknowledgments

We would like to thank Lynn Schrader, Andy Austin, and Brian Todd for their thorough review of the manuscript. We would also like to thank Ivan Vining for assistance with statistical analysis and Alice Uhrhan, Dominique Priester, Gaye Puchbauer, and Eileen Meier for their help processing tag returns. This study was funded by Federal Aid in Sportfish Restoration Project through the F50D grant administered by the U.S. Fish and Wildlife Service.

References

Ackerson, J. R., C. J. Fuller, J. A. Girondo, K. J. Meneau, and R. R. Reitz. 2012. Smallmouth Bass and Rock Bass fishing in Missouri, results from the 2011 survey. Final report. Missouri Department of Conservation, Jefferson City.

Allen, M. S., C. J. Walters, and R. Myers. 2008. Temporal trends in Largemouth Bass mortality, with fishery implications. North American Journal of Fisheries Management 28:418–427.

Allen, M. S., and J. E. Hightower. 2010. Fish population dynamics: mortality, growth, and recruitment. Pages 43–79 *in* W. A. Hubert and M. C. Quist, editors. Inland Fisheries Management in North America, 3rd edition. American Fisheries Society, Bethesda, Maryland.

Beamesderfer, R. C. P., and J. A. North. 1995. Growth, natural mortality, and predicted response to fishing for Largemouth Bass and Smallmouth Bass populations in North America. North American Journal of Fisheries Management 15:688–704.

Beaverton, R. J. H., and S. J. Holt. 1957. On the dynamics of exploited fish populations. Chapman and Hall, London.

Boucher, D. P. 1998. Precision and relative accuracy of Maine black bass age estimates from scales and otolith sections. Final report. Maine Department of Inland Fisheries and Wildlife, Fisheries Progress Report Series No. 98–1, Augusta.

Carlson, A. J., and D. A. Isermann. 2010. Mandatory catch and release and maximum length limits for Largemouth Bass in Minnesota: is exploitation still a relevant concern? North American Journal of Fisheries Management 30:209–220.

Fajen, O. F. 1975a. The standing crop of Smallmouth Bass and associated species in Courtois Creek. Pages 240–249 *in* R. H. Stroud and H. Clepper, editors. Black bass biology and management. Sport Fishing Institute, Washington, D.C.

Fajen, O. F. 1975b. The effect of a "fish-for-fun" regulation on black bass upon the standing crop of fish in Courtois Creek. Final report.

Missouri Department of Conservation, Jefferson City.

Fajen, O. F. 1981. An evaluation of the 12-inch minimum length limit on black bass in streams. Final report. Missouri Department of Conservation, Jefferson City.

Fiss, F. C., T. A. Clevland, B. D. Carter, R. D. Bivens, and J. M. Swearengin. 2001. Population characteristics of riverine Smallmouth Bass in Tennessee, simulated effects of length limits, and management recommendations. Tennessee Wildlife Resources Agency, Fisheries Report 01–19.

Fleener, G. G. 1975. Harvest of Smallmouth Bass and associated species in Courtois Creek. Pages 250–256 *in* R. H. Stroud and H. Clepper, editors. Black bass biology and management. Sport Fishing Institute, Washington, D.C.

Funk, J. L., and G. G. Fleener. 1974. The fishery of a Missouri Ozark stream, Big Piney River, and the effects of stocking fingerling Smallmouth Bass. Transactions of the American Fisheries Society 103:751–771.

Hayes, M. C., L. F. Gates, and S. A. Hirsch. 1997. Multiple catches of Smallmouth Bass in a special regulation fishery. North American Journal of Fisheries Management 17:182–187.

Heidinger, R. C., and K. Clodfelter. 1987. Validity of the otolith for determining age and growth of Walleye, Striped Bass, and Smallmouth Bass in power plant cooling ponds. Pages 241–251 *in* R. C. Summerfelt and G. E. Hall, editors. Age and growth of fish. Iowa State University Press, Ames.

Herlihy, A. T., and J. C. Sifneos. 2008. Developing nutrient criteria and classification schemes for wadeable streams in conterminous United States. Journal of the North American Benthological Society 27:932–948.

Isermann, D. A., and C. P. Paukert. 2010. Regulating harvest. Pages 185–212 *in* W. A. Hubert and M. C. Quist, editors. Inland fisheries management in North America, 3rd Edition. American Fisheries Society, Bethesda, Maryland.

Jackson, Z. L., M. C. Quist, and J. G. Larscheid. 2008. Growth standards for nine North American fish species. Fisheries Management and Ecology 15:107–118.

Jones, R. 1957. A much simplified version of the full yield equation. International Commission for Northeast Atlantic Fisheries, International Council for the Exploration of the Sea, and Food and Agricultural Organization of the United Nations. Document P.21, Lisbon, Portugal.

Kolander, T. D., D. W. Willis, and B. R. Murphy. 1993. Proposed revision of the standard weight (W_s) equation for Smallmouth Bass. North American Journal of Fisheries Management 13:398–400.

Kruse, M., and K. DeiSanti. 2002. Evaluation of strategies for quality management of Smallmouth Bass in Missouri streams. Final report. Missouri Department of Conservation, Jefferson City.

Long, J. M., and W. L. Fisher. 2001. Precision and bias of Largemouth, Smallmouth, and Spotted Bass ages estimated from scales, whole otoliths, and sectioned otoliths. North American Journal of Fisheries Management 21:636–645.

Maceina, M. J., P. W. Bettoli, S. D. Finely, and V. J. DiCenzo. 1998a. Analyses of the Sauger fishery with simulated effects of a minimum size limit in the Tennessee River of Alabama. North American Journal of Fisheries Management 18:66–75.

Maceina, M. J., O. Ozen, M. S. Allen, and S. M. Smith. 1998b. Use of equilibrium yield modeling to evaluate length limits for crappie in Weiss Lake, Alabama. North American Journal of Fisheries Management 18:854–863.

Maceina, M. J., and S. M. Sammons. 2006. An evaluation of different structures to age freshwater fish from a northeastern US river. Fisheries Management and Ecology 13:237–242.

Maceina, M. J., J. Boxrucker, D. L. Buckmeier, R. S. Gangl, D. O. Lucchesi, D. A. Isermann, J. R. Jackson, and P. J. Martinez. 2007. Current status and review of freshwater fish aging procedures used by state and provincial fisheries agencies with recommendations for future directions. Fisheries 32:329–340.

Meneau, K. J. 1998. Smallmouth Bass special management in Missouri streams. Final report. Missouri Department of Conservation, Jefferson City.

Meneau, K. J. 2009. Smallmouth Bass white paper summary. Final report. Missouri Department of Conservation, Jefferson City.

Meyer, K. A., F. S. Elle, J. A. Lamansky Jr., E. R. J. M. Mamer, and A. E. Butts. 2012. A reward-recovery study to estimate tagged-fish reporting rates by Idaho anglers. North American Journal of Fisheries Management 32:696–703.

Meyer, K. A., and D. Schill. 2014. Use of a statewide angler tag reporting system to estimate rates of exploitation and total mortality for Idaho sport fisheries. North American Journal of Fisheries Management 34:1145–1158.

Miranda, L. E., R. E. Brock, and B. S. Dorr. 2002. Uncertainty of exploitation estimates made from tag returns. North American Journal of Fisheries Management 22:1358–1363.

Nigh, T. A., and W. A. Schroder. 2002. Atlas of Missouri ecoregions. Missouri Department of Conservation, Jefferson City.

Paragamian, V. L. 1984. Evaluation of a 12.0-inch minimum length limit on Smallmouth Bass in the Maquoketa River, Iowa. North American Journal of Fisheries Management 4:507–513.

Paukert, C., M. McInerny, and R. Schultz. 2007. Historical trends in creel limits, length-based limits, and season restrictions for black basses in the United States and Canada. Fisheries 32:62–72.

Pollock, K. H., J. M. Hoenig, W. S. Hearn, and B. Calingaert. 2001. Tag reporting rate estimation: 1. An evaluation of the high-reward tagging method. North American Journal of Fisheries Management 21:521–532.

Purkett, C. A. 1958. Growth rates of Missouri stream fishes. Final report. Missouri Department of Conservation, Jefferson City.

Sammons, S. M. 2016. Catch and exploitation of Shoal Bass in the Flint River, Georgia, USA: implications for harvest restrictions. North American Journal of Fisheries Management 36:606–620.

Smith, M. 1998. Bourbeuse River Smallmouth Bass population evaluation. Final report. Missouri Department of Conservation, Jefferson City.

Slipke, J. W., M. J. Macenia, V. H. Travnicheck, and K. C. Weathers. 1998. Effects of a 356-mm minimum-length limit on the population characteristics and sport fishery of Smallmouth Bass in the Shoals Reach of the Tennessee River, Alabama. North American Journal of Fisheries Management 18:76–84.

Slipke, J. W., and M. J. Maceina. 2010. Fishery analyses and modeling simulator (FAMS 1.0). American Fisheries Society, Bethesda, Maryland.

Taylor, R. G., J. A. Whittington, W. E. Pine, III, and K. H. Pollock. 2006. Effect of different reward levels on tag reporting rates and behavior of common snook anglers in southeast Florida. North American Journal of Fisheries Management 26:645–651.

von Bertalanffy, L. 1938. A quantitative theory of organic growth. Human Biology 10:181–213.

Weathers, K. C., and M. B. Bain. 1992. Smallmouth Bass in the Shoals Reach of the Tennessee River: population characteristics and sport fishery. North American Journal of Fisheries Management 12:528–537.

Westoff, J. T., C. Paukert, S. Ettinger-Dietzel, H. Dodd, and M. Siepker. 2016. Behavioral thermoregulation and bioenergetics of riverine Smallmouth Bass associated with ambient cold-period thermal refuge. Ecology of Freshwater Fish 25:72–85.

Wilde, G. R. 1997. Largemouth Bass fishery responses to length limits. Fisheries 22:14–23.

Williamson, C. J. 2012. Salt River Smallmouth Bass evaluation. Final report. Missouri Department of Conservation, Jefferson City.

United States Fish and Wildlife Service. 2011. A national survey of fishing, hunting, and wildlife-associated recreation. United States Fish and Wildlife Service FHW/11-NAT (RV), Washington, D.C.

American Fisheries Society Symposium 87:129–145, 2019

Factors Affecting Black Bass Species Composition in Louisiana Gulf Slope and Lower Mississippi Alluvial Valley Streams

Michael D. Kaller*, Tiffany E. Pasco, William E. Kelso, and Debra G. Kelly

School of Renewable Natural Resources, Louisiana State University Agricultural Center, 227 Renewable Natural Resources Building Baton Rouge, Louisiana, 70803, USA

Abstract.—Historic reports and museum records describe a distinct spatial distribution pattern and habitat partitioning between Largemouth Bass *Micropertus salmoides salmoides* and Spotted Bass *Micropterus punctulatus*. However, beginning in the mid-20th Century, the literature describes more co-occurrences and less species specificity among habitats and distributions. Herein, we conducted a retrospective analysis of recently collected data (2005–2014) to assess: (1) how current distributional patterns of Spotted Bass and Largemouth Bass compare with historic distributions in Louisiana; (2) evidence of species-specific habitat associations based on abundance and habitat data collected; and (3) whether the Florida Largemouth Bass *M. salmoides floridanus* stocking program has affected Spotted Bass populations in adjacent river systems. Analyses of 68 wadeable streams and rivers indicated that current spatial distributions matched historic descriptions for 6 of 9 major drainages, with evidence still supporting distinct habitat associations for each species. Importantly, Spotted Bass did appear to be negatively influenced by widespread Largemouth Bass stocking. Further, construction of reservoirs across the state and landscape conversion from wet prairie and forest to intensive agriculture may have negatively impacted both species. Overall, habitat alteration and loss was the most widespread explanation for declines in Spotted Bass, with Largemouth Bass stocking of secondary and more localized importance.

*Corresponding author: mkalle1@lsu.edu

Introduction

Unlike other southern U.S. states, Louisiana streams, rivers, lakes, and reservoirs support only two native Black Bass species, Largemouth Bass *Micropterus salmoides salmoides* and Spotted Bass *Micropterus punctatus*. These popular sportfishes are socio-culturally important (e.g., Louisiana hosted six Bassmaster events during 2016), and largely support the U.S.$953 million recreational fishery within the State (Southwick Associates 2008). Writing shortly after the determination that Spotted Bass in Louisiana were not a form of Smallmouth Bass (Viosca 1931), Gowanloch (1933) reported strong differences in habitat associations and geographic distributions of each species, which has been subsequently supported by natural history museum records (e.g., FishBase hosted by Tulane University, New Orleans, Louisiana). Spotted Bass were most often associated with swifter current velocities and sand and gravel substrates, whereas Largemouth Bass were associated with slower current velocities and substrate dominated by silt (Gowanloch 1933; Carver 1975). Consistently, the earliest literature, reports, and popular press described Spotted Bass occurring in quite different habitats and regions within the state than Largemouth Bass (Viosca 1931; Gowanloch 1933; Gresham 1958). Biogeographic studies of museum specimens and sampling reports of the late 19th and early 20th Century supported these observations and suggested that habitats and spatial distributions exploited by the two species differed (Connor and Suttkus 1986). Moreover, during this time period, data indicated only one of the two species should have been present in a given stream, or if both were present, one should have been dominant (Swift et al. 1986). In contrast, later publications reported statewide distributions of both species (Douglas 1974; Felley 1992) and co-occurrences within stream habitats (Watson et al. 1981), without comment on the unexpected nature of species co-occurrence, given the previous widely accepted delineations. The statewide occurrence described by Douglas (1974) could have represented a better and more nuanced understanding of distribution and habitat associations based on a greater number of samples and specimens, or it could have meant that the species were common throughout the state but did not necessarily co-occur. Similarly, the lack of comment on the co-occurrence of both bass species by Watson et al. (1981) could be attributable to the fact that the study was not focused on natural history of fishes in the Blind River. However, we believe it more likely that information presented in these sources represent recognition of a change in occurrence and distribution of these two species. Further, we propose that the lack of distinct geographic dominance and habitat partitioning between Largemouth Bass and Spotted Bass could be the result of anthropogenic alteration of streams and rivers across the landscape (including stocking), as noted by Felley (1992) for fishes in general and Carver (1975) for Spotted Bass specifically.

Since the publication of Gowanloch (1933) and collection of early museum records from Louisiana streams, two important anthropogenic changes in aquatic system structure and function have occurred within Louisiana. First, 73 reservoirs were constructed throughout the state between 1900 and 2000, which may have had pervasive impacts on native fishes, as observed elsewhere (e.g., Schlosser et al. 2000; Gido et al. 2002; Herbert and Gelwick 2003; Falke and Gido 2006; Guenther and Spacie 2006; Freeman and Marcinek 2006; Matthews and Marsh-Mathews 2007; Franssen and Tobler 2013; Freedman et al. 2014). Southern Louisiana lacks rugged topography and large elevation changes, hence, reservoirs in these landscapes tend to be shallow and characterized by extensive emergent vegetation and trees (Fries 2010; Reuter 2015). In contrast, hilly and mountainous regions support deeper and more complex reservoir habitats that characteristically support robust populations of Spotted Bass and Largemouth Bass (e.g., Novinger 1987; Sammons and Bettoli 1999; Greene and Maceina 2000; Hunter and Maceina 2008). Viosca (1953) reported that Louisiana reservoirs supported both species, although Largemouth Bass dominated most samples. After reservoir construction, Carver (1975) reported upstream population reductions and localized extirpations of Spotted Bass in previously productive habitats, similar to Reuter et al. (in press), who reported few Spotted Bass and numerous Largemouth bass in many reservoir tributaries.

The second factor that may be strongly influencing interspecific interactions between these two *Micropterus* species is the Louisiana Department of Wildlife and Fisheries Largemouth Bass stocking program. Since 1981, approximately 104 million predominantly Florida Largemouth Bass *M. salmoides floridanus* have been stocked in large rivers and reservoirs throughout the state (this total includes a small number of hybrid Florida/northern Largemouth Bass spawned and stocked at the request of state biologists based on management objectives for a few waterbodies in the state). To date, impacts of the Largemouth Bass stocking program on Spotted Bass populations in reservoirs or adjacent tributary streams have not been evaluated.

Louisiana is similar to Texas in that the saltwater termini of many rivers limits fish dispersal among drainages (Hoeinghaus et al. 2007), resulting in taxonomically distinct fish assemblages within drainages, with anthropogenic alterations to stream habitat and natural and anthropogenic barriers to connectivity reducing potential recolonization of Spotted Bass (Kaller et al. 2013). Because of observed reductions in Spotted Bass populations in streams where the species was historically known and, in places, dominant (Gowanloch 1933; Lambou 1959; Carver 1975; Reuter 2015), we retrospectively analyzed existing data to examine: (1) how current distributional patterns of Spotted Bass and Largemouth Bass compare with historic distributions in Louisiana; (2) evidence of species-specific habitat associations based on abundance and habitat data; and (3) potential effects of the Largemouth Bass stocking program on Spotted Bass populations in adjacent riverine systems. We anticipate results of these analyses will be helpful in conservation and management of Spotted Bass and Largemouth Bass in coastal plain streams.

Methods

We determined historic distributions from natural history museum, state, and federal field survey data that included complete spatial information (i.e., GPS coordinates or clear descriptions with a geographic reference point, such as a bridge), and from literature and agency reports lacking complete spatial information, but including named water bodies. We were specifically interested in historic data that were collected prior to or during the impoundment construction period (1940–2000) and the first Largemouth Bass stocking event in these drainages (first stocking events ranged from 1981 to 1986). Historic field survey data sources were described in detail in Kaller et al. (2013) and included data collected by Louisiana State University (LSU), Louisiana Department of Wildlife and Fisheries, Louisiana Department of Environmental Quality, U.S. Geological Survey (National Water Quality Data Assessment), U.S. Department of Agriculture Forest Service, and U.S. Environmental Protection Agency (Regional Environmental Monitoring and Assessment). Museum data were obtained from examination of voucher and teaching collections at Louisiana State University and by accessing FishBase, hosted by Tulane University, New Orleans, Louisiana. All data included definitive and complete spatial information, although they varied in collection methodology (e.g., seining, single backpack or dual backpack electrofishing, barge electrofishing, or sometimes by both seining and electrofishing), so all data were simplified to the presence of one or both species. Literature (e.g., Gowanloch 1933) and agency reports frequently reported only presence of a species in a stream or river system, and these data were considered presence reports. Dominance of Spotted Bass or Largemouth Bass within a drainage was determined based on which species occurred in the largest number of sampling events during the historic period (Figure 1a). The Atchafalya River was excluded from these analyses because of unique management actions, including harvest restrictions and extensive stocking following Hurricane Andrew in 1992 (Perret et al. 2010), which we believed would complicate comparisons.

From 2005 through 2014, we collected Spotted Bass and Largemouth Bass from 68 wadeable rivers and streams in the lower Mississippi Alluvial Valley and Louisiana Gulf Slope that either were stocked or were connected to locations stocked with Florida Largemouth Bass. Each stream or river was sampled 2–4 times (162 total sampling events), depending on the nature of the project active at the time. Although projects varied in purpose and funding source (e.g., State Wildlife Grant Program through Louisiana Department of Wildlife and Fisheries and stream surveys for the Louisiana Department of Environmental Quality), the sampling methodology was consistent across projects. At each sampling location, a 150–200 m reach was isolated by block nets. Within the reach, 2–3 fish removal passes were conducted with backpack DC electrofishing (Halltech Aquatic Research, HT-2000, or Smith-Root, LR24, 150–200 V, a minimum output of 1.8 amps), followed by 2–3 hauls with a 10 m seine (6.35 mm bar mesh). Most fishes were identified, enumerated and released on site, with unidentifiable individuals placed in an ice slurry and returned to LSU for identification. Abundance was estimated by the Zippin estimator (2 passes) or the three pass estimation method described in Hayes et al. (2007), with sample abundance expressed as *n*/ha. All fish handling following LSU Institutional Animal Care and Use Committee (IACUC) A2011–16 and earlier approved IACUC protocols.

At each sampling site, we also used a multi-probe meter (model 650, YSI, Inc.) to determine temperature (°C), specific conductance (mS/cm), dissolved oxygen (DO; mg/L), pH, and turbidity (NTU). Ten cross-sectional transects, 15 m apart, were spaced along each sample reach. Along each transect, wetted width, stream bank stability, riparian cover, and percent canopy cover were recorded. Additionally, we recorded depth (m), flow velocity at 0.6 of depth (m/s; SONTEK, YSI, Inc.), and the abundance of woody debris (recorded as count) at 25%, 50%, and 75% of stream width. Substrate was visually categorized as silt, clay, sand, and gravel along transects following the frequency of size classes method in Bain (1999).

Comparison of historic and current distributions involved a sequence of analyses. First, differences in total black bass abundance (*n*/ha), Spotted Bass abundance (*n*/ha), and Largemouth Bass abundance (*n*/ha) among drainages were assessed by generalized linear mixed models (GLMM; PROC GLIMMIX, SAS vers. 9.4, SAS, Institute, Inc., Cary, North Carolina), with drainage as the fixed explanatory variable for among-drainage comparisons, an interaction of drainage with species for within-drainage comparisons, and sampling site nested within drainage as a random variable. Estimation was based on Laplace approximation. Multiple candidate GLMMs (e.g., log link and

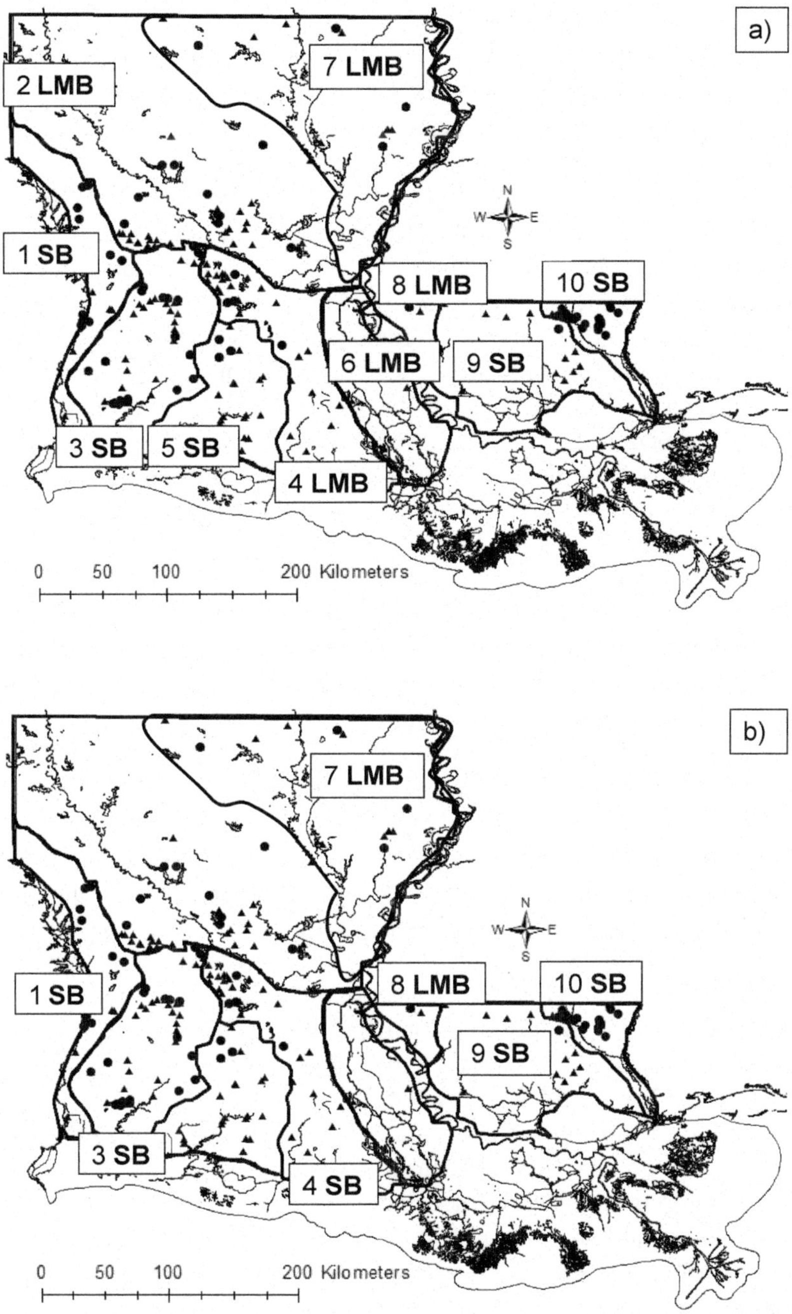

Figure 1. Historic Largemouth Bass (LMB) and Spotted Bass (SB) distributions based upon reports and (a) museum records and current dominance based on 2005–2014 sampling (b). Triangles represent historic records, and dots indicate current locations. No labels indicates no dominance. From west to east, drainages are: (1) Sabine River; (2) Red River; (3) Calcasieu River; (4) Vermillion/Teche River; (5) Mermentau River; (6) Atchafalaya River (not included in study); (7) Ouachita River; (8) Mississippi River; (9) Lake Ponchartrain; and (10) Pearl River.

Poisson distribution, log link and negative binomial distribution, and inverse link and gamma distribution) were assessed for fit to the data and appropriateness for interpretation by examining AIC and ĉ (χ^2/degree of freedom). The GLMM that exhibited the lowest AIC and a ĉ closest to but not less than 1 was selected for interpretation. Differences among drainages were evaluated posthoc by least square means adjusted by the Tukey-Kramer method. Dominance for these collections was based on which species had a significantly greater abundance (*n*/ha) within a drainage (i.e., the magnitude of the difference was substantial and statistically significant). At the drainage scale, we then compared species dominance from the historic (greater number of sites with species) and current (statistically greater abundance) data sets to assess differences in species dominance through time.

Habitat data were reorganized by ordination to evaluate relationships between habitat characteristics with current Spotted Bass and Largemouth Bass abundance data, and bass abundances were analyzed with GLMMs as a function of linear combinations of habitat components (Table 1). Several candidate ordinations were attempted (e.g., principal component analysis, detrended correspondence analysis, and nonmetric multidimensional scaling; Program R, vers.3.3. R Core Team 2015; Oksanen et al. 2017 and SAS vers. 9.4, SAS, Institute, Inc., Cary, North Carolina). Following STRESS2 and axis length criteria (ter Braak and Verdonschot 1995; Hirst and Jackson 2007), principal component analysis with varimax rotation was selected to reorganize the habitat variables into principal components for GLMM analyses. The number of principal components retained for analysis was determined with the broken stick method (Horn's Test; Hirst and Jackson 2007). Similar to the GLMMs previously described, total black bass abundance (*n*/ha), Spotted Bass abundance (*n*/ha), and Largemouth Bass abundance (*n*/ha) were the response variables with habitat principal components as fixed explanatory variables and sampling site nested within drainage as a random variable. Estimation and GLMM assessment were as described above.

We used a GLMM with Spotted Bass abundance (*n*/ha) as the response variable, and number of Largemouth Bass stocked in study streams or directly-connected, adjacent waters as fixed explanatory variables, and site nested within drainage as a random variable, to assess the impact of Largemouth Bass stocking on Spotted Bass populations. Of the approximately 104 million Largemouth Bass stocked in Louisiana since 1981, 32,808,221 Largemouth Bass were stocked in study streams or directly-connected, adjacent waters (Table 1). Because previous analyses of the stocking program indicated the cumulative number of fish stocked in a location was a good predictor of establishment of stocked fish (Kaller et al. 2006), we used this number in each drainage as the fixed explanatory variable in the analysis. Estimation and GLMM assessment were as described above. Lake Ponchartrain drainage sites were not included in this analysis because none of the current samples were collected in streams or connected to water bodies that were stocked.

Results

Current and Historic Distributions of Spotted Bass and Largemouth Bass

Spotted Bass and Largemouth Bass exhibited patterns of dominance by drainage (Figure 2). Spotted Bass density was greater ($F_{7,62} = 3.21$, $P < 0.01$) in Lake Ponchartrain, Pearl River, and Vermillion River drainages, whereas Largemouth Bass density was greater ($F_{7,62} = 3.75$, $P < 0.01$) in the Ouachita River (a tributary of the Red River) and Sabine River drainages. Analyses of total black bass did not detect any statistically significant dif-

Table 1. Cumulative Florida Largemouth Bass and Largemouth Bass (LMB) stocking from 1982–2016 in areas either directly sampled or connected to sampled rivers and streams in Louisiana. Sabine River drainage stocking data includes Louisiana stocking data only.

River Drainage	Florida LMB	LMB	All LMB
Calcasieu	537,600	13,650	551,160
Mermentau	249,435	0	249,435
Ouachita	10,089,101	72,235	10,161,336
Pearl	610,359	66,190	676,549
Red	1,540,511	0	1,540,511
Sabine	18,421,813	0	18,421,813
Vermillion/Teche	68,226	0	68,226

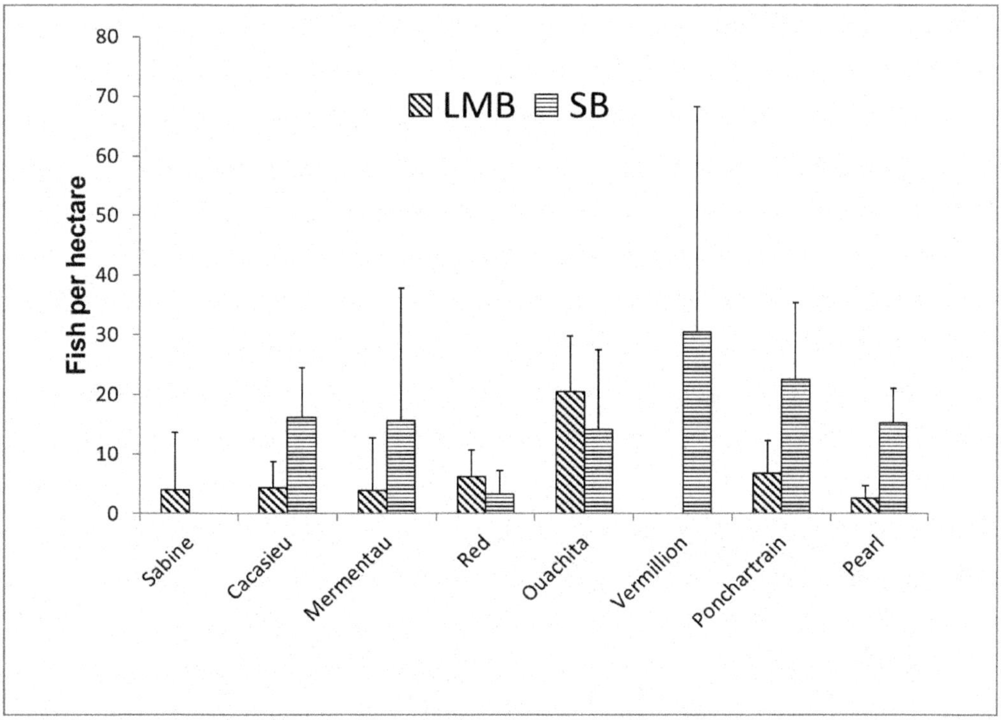

Figure 2. Mean Largemouth Bass and Spotted Bass abundance (*n*/ha) with standard error bars across the studied river drainages in Louisiana.

ferences among drainages. Comparisons with reported historic distributions and dominance patterns indicated that Largemouth Bass remained dominant in the Ouachita River drainage (i.e., were reported at a majority of historical sampling sites and the difference in abundance (*n*/ha) was greater than Spotted Bass) but were no longer dominant in other Red River drainages and the Vermillion River drainage. Spotted Bass dominance remained consistent with historic data in the Calcasieu River, Lake Ponchartrain, and Pearl River drainages, but this species was no longer detected in the historically-documented Sabine River drainage (Figure 1b).

Stream Habitat Characteristics and Density of Spotted Bass and Largemouth Bass

Five principal components selected by broken stick analysis explained 76% of the variation in habitat and water quality data in the study streams (Tables 2 and 3). Spotted Bass densities were positively associated with greater current velocity and higher dissolved oxygen (principal component 3; $F_{1,129} = 17.77$, $P < 0.01$; Figure 3). In contrast, Largemouth Bass densities were positively associated with silt substrates and greater accumulations of woody debris (principal

Table 2. Habitat and water quality variables in each drainage from 2005–2014 stream surveys in Louisiana. Values are means (SE) based on *n* sampling events.

Variable	Sabine (*n* =14)	Calcasieu (*n* = 31)	Mermentau (*n* = 9)	Red (*n* = 34)
Woody debris (*n*/150m)	4.9 (1.6)	12.6 (6.2)	16.8 (2.4)	2.3 (0.4)
Dissolved oxygen (mg/L)	5.6 (0.5)	5.6 (0.5)	4.0 (0.8)	3.3 (0.5)
Current velocity (m/s)	0.1 (0.05)	75.0 (20.5)	16.9 (6.8)	0.5 (0.8)
Turbidity (NTU)	36.8 (5.2)	42.3 (10.9)	119.6 (59.4)	23.3 (6.5)
pH	6.6 (0.1)	6.6 (0.1)	6.8 (0.2)	6.7 (0.09)
Specific conductance (mS/cm)	0.4 (0.1)	0.1 (0.2)	0.1 (0.03)	0.1 (0.02)
Silt substrate (%)	43.6 (13)	22.6 (7.6)	77.8 (14.7)	5.9 (4.1)
Clay substrate (%)	0	0	22.2 (14.7)	5.9 (4.1)
Sand substrate (%)	43.6 (13)	74.2 (8.0)	0	88.2 (5.6)
Variable	**Ouachita (*n* = 14)**	**Vermillion (*n* = 31)**	**Ponchartrain (*n* = 9)**	**Pearl (*n* = 34)**
Woody debris (*n*/150m)	25.6 (10.3)	16.6 (9.8)	7.2 (1.4)	3.5 (0.6)
Dissolved oxygen (mg/L)	2.9 (0.4)	6.5 (1.1)	6.5 (0.3)	7.1 (0.2)
Current velocity (m/s)	21.3 (15.7)	9.5 (2.8)	31.1 (10.6)	99.3 (12.5)
Turbidity (NTU)	49.1 (10.5)	12.0 (2.8)	38.3 (11.5)	17.0 (5.1)
pH	6.9 (0.1)	7.5 (0.08)	6.3 (0.2)	6.6 (0.1)
Specific conductance (mS/cm)	0.2 (0.03)	0.1 (0.03)	0.1 (0.03)	0.1 (0.01)
Silt substrate (%)	100 (0)	42.9 (20.2)	33.3 (14.2)	0
Clay substrate (%)	0	0	0	0
Sand substrate (%)	0	57.1 (20.2)	66.7 (14.2)	100 (0)

Table 3. Correlations between measured habitat and water quality variables with the five principal components (PCs). Bolded correlations met sample size adjusted criteria for interpretation (Stevens 2002).

	PC 1	PC 2	PC 3	PC 4	PC 5
Current velocity (m/s)	–0.01	0.0	**0.80**	–0.17	0.08
Dissolved oxygen (mg/L)	–0.22	–0.16	**0.77**	0.05	0.04
Turbidity (NTU)	0.25	**0.83**	–0.07	0.04	–0.03
pH	0.0	0.06	0.08	**0.84**	0.07
Specific conductance (mS/cm)	0.10	–0.04	–0.22	**0.76**	–0.04
Silt substrate	**0.90**	–0.06	–0.27	0.07	0.02
Clay substrate	–0.10	**0.90**	–0.07	–0.01	0.01
Sand substrate	**–0.83**	–0.24	0.23	–0.08	–0.37
Gravel substrate	–0.01	–0.02	0.14	0.03	**0.93**
Woody debris count	**0.65**	0.06	0.12	0.01	–0.31
% Variation explained	26.4	14.4	12.8	12.5	9.3

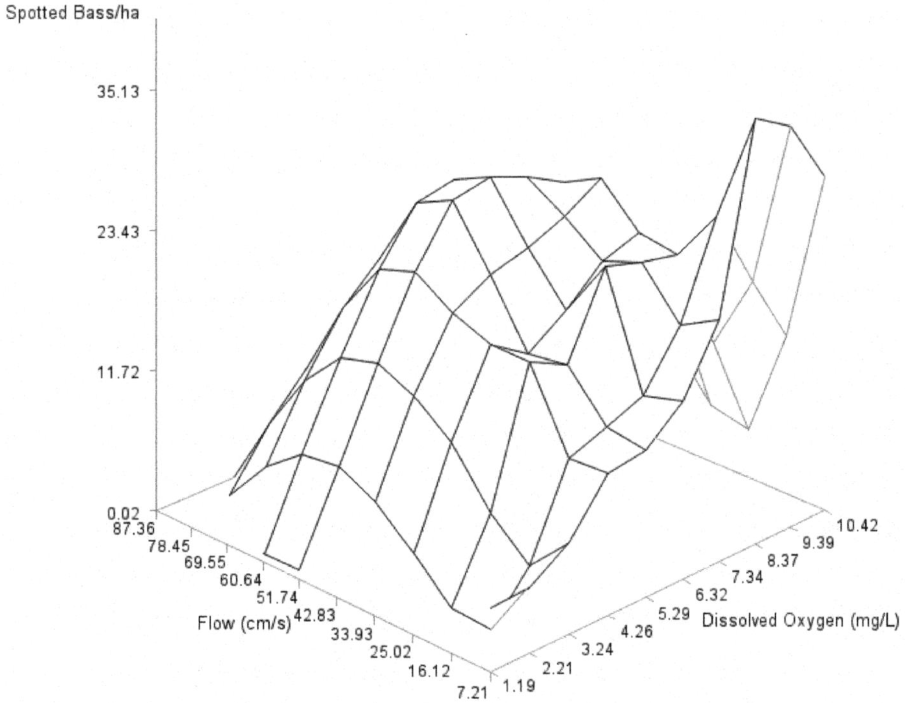

Figure 3. Spotted Bass abundance (*n*/ha) projected with current velocity (m/s) and dissolved oxygen (mg/L) among Louisiana streams. Surface produced by a thin-plate spline with a 0.05 smoothing coefficient (PROC G3GRID, SAS Vers. 9.4).

component 1; $F_{1,78}$ = 4.29, P = 0.04; Figure 4a) and negatively associated with higher current velocities and higher dissolved oxygen (principal component 3; $F_{1,78}$ = 8.98, P < 0.01; Figure 4b). Total bass densities were not significantly related (P > 0.05) to any of the measured habitat and water quality components.

Impacts of Largemouth Bass Stocking on Spotted Bass

Analyses of the impacts of Largemouth Bass stocking on Spotted Bass populations were complicated by habitat changes and changes in stocking site and stocking parameters (e.g., initially based on total surface area and later based on littoral zone area) over time. Moreover, because generalized linear models were employed, inference was based on link function (natural log) transformed estimates. For interpretation, the estimates presented herein were inverse link transformed to abundance (*n*/ha). However, when inverse link transformed, standard errors of the estimates may appear to contradict determinations of statistical significance and inference. It is important to note that given the reported statistically significant F and t values: (1) the link transformed parameter estimates accounted for significantly more variation than the null model; and (2) the link transformed estimates were statistically different from zero. Therefore, by generalized linear model theory and logarithm rules, inference about the impacts of Largemouth Bass stocking drawn from transformed estimates can inform interpretation of the estimates that follow.

The cumulative number of introduced Largemouth Bass had a detectable influence ($F_{1,87}$ = 6.51, P = 0.01) on Spotted Bass densities in streams that had been either directly or indirectly (adjacent, connected rivers and streams) stocked. Overall, Spotted Bass density was lower by –0.99 ± 1.00 SE fish/ha per Largemouth Bass stocked (t_1 = –2.73, P < 0.01). By drainage, reductions in mean density ± SE of Spotted Bass per Largemouth Bass stocked were: Calcasieu River –0.84 ± 1.9 fish/ha; Mermentau River –1.44 ± 1.60 fish/ha; Ouachita River –2.05 ± 1.81 fish/ha; Pearl River –1.30 ± 1.99 fish/ha; Red River –0.52 + –1.74 fish/ha; Sabine River –0.78 ± 1.91 fish/ha; and Vermillion River –1.06 ± 2.04 fish/ha.

Conversely, Largemouth Bass exhibited an overall pattern of significantly higher abundance (mean fish/ha 1.00 ± 1.00 SE) per Largemouth Bass stocked, ($F_{1,87}$ = 5.37, P = 0.02, t_1 = –1.86, P = 0.06). By drainage, increases in mean density ± SE of Largemouth Bass per Largemouth Bass stocked were: Calcasieu River 1.0 ± 1.7 fish/ha; Mermentau River 1.02 ± 1.60 fish/ha; Ouachita River 1.93 ± 1.64 fish/ha; Pearl River 0.98 ± 1.71 fish/ha; Red River 0.96 ± 1.60 fish/ha; Sabine River 0.59 ± 1.67 fish/ha; and Vermillion River 0.98 ± 1.71 fish/ha.

Discussion

Our retrospective analysis of wadeable stream data suggested that historic biogeography remains an important factor in prevailing distributions of Largemouth Bass and Spotted Bass in the lower Mississippi River Alluvial Valley and Louisiana Gulf Slope. Historic records of distribution and dominance matched current data in six of nine evaluated drainages. Habitat associations of both species were also consistent with historic records and with contemporary literature (e.g., Horton and Guy 2002; Warren 2009; Churchill and Bettoli 2015; Claussen 2015). Overall, it appears that habitat alteration and loss was the most widespread explanation for declines in Spotted Bass populations, with Largemouth Bass stocking of secondary and more localized importance. Statistical analyses and projections of observed data suggest that an optimal combination of current

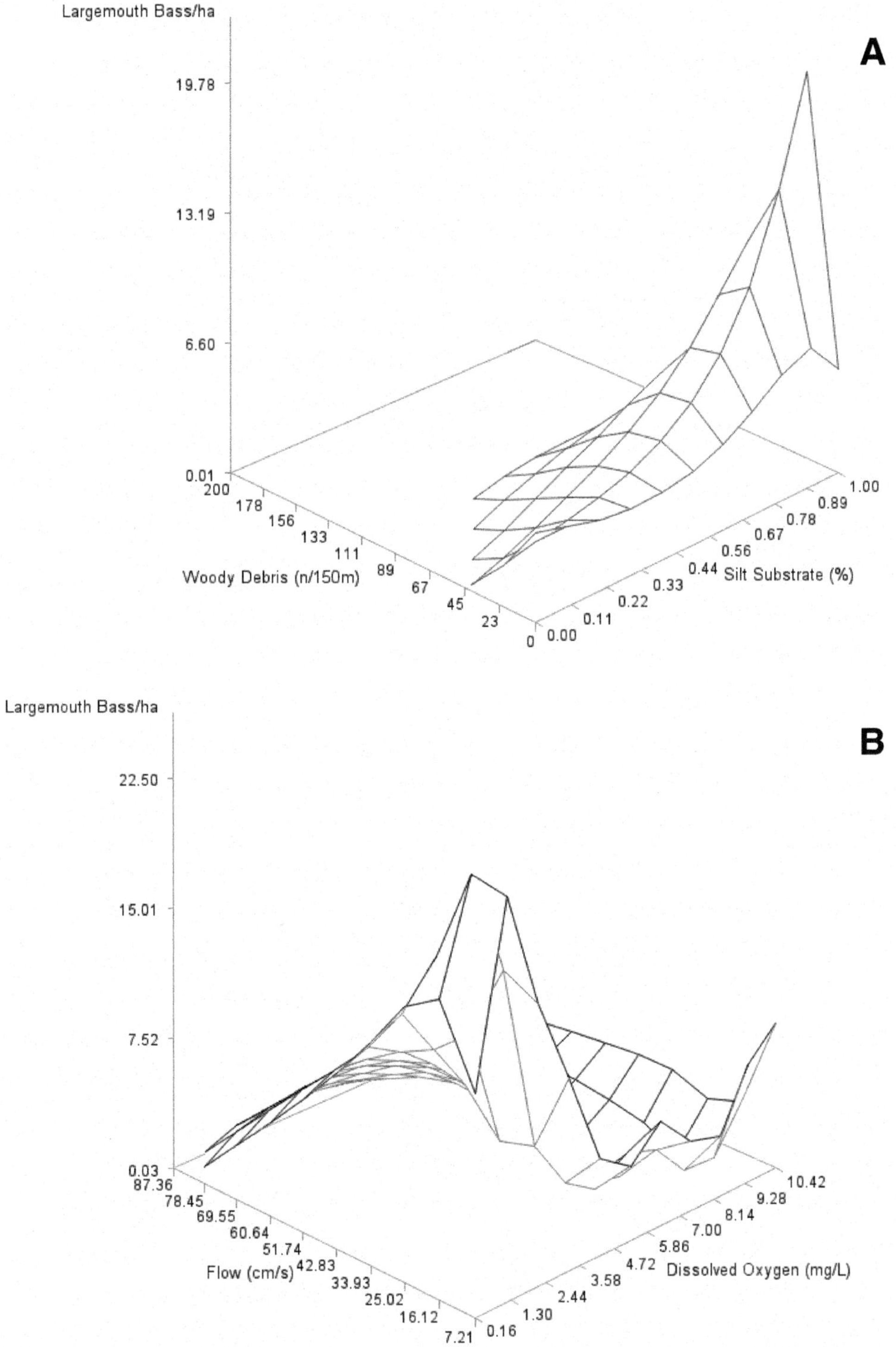

Figure 4. Largemouth Bass abundance projected with (a) silt substrate and woody debris and (b) current velocity (m/s) and dissolved oxygen (mg/L). Surface produced by a thin-plate spline with a 0.05 smoothing coefficient (PROC G3GRID, SAS Vers. 9.4).

velocity (30–70 cm/s) and dissolved oxygen (>3 mg/L) appeared to support the greatest Spotted Bass densities. Although many historic species accounts did not measure stream physicochemistry, one can infer that the common description of Spotted Bass streams as free-flowing and turbulent (Gowanloch 1933; Carver 1975) provided these conditions. Conversely, Largemouth Bass have become dominant in habitats characterized by silt substrates, low current velocities (<30 cm/s), and lower dissolved oxygen levels. Future management efforts designed to enhance and restore Spotted Bass populations in Louisiana should be focused on habitat improvement, specifically the maintenance of currently velocity and turbulence to promote oxygenation and reduce siltation.

The reduction in Spotted Bass in the Red River drainage may be related to anthropogenic changes to this large river system. In the historic data, although not the most frequently reported of the two species, Spotted Bass were reported from many sites, which was not the case in the current data. The topography and geology of the Red River region in Louisiana is particularly suitable for reservoir construction, and the river contains a disproportionate number of reservoirs relative to other drainages in Louisiana. Potentially, as described by Carver (1975), reservoir construction inundated productive Spotted Bass habitats, and/or resulted in the introduction of piscivores and competitors into historic Spotted Bass habitats. A recent study in the Red River region indicated considerable similarities in fish assemblages between reservoir tributaries and nearby nontributary streams (Reuter et al., in press), suggesting changes in fish assemblages in these streams through time were not related to invasion of more lentic species (with the possible exception of Florida and hybrid Largemouth Bass). More likely, widespread inundation and alteration of habitats historically supporting viable Spotted Bass populations has decreased the distribution and abundance of Spotted Bass in this watershed.

More significantly, Spotted Bass were not found in recent collections in Sabine River streams, although they were historically abundant in the watershed. Completion of the 75,000-ha Toledo Bend Reservoir in 1969 inundated 13% of the land and water area within the Sabine River drainage in Louisiana and completely drowned numerous low lying, low gradient streams and bayous. Additionally, the reservoir has been extensively stocked with Largemouth Bass by both Louisiana (Table 1) and Texas (Forshage and Fries 1995; Ashe et al. 2016). It is likely that habitat loss from reservoir inundation and biotic effects from Florida Largemouth Bass stocking have combined to decrease Spotted Bass populations in streams within this system.

Spotted Bass were also no longer dominant in streams within the Mermentau River basin, which has experienced profound changes in land use since the early 1900s, particularly the adoption of intensive rice-crayfish production over the last 40 years (Vidrine 1993; Allen and Vidrine 2003; Vidrine et al. 2004). Wet prairie draining, stream dredging and reconfiguration projects, and extensive urbanization have had tremendous negative impacts on the native aquatic fauna, including freshwater mussels (Vidrine et al. 2004), fish (Fitzgerald 2012), and crayfish (Budnick 2015), which is probably related to widespread hypoxia that has become characteristic of streams throughout this drainage (Felley 1992; Louisiana Department of Environmental Quality 2016). Given the positive association between Spotted Bass density and higher current velocities and dissolved oxygen levels, declines in density have presumably been directly related to changes in water quality associated with agricultural and urban development in this region.

Cumulative Florida Largemouth Bass stocking (1986–2014) may have had a negative effect on Spotted Bass density at the drainage level and may be a contributing factor to the differences noted in current and historic Spotted Bass distributions. Importantly, this study is retrospective and lacks the greater certainty of a stocking experiment or concurrent monitoring data, and given the habitat changes over time and shifting stocking strategies, we caution that declines in Spotted Bass populations may also have been the result of modified habitat facilitating successful stocking, rather than direct competition between the species. Although Florida Largemouth Bass stocking has been reported to have potentially deleterious impacts on endemic Largemouth Bass (e.g., Philipp et al. 2002) and other native fishes (e.g., Jackson 2002), the program is regarded as successful by the Louisiana Department of Wildlife and Fisheries (Hughes and Wood 1995) and anglers (Louisiana Sportsman 2015). As such, the stocking program is an important component of freshwater fisheries management in the state. Our results suggest that potential impacts to Spotted Bass populations should be considered when evaluating future stocking options in rivers and connected systems.

Management implications of our analyses are relatively straightforward. For 67% of the study region, fisheries managers can continue programs for Largemouth Bass and Spotted Bass in systems where they have maintained populations through time, as habitat still appears to be adequate to maintain both species. Where past distributions have changed, however, managers may need to consider actions to conserve and restore lotic habitat. Foremost among recommendations would be to prevent further Spotted Bass habitat loss by enhancing stream gradient to promote year-round flow and reduce summer stagnation and the concomitant depression of dissolved oxygen (Ice and Sugden 2003). Gradient enhancement activities should target erosion prevention and mitigation of sedimentation to prevent infilling and loss of pool-riffle channel structure (Walser and Bart 1999). Additional management includes conservation and restoration of riparian forests to reduce stream temperatures and increase dissolved oxygen levels (Binkley and Brown 1993) and development/improvement of streamside management zones and other agricultural and urban aquatic conservation practices. In this region, dissolved oxygen levels can be enhanced by careful integration of best management practices (DaSilva et al. 2013) into forest management activities, as organic inputs related to harvesting can reduce dissolved oxygen to very low levels (Ice and Sugden 2003). Agricultural development has also been associated with reduced stream dissolved oxygen levels in Louisiana (Fitzgerald 2012; Murphy 2017), and programs to control field runoff and conserve riparian cover could help mitigate hypoxic conditions.

The Largemouth Bass stocking program in Louisiana has been popular among anglers and fisheries managers and has served to increase harvest of trophy fish (>4 kg) in many water bodies throughout the state (Hughes and Wood 1995; Louisiana State Fish Records 2016). Relative to the Sabine River and Ouachita River drainages, fewer fish have been stocked in the Pearl River, Calcasieu River, Mermentau River, and Vermillion/Teche River drainages. Notably, many streams in these drainages continue to support high density Spotted Bass populations [e.g., Hurricane Creek (Vermillion), Calcasieu and Whisky Chitto rivers, Six Mile Creek (Calcasieu), West Fork Caney Creek (Mermentau), and Bogalusa Creek (Pearl)]. Additionally, in free-flowing streams that drain into Lake Pontchartrain, Spotted Bass exhibit growth and mortality rates similar to other riverine Spotted Bass populations in the U.S. that contain adequate habitat (Alford and Heimann 2016). With the exception of infrequent demands from the public to stock Florida Largemouth Bass in some river sections after hurricane-induced fish kills, Lake

Pontchartrain drainage Spotted Bass typically do not co-occur with Florida Largemouth Bass. Future decisions to perennially stock additional Florida Largemouth Bass in these systems should carefully weigh perceived benefits to bass anglers against the potential negative effects on resident Spotted Bass populations.

Acknowledgments

The authors thank Michael Campbell, Anna Evans, Melissa Gray, Brad Hester, A. Raynie Harlan, Ann Hijuelos, Kayla Kimmel, Claire LaBarbera, Tyler Loeb, Samantha Lott, Peter Markos, Allison Martin, Brett Miller, Catherine Murphy, Aaron Podey, Will Sheftall, Adam Piehler, Mason Piehler, Catherine Reuter, Jesse Sabo, Matt Songy, Chad Thomas, Brad Trumbo, Alexandra Trumbo, Rachel Walley, Brian Ward, Jonathan West, Angela Williamson, and Jose Vazquez. Portions of the data used in this study were funded by the Louisiana Department of Environmental Quality and the State Wildlife Grant Program through the Louisiana Department of Wildlife and Fisheries and U.S. Fish and Wildlife Service Division of Federal Aid. This material is based upon work that also was partially supported by the National Institute of Food and Agriculture, U.S. Department of Agriculture, under the McIntire-Stennis Cooperative Forestry Program as project number LAB-94171. This manuscript was approved for publication by the Director of the Louisiana Agricultural Experiment Station as MS 2017–241–30834.

References

Alford, J. B., and B. Heimann. 2016. Spotted bass population structure and diet inwadeable and non-wadeable streams draining the Lake Pontchartrain Basin. Journal of the Southeastern Association of Fish and Wildlife Agencies 3(2015):17–24.

Allen, C. M., and M. F. Vidrine. 2003. Flowering phenology of Cajun Prairie plants. Pages 174–177 *in* S. Foré, editor. Proceedings of the 18th North American Prairie Conference: Promoting Prairie. Truman State University Press, Kirksville, Missouri.

Ashe, D. E., M. T. Driscoll, and J. W. Schlechte. 2016. Stocking contribution of fingerling Largemouth Bass in three aquatic vegetation types in Toledo Bend Reservoir, Texas-Louisiana. Journal of the Southeastern Association of Fish and Wildlife Agencies 3(2015):49–49.

Bain, M. B. 1999. Substrate. Pages 95–103 *in* M. B. Bain and N. J. Stevenson, editors. Aquatic Habitat Assessment: Common Methods. American Fisheries Society, Bethesda, Maryland.

Binkley, D., and T. C. Brown. 1993. Forest practices as nonpoint sources of pollution in North America. Journal of the North American Water Resources Association 29:729–740.

Budnick, W. R. 2015. Metacommunity dynamics and the biogeography of central Louisiana crayfish. Master's thesis. Louisiana State University, Baton Rouge.

Carver, D. C. 1975. Life history study of the Spotted Bass, *Micropterus punctulatus* (Rafinesque) in Six-Mile Creek, Louisiana. Fisheries Bulletin 15. Louisiana Department of Wildlife and Fisheries, Baton Rouge.

Churchill, T. N. and P. W. Bettoli. 2015. Spotted Bass *Micropterus punctulatus* (Rafinesque, 1819). Pages 34–41 *in* M. D. Tringali, J. M. Long, T. W. Birdsong, and M. S. Allen, editors. Black bass diversity: multidisciplinary science for conservation. American Fisheries Society Symposium 82, Bethesda, Maryland.

Claussen, J. E. 2015. Largemouth Bass *Micropterus salmoides* (Lacepede, 1802). Pages 27–34 *in* M. D. Tringali, J. M. Long, T. W. Birdsong, and M. S. Allen, editors. Black bass diversity: multidisciplinary science for conservation. American Fisheries Society Symposium 82, Bethesda, Maryland.

Connor, J. V., and R. D. Suttkus. 1986. Zoogeography of freshwater fishes of the western Gulf Slope. Pages 413–456 *in* C. H. Holcott and E. O. Wiley, editors. The zoogeography of North American freshwater fishes. Wiley, New York.

DaSilva, A., Y. J. Xu, J. Beebe, and G. Ice. 2013. Effects of timber harvesting on dissolved oxygen in a northern Louisiana headwater stream. Forest Science 59:127–138.

Douglas, N. E. 1974. Freshwater fishes of Louisiana. Claitor's Publication Division, Baton Rouge, Louisiana.

Falke, J. A., and K. S. Gido. 2006. Effects of reservoir connectivity on stream fish assemblages in the Great Plains. Canadian Journal of Fisheries and Aquatic Sciences 63:480–493.

Felley, J. D. 1992. Medium–low–gradient streams of the Gulf coastal plain. Pages 233–270 *in* C. T. Hackney, S. M. Adams, and W. H. Martin, editors. Biodiversity of the southeastern United States: aquatic communities. Wiley, New York.

Fitzgerald, A. M. 2012. Effects of varying land use on headwater stream fish assemblages and in-stream habitats in southwestern Louisiana. Master's thesis, Louisiana State University, Baton Rouge.

Forshage, A. A., and L. T. Fries. 1995. Evaluation of the Florida Largemouth Bass in Texas, 1972–1993. Pages 484–491 *in* H. L Schramm and R. G. Piper, editors. Uses and effects of cultured fishes in aquatic ecosystems. American Fisheries Society, Symposium 15, Bethesda, Maryland.

Franssen, N. R., and M. Tobler. 2013. Upstream effects of a reservoir on fish assemblages 45-years following impoundment. Journal of Fish Biology 82:1659–1670.

Freedman, J. A., B. D. Lorson, R. B. Taylor, R. F. Carline, and J. R. Stauffer, Jr. 2014. River of the dammed: longitudinal changes in fish assemblages in response to dams. Hydrobiologia 727:19–33.

Freeman, M. C., and P. A. Marcinek. 2006. Fish assemblage responses to water withdrawals and water supply reservoirs in piedmont streams. Environmental Management 38:435–450.

Fries, M. A. 2010. Introgression, health, and condition of Florida, northern, and Fx hybrid Largemouth Bass in Louisiana water bodies. Master's thesis, Louisiana State University, Baton Rouge.

Gido, K. B., C. W. Hargrave, W. J. Matthews, G. D. Schnell, D. W. Pogue, and G. W. Sewell. 2002. Structure of littoral-zone fish communities in relation to habitat, physical, and chemical gradients in a southern reservoir. Environmental Biology of Fishes 63:253–263.

Gowanloch, J. N. 1933. Fishes and fishing in Louisiana. Bulletin 23. Louisiana Department of Conservation, New Orleans.

Greene, J. C., and M. J. Maceina. 2000. Influence of trophic state on Spotted Bass and Largemouth Bass spawning time and age-0 population characteristics in Alabama reservoirs. North American Journal of Fisheries Management 20:100–108.

Gresham, G. 1958. Fish and Boating in Louisiana. TAM Publications, Shreveport, Louisiana.

Guenther, C. B., and A. Spacie. 2006. Changes in fish assemblage structure upstream of impoundments within the Upper Wabash River basin, Indiana. Transactions of the American Fisheries Society 135:570–583.

Hayes, D. B., J. R. Bence, T. J. Kwak, and B.E. Thompson. 2007. Abundance, biomass, and production. Pages 327–374 *in* C. S. Guy and M. L. Brown, editors. Analysis and interpretation of freshwater fisheries data. American Fisheries Society, Bethesda, Maryland.

Herbert, M. E. and F. P. Gelwick. 2003. Spatial variation of headwater fish assemblages explained by hydrologic variability and upstream effects of impoundment. Copeia 2:273–284.

Hirst, C. N., and D. A. Jackson. 2007. Reconstructing community relationships: the impact of sampling error, ordination approach, and gradient length. Diversity and Distributions 13:361–371.

Hoeinghaus, D. J., K. O. Winemiller, and J. S. Birnbaum. 2007. Local and regional determinants of stream fish assemblage structure: inferences based on taxonomic vs. functional groups. Journal of Biogeography 34:324–338.

Horton, T. B., and C. S. Guy. 2002. Habitat use and movement of Spotted Bass in Otter Creek, Kansas. Pages 161–171 *in* D. P. Phillip and M. S. Ridgway, editors. Black bass: ecology, conservation, and management.

American Fisheries Society, Symposium 31, Bethesda, Maryland.

Hughes, J. S., and M. G. Wood. 1995. Development of a trophy Largemouth Bass fishery in Louisiana. Proceedings of the Southeastern Association of Fish and Wildlife Agencies 49:58–68.

Hunter, R. W., and M. J. Maceina. 2008. Movements and home ranges of Largemouth Bass and Alabama Spotted Bass in Lake Martin, Alabama. Journal of Freshwater Ecology 23:599–606.

Ice, G., and B. Sugden. 2003. Summer dissolved oxygen concentrations in forested streams of northern Louisiana. Southern Journal of Applied Forestry 27:92–99.

Jackson, D. A. 2002. Ecological effects of *Micropterus* introductions: the dark side of Black Bass. Pages 221–232 *in* D. P. Phillip and M. S. Ridgway, editors. Black bass: ecology, conservation, and management. American Fisheries Society, Symposium 31, Bethesda, Maryland.

Kaller, M. D., C. E. Murphy, W. E. Kelso, and M. R. Stead. 2013. Basins for fish and ecoregions for macroinvertebrates: different spatial scales are needed to assess Louisiana wadeable streams. Transactions of the American Fisheries Society 142:767–782.

Kaller, M. D., D. G. Kelly, and W. E. Kelso. 2006. Introgression of *Micropterus salmoides floridanus* into Louisiana Largemouth Bass populations, and the associated effects on length-weight relationships. Final Report, Louisiana Department of Wildlife and Fisheries, Baton Rouge.

Lambou, V. W. 1959. Fish populations of blackwater lakes in Louisiana. Transactions of the American Fisheries Society 88:7–15.

Louisiana Department of Environmental Quality. 2016. LDEQ Water Quality/Integrated Report/ 305(b)/ 303(d)/ 2016 Integrated Report. Louisiana Department of Environmental Quality, Baton Rouge.

Louisiana Sportsman. 2015. The making of America's No. 1 bass fishery: How Toledo Bend became a true trophy destination. Available: http://louisianasportsman.com/details.php?id=8512. (September 2018).

Louisiana State Fish Records. 2016. Available: http://laoutdoorwriters.com/wp-content/uploads/2016/02/LOWALouisianaStateFishRecords-7.pdf. (September 2018).

Matthews, W. J., and E. Marsh-Mathews. 2007. Extirpation of red shiner in direct tributaries of Lake Texoma (Oklahoma-Texas): a cautionary case history from a fragmented river-reservoir system. Transactions of the American Fisheries Society 136:1041–1062.

Murphy, C. E. 2017. Temporal dynamics of benthic response to habitat disturbance in coastal plain headwaters of southwestern Louisiana. Doctoral dissertation. Louisiana State University, Baton Rouge.

Novinger, G. D. 1987. Evaluation of a 15.0-inch minimum length limit on Largemouth Bass and Spotted Bass catches at Table Rock Lake, Missouri. North American Journal of Fisheries Management 7:260–272.

Oksanen, J., F. G. Blanchet, M. Friendly, R. Kindt, P. Legendre, D. McGlinn, P. R. Minchin, R. B. O'Hara, G. L. Simpson, P. Solymos, M. H. H. Stevens, E. Szoecs, and H. Wagner. 2017. Vegan: Community ecology package. R package version 2.4–3.

Perret, A. J., M. D. Kaller, W. E. Kelso, and D. A. Rutherford. 2010. Effects of Hurricanes Katrina and Rita on sport fish abundance in the eastern Atchafalaya River basin, Louisiana. North American Journal of Fisheries Management 30: 511–517.

Philipp, D. P., J. E. Claussen, T. W. Kassler, and J. M. Epifiano. 2002. Mixing stocks of Largemouth Bass reduces fitness through outbreeding depression. Pages 349–363 *in* D. P. Phillip and M. S. Ridgway, editors. Black bass: ecology, conservation, and management. American Fisheries Society, Symposium 31, Bethesda, Maryland.

R Core Team. 2015. R: A language and environment for statistical computing. R Foundation for Statistical Computing. Vienna, Austria. http://www.R-project.org/

Reuter, C. N., M. D. Kaller, C. E. Walsh, and W. E. Kelso. In press. Fish assemblage response to altered dendritic connectivity in the Red River basin, central Louisiana. American Midland Naturalist 180.

Sammons, S. M., and P. W. Bettoli.1999. Spatial and temporal variation in electrofishing catch rates of three species of black bass (*Micropterus* spp.) from Normandy Reservoir, Tennessee. North American Journal of Fisheries Management 19:454–461.

Schlosser, I. J., J. D. Johnson, W. L. Knotek, and M. Lapinska. 2000. Climate variability and-size-structured interactions among juvenile fish along a lake-stream gradient. Ecology 81:1046–1057.

Southwick Associates. 2008. Sportfishing in America: An Economic Engine and Conservation Powerhouse. American Sportfishing Association, Alexandria, Virginia.

Stevens, J. P. 2002. Applied multivariate statistics for the social sciences, 4th Edition. Lawrence Erlbaum Associates, Inc., Mahwah, New Jersey

Swift, C. C., C. R. Gilbert, S. A Bortone, G. H. Burgess, and R. W. Yerger. 1986. Zoogeography of the freshwater fishes of the southeastern United States: Savannah River to Lake Ponchartrain. Pages 213–266 *in* C. H. Holcott and E. O. Wiley, editors. The Zoogeography of North American freshwater fishes. Wiley, New York.

ter Braak, C. J. F., and P. F. M. Verdonschot. 1995. Canonical correspondence-analysis andrelated multivariate methods in aquatic ecology. Aquatic Sciences 57:255–289.

Vidrine, M. E 1993. The historical distributions of freshwater mussels in Louisiana. Gail Q. Vidrine Collectables, Eunice, Louisiana.

Vidrine, M. F., G. J. Quilmman-VIdrine, M. F. Vidrine II, D. J. Vidrine, and C. E. Vidrine. 2004. Freshwater mussels (Bivalvia: Unionidae) in the Cajun Prairie ecosystem in southwestern Louisiana. Pages 133–136 *in* D. Egan and J. A. Harrington, editors, Proceedings of the 19th North American Prairie Conference: The conservation legacy lives on..., University of Wisconsin-Madison, August 8–12 2004. University of Wisconsin-Madison, Madison, Wisconsin.

Viosca Jr., P. 1931. The Southern Small-mouth Black Bass *Micropterus pseudaplites* Hubbs. Transactions of the American Fisheries Society 61:95–105.

Viosca, Jr., P. 1953. Growth rates of black basses and crappie in an impoundment of northwestern Louisiana. Transactions of the American Fisheries Society 82:255–264.

Walser, C. A., and H. L. Bart, Jr. 1999. Influence of agriculture on in-stream habitat and fish community structure in Piedmont watersheds of the Chattahoochee River system. Ecology of Freshwater Fish 8:237–246.

Warren, Jr., M. L. 2009. Centrarchid identification and natural history. Pages 375–530 *in* S. Cooke and D. P. Philipp, editors, Centrarchid fishes: diversity, biology, and conservation. Wiley-Blackwell Scientific Publications, Chichester, UK.

Watson, M. B., C. J. Killebrew, M. H. Schultz, and J .L. Landry. 1981. A preliminary survey of the Blind River, Louisiana. Pages 303–319 *in* L. A. Krumholtz, editor, The Warmwater Streams Symposium: a national symposium on fisheries aspects of warmwater streams: Proceedings of a symposium held at the University of Tennessee, Knoxville, Knoxville, Tennessee, 9–11 March 1980. American Fisheries Society, Southern Division, Knoxville, Tennessee.

American Fisheries Society Symposium 87:147–166, 2019

Genetic Relationships of Florida Bass and Largemouth Bass Inhabiting Salt Marsh and Freshwater Habitats in Northwest Florida Coastal Streams

Matthew G. Wegener*
Florida Fish and Wildlife Conservation Commission
Fish and Wildlife Research Institute, Holt, Florida 32514, USA

Brandon L. Barthel
Florida Fish and Wildlife Conservation Commission
Fish and Wildlife Research Institute, St. Petersburg, Florida 33701, USA

Abstract.—Coastal rivers provide important fishing opportunities throughout the southeastern United States. Black bass *Micropterus* spp. are major components of these fisheries and are highly valued by anglers, especially in northwest Florida, where few impoundments exist. In the Escambia and Yellow rivers, as in most other rivers on the northern Gulf Coast, salinity is low in the upper segments but may exceed 13‰ in the lower, mesohaline sections. Both rivers are located within the natural intergrade zone between Largemouth Bass *M. salmoides* and Florida Bass *M. floridanus*, and the age and size structure of fish inhabiting riverine reaches differ considerably from those in salt marsh habitats. We collected tissue samples from fish in both the riverine and salt marsh habitats in the Escambia and Yellow rivers and conducted genetic analyses to determine whether individuals from the two habitat types exist as a single randomly mating (panmictic) population in each system. Analysis of 15 microsatellite loci identified significant differences in the allele frequency distributions of samples collected from riverine versus salt marsh habitats in each river. Both rivers were found to be inhabited by intergrade populations that included pure Largemouth Bass and hybrids between Largemouth Bass and Florida Bass. Fish in the salt marsh habitats tended to have more Largemouth Bass alleles than fish in riverine habitats in both systems. Together, these findings indicate fish in the salt marsh and river habitats do not

*Corresponding author: Matt.Wegener@myfwc.com

exist in panmixia and have different genetic compositions. Environmental differences between the riverine and estuarine habitats may have led to divergent selection and limited mating between subpopulations within the same river system. Largemouth Bass alleles may confer greater fitness than Florida Bass alleles in estuarine environments of the northern Gulf Coast, which could explain the greater frequencies of Largemouth Bass alleles in these unique habitats.

Introduction

For decades, managers and researchers have recognized that the life histories of Largemouth Bass *Micropterus salmoides* and intergrade Largemouth Bass × Florida Bass *M. floridanus* (hereafter, coastal bass) inhabiting mesohaline habitats in many U.S. coastal drainages differ from upriver freshwater habitats (Meador and Kelso 1990a; Krause 2002; Markham et al. 2002; DeVries et al. 2015). Coastal bass in these mesohaline environments provide popular recreational fishing opportunities for anglers, but the catch rates of large individuals (>2.3 kg) tend to be lower than in upstream, freshwater reaches (Nack et al. 1993; Haffner et al. 2003). The most extensive studies of coastal bass have been done on the Mobile-Tensaw River delta, where researchers have proposed that the life history of fish in the salt marsh habitats results from the complex interaction of salinity, temperature, amount of freshwater inflow, differences in prey availability, and possibly even local adaptation to the salt marsh environment (DeVries et al. 2015). While there has been limited research on the fish that inhabit these systems, mesohaline salt marsh-dwelling individuals are known to be smaller, have greater condition, shorter lifespans, greater lipid storage, and occur at greater densities than individuals from freshwater habitats (DeVries et al. 2015). The full geographic extent of these life history characteristics has not been determined, although it has been suggested that it is present in many parts of the range in which coastal bass occupy mesohaline environments (Meador and Kelso 1990a).

Largemouth Bass and Florida Bass are generalist species well adapted to a wide range of freshwater environments, but exposure to salinity is known to be a source of stress that impedes growth (Meador and Kelso 1990a; Peterson 1991), osmoregulation (Peterson 1988; Meador and Kelso 1990b), and reproduction (Tebo and McCoy 1964). Early work on coastal bass populations by Swingle and Bland (1974) proposed that reduced growth rates among Largemouth Bass in salt marsh habitats resulted from crowding when fish moved upstream to escape elevated salinity levels. Initially, a telemetry project in the Mobile-Tensaw Delta (Meador and Kelso 1989) appeared to support the hypothesis that fish moved upstream as salinity increased, but more extensive research determined that there was no large-scale dispersal of Largemouth Bass out of salt marsh habitats for adults (Norris et al. 2005) or age-0 fish (Lowe et al. 2009). Laboratory trials found that fish from freshwater habitats had decreased growth when exposed to salinity levels of 4 and 12‰, while fish from the salt marsh continued to grow under elevated salinity (Meador and Kelso 1990a). Salinity can reduce the reproductive success and juvenile survival rates of Largemouth Bass in the salt marsh habitat (DeVries and Wright 2010); conversely, age-0 fish have been shown to grow faster in the salt marsh than in freshwater habitats, most likely because more energy-rich prey is available there (Peer et al. 2006). This suggests there may be trade-offs in terms of increased mortality versus faster growth for juvenile fish in salt marsh habitat.

Differences in salinity levels have been associated with genetic divergence and local adaptation in a number of fish species (Whitehead et al. 2011; Jones et al. 2012; Papakostas et al. 2012; Ferchaud et al. 2014; Berg et al. 2015; Dennenmoser et al. 2017), including Largemouth Bass (Hallerman et al. 1986). For instance, researchers found that Atlantic Cod *Gadus morhua* captured in low-salinity and high-salinity areas of the North Atlantic Ocean exhibited different allele frequencies at markers associated with genes involved in osmoregulation and oocyte development (Berg et al. 2015). They concluded that these differences likely contributed to reproductive isolation between cod in the North Sea (35‰ salinity) and the Baltic Sea (7‰ salinity). Other studies have shown that fish living in brackish habitats have unique adaptations to mesohaline environments that inhabitants of upriver freshwater habitats do not have (e.g., Threespine Stickleback *Gasterosteus aculeatus*, Jones et al. 2012; European Whitefish *Coregonus lavaretus*, Papakostas et al. 2012; Mummichog *Fundulus heteroclitus*, Whitehead et al. 2013; American Eel *Anguilla rostrata*, Pavey et al. 2015; Prickly Sculpin *Cottus asper*, Dennenmoser et al. 2017).

The only experimental evaluation of salinity adaptation in coastal bass found that when fish from the Mobile-Tensaw Delta were released into a freshwater impoundment they grew faster than fish in salt marsh habitat, but continued to have lower survival than typical freshwater populations (Norris et al. 2010). This led the authors to conclude that growth was impacted by environmental differences between the salt marsh habitat and the impoundment, but there was a strong genetic component to survival (Norris et al. 2010). This suggests some of the unique characteristics of coastal bass populations may be adaptive responses to mesohaline environments.

Fish inhabiting the Escambia River marsh are known to exhibit very similar growth patterns and longevities to coastal Largemouth Bass in the Mobile-Tensaw Delta (Krause 2002; Norris et al. 2010; FWC 2012). This system is on the western edge of the natural intergrade zone between Largemouth Bass and Florida Bass (Barthel et al. 2010) and individuals carrying alleles from each of the species are likely to have coexisted within this system since the taxa came into secondary contact during the Pleistocene epoch (Near et al. 2003). Recent otolith analysis indicated growth patterns differed between individuals in the main-stem freshwater river and the mesohaline salt marsh habitats (FWC 2012). Fish in the salt marsh exhibited faster growth than fish in the riverine habitat at ages <2; but at ages ≥ 2, riverine fish grew faster than did individuals in the salt marsh (FWC 2012). Only a few individuals $\geq$ age-3 were observed in the salt marsh sample, but the river sample included individuals $\geq$ age 7 (FWC 2012).

The extent to which the fish in the two habitat types are demographically connected by gene flow is poorly understood. Hallerman et al. (1986) found that fish from the Mobile Delta had different allele frequencies than fish upriver in freshwater habitats at one of 12 polymorphic allozyme loci. However, the low polymorphism of allozyme loci may not have been adequate to resolve the fine-scale divergence between groups of fish inhabiting the two interconnected habitats. There are many reasons why fish in the two habitats may interbreed less frequently than would be expected under panmixia. For instance, restricted movement between the two habitats (i.e., fish that hatch in the salt marsh tend to spawn in the salt marsh, and fish that hatch in the river tend to spawn in the river) and assortative mating (i.e., individuals preferentially spawn with fish with similar phenotypes) could both lead to low gene flow between two subpopulations. Studies have found most adult and juvenile fish remained in the marsh habitat of the Mobile-Tensaw system even when exposed to rising salinities (Norris et al. 2005; Lowe et al. 2009). Alternatively, habitat-specific adaptations might result in fish having depressed fitness outside of their natal habitat type. This could lead to migration-selection equilibrium between the groups inhabiting the two habitats (Nagylaki 1975). The present study investigates whether fish captured in riverine and salt marsh habitats of two northwest Florida coastal streams exhibited genetic differentiation, as would be expected if fish in the two habitat types were not mating randomly. The Escambia and Yellow river systems were selected because both were known to be inhabited by fish exhibiting the coastal life history in salt marsh habitats and typical phenotypes in upriver sections of these streams. The amount of salt marsh differed greatly between the two systems; the Escambia system includes 1,669 ha of salt marsh, while the Yellow River has 855 ha. Although salinity varies seasonally in both systems (Krause 2002), it tends to be higher in the salt marshes of the Escambia River (0.2–18.0 ppt) than the Yellow River (0.1–12.0 ppt; Olinger et al. 1975).

Phenotypic plasticity and genetic adaptation may have played a role in the development and persistence of life history divergence between the fish in the salt marsh and river, but the relative influence of these two forces is not known. In a scenario involving only plasticity, the two groups would be genetically identical but exhibit different phenotypes due to environmental differences between the salt marsh and river habitats. In contrast, an adaptive explanation would involve there being a genetic basis for differences between groups of fish that either return to their natal habitats to spawn (i.e., homing) or do not disperse between habitats. This study investigated whether there were genetic differences between fish inhabiting the river and salt marsh habitats. The study objectives were: (1) to determine whether the fish inhabiting each river system were panmictic, (2) to assess the ancestry of fish inhabiting the two habitats in each river system (identify fish as either Florida Bass [FLB], Largemouth Bass [LMB], or hybrids between the two), and (3) to estimate the effective population sizes of fish in the salt marsh and river habitats in each river system. The results of this study are expected to be used to benefit the management and conservation of these species.

Methods

Sampling Design

The Escambia River is a 415-km-long stream that flows through Florida and Alabama (where it is known as the Conecuh River). Eighty-six of the river kilometers are in Florida, where they made up the border between Escambia and Santa Rosa counties. The Yellow River also runs through Florida and Alabama. Ninety-six of its 190 km are located in Florida. Each river was divided into riverine and salt marsh segments based on available habitat, tidal influence, and salinity levels (Figure 1).

Riverine segments are characterized by sand and gravel substrate. The water is generally fresh with little tidal influence except during dry periods in late summer. In-stream cover is mainly large woody structure, specifically log jams and fallen trees. Aquatic macrophyte coverage is sparse and dominated by Spatterdock *Nuphar lutea*. In salt marsh habitats, the water becomes brackish (1.0–16 ‰), and the substrate is silt. Tidal fluctuations are diurnal with a range of 0.5 m. Coverage by aquatic macrophytes increases in a downstream direction and primarily includes Eelgrass *Vallisneria americana* and Eur-

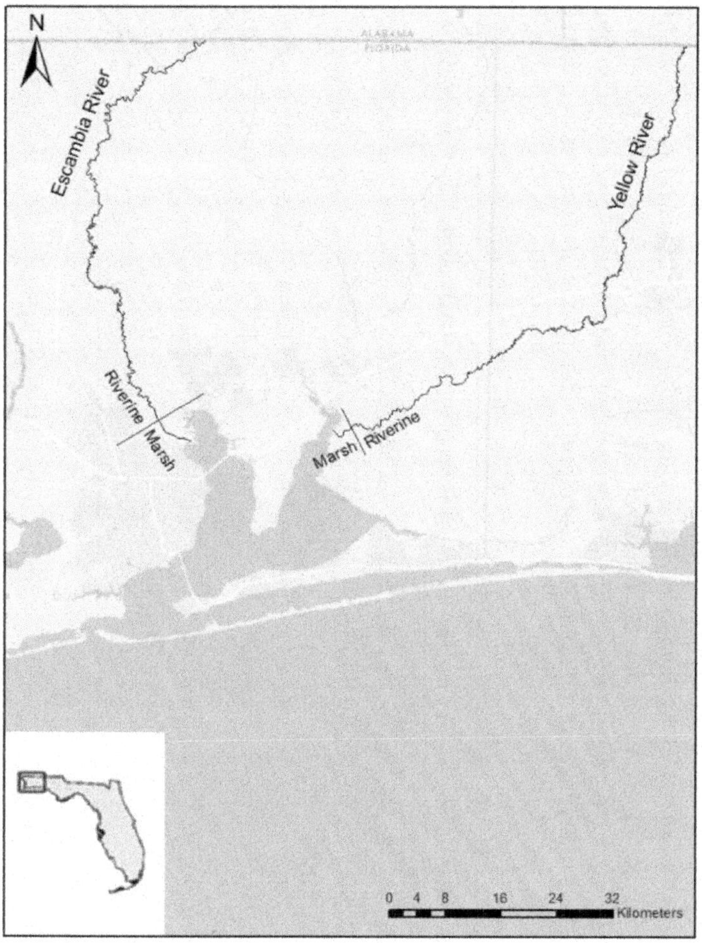

Figure 1. Riverine and salt marsh habitat boundaries on the Escambia and Yellow rivers, Florida. Habitat divisions were determined based on available habitat, tidal influence, and salinity levels.

asian Water Milfoil *Myriophyllum spicatum*. Large-woody structure becomes less abundant in a downstream direction, as heavily wooded shorelines are replaced by emergent vegetation, specifically Black Needlerush *Juncus roemerianus*, Bulrush *Scirpus* spp., and Sawgrass *Cladium jamaicense*.

Fish were collected using a boat-mounted, 7.5-GPP Smith-Root electrofisher (Smith-Root, Vancouver, Washington) in riverine segments (salinity <1.0 ‰) and a 9.0-GPP Smith-Root electrofisher in salt marsh segments and were operated at 60 Hz pulsed DC. Both rivers were sampled with FWC long-term monitoring protocols, where 30–35 sampling locations are randomly selected at a minimum distance of 200-m apart to ensure collection of a representative sample from riverine and salt marsh segments with an effort to have complete coverage of both habitats. However, a 5-km area near the interface between segments was excluded from sampling to avoid collecting fish that either may have recently moved from one segment to the other. The goal was to collect caudal fin clips from as many as 70 individuals from each habitat in each river per sampling year. Fin-clip collections were initially collected

Table 1. Genetic composition of fish sampled from the salt marsh and freshwater river habitats of the Escambia River (2012, 2013, 2015, and 2016) and Yellow River (2013 only) drainages, Florida. The average proportional assignment of fish from each collection to Largemouth Bass (LMB prop.) and Florida Bass (FLB prop.) are presented, along with the number and percentage of the fish that were identified as hybrids (Hybrids *N* and Hybrids %) based on multilocus microsatellite genotypes. The means and ranges of the total lengths (TL) of fish from each collection are also provided.

River	Collection year	Habitat	*N*	Mean TL (mm)	TL range (mm)	FLB prop.	LMB prop.	Hybrids *N*	Hybrids %
Escambia	2012	salt marsh	30	208.1	105–230	0.032	0.968	6	20%
		river	60	245.6	79–353	0.094	0.906	39	65%
	2013	salt marsh	58	121.0	84–177	0.056	0.944	35	60%
		river	54	116.3	62–220	0.091	0.909	34	63%
	2015	salt marsh	58	136.6	87–179	0.030	0.970	19	33%
	2016	salt marsh	68	128.6	93–179	0.024	0.976	24	35%
Yellow	2013	salt marsh	64	112.5	35–198	0.146	0.854	51	82%
		river	60	114.3	60–163	0.340	0.660	55	92%

from Largemouth Bass of all sizes in 2012 but then were restricted to fish ≤180 mm TL (total length) from 2013 through 2016 to reduce the chance of collecting genetic tissues from individuals older than age-1 (FWC 2012) since those fish would have had greater time to migrate out of their natal habitat. Collections were limited to 2–3 fish per transect to reduce the possibility of collecting siblings, which could bias genetic differences. Tissue samples were collected from fish captured in the salt marsh and river habitats of the Escambia River system in October–November 2012 and October–November 2013 and only in the salt marsh habitat in March–April 2015 and March–April 2016 (Table 1). Samples were collected from the salt marsh and river habitats of the Yellow River system in October–November 2013 (Table 1). The clips were stored in 95% ethanol and sent for genetic analysis to the Fisheries Genetics Laboratory of the FWC's Fish and Wildlife Research Institute in St. Petersburg, Florida.

Microsatellite DNA Genotyping

Genomic DNA was extracted from small amounts of fin tissue following the protocols from the PUREGENE DNA purification kits (Gentra Systems, Minneapolis, Minnesota). Extracted DNA was adjusted to a concentration of 50 ng/μL with low tris-EDTA buffer (10 mM tris [ph 7–8], 1 mM EDTA; Qiagen). Six polymerase chain reactions (PCR) were used to amplify 15 microsatellite markers (Seyoum et al. 2013): Msaf1, Msaf5, Msaf6, Msaf9, Msaf10, Msaf14, Msaf17, Msaf22, Msaf24, Msaf25, Msaf27, Msaf28, Msaf29, Msaf31, and Msaf32. The PCR reaction profile was: (1) 94°C for 2 min, (2) 35 cycles of 94°C for 30 s, 58°C for 30 s, and 72°C for 30 s, and (3) a 7-min extension at 72°C. Reactions were performed in 13-μL volumes that included 50 ng of template DNA, 1× buffer (20-mM tris-HCl [ph 8.4], 50-mM KCl; Invitrogen), 1.5–2.5 μM of both the forward and reverse primers for each locus (one having a 5' label of FAM, HEX, or NED), and 0.5 units of Taq DNA polymerase (Promega, Madison, Wisconsin). The Genescan 500 ROX size standard (ThermoFisher Scientific) was used to visualize fragment sizes on an ABI 3130 genetic analyzer (Applied Biosystems). Genotype data were evaluated with GENEMAPPER version 3.7 (Applied Biosystems).

The ARLEQUIN software package version 3.5.1 (Excoffier and Lischer 2010) was used to test whether the genotype data from each collection conformed to Hardy-Weinberg (HWE) and linkage equilibria. The contingency table analysis described by Guo and Thompson (1992) was used to determine whether allelic associations met HWE expectations for each locus in each collection (100,000 Markov chain steps). Likelihood-ratio tests were conducted to evaluate whether allelic patterns were indicative of linkage disequilibria, wherein some combination of alleles at different loci were observed together more often than would be expected by chance (10,000 permutations; Excoffier and Slatkin 1998).

Genetic Composition Analyses

Individual fish were identified as Florida Bass (FLB), Largemouth Bass (LMB), or hybrids via the Bayesian assignment protocols in the program STRUCTURE (Pritchard et al. 2000). The genotype of each fish collected for the study was included in a STRUCTURE analysis with previously identified reference samples of pure LMB and FLB. The LMB reference samples were collected from Ontario (Warner Lake, n = 25), Wisconsin (Minocqua Lake and Lake Chippewa, n = 34), Illinois (Lake Shelbyville, n = 13), Maryland (Northeast River and Potomac River, n = 47), Tennessee (Reelfoot Lake and Lake Norris, n = 35), Kentucky (Reformatory Lake and Cedar Creek Lake, n = 91), and Virginia (Holston Lake and Claytor Lake, n = 50). The FLB reference samples were collected from 21 water bodies in Florida (sample sizes ranged from 7 to 45). The program proportionally assigned each individual genotype to two groups associated with the LMB versus FLB reference samples (e.g., a first-generation hybrid would be expected to have equal assignment to each of the groups). The two taxa have overlapping allele distributions at the microsatellite markers utilized for this study, but the distributions differ between taxa and a number of alleles are only found in one taxon (Seyoum et al. 2013). We used the admixture ancestry model with independent allele frequencies among collections to run 50 STRUCTURE replicates at $K = 2$. If an individual fish had a ≥99.2% assignment to one of the groups, it was considered a pure member of that group. Fish were considered hybrids if assignments to a single group were <99.2%. This threshold was determined by Tringali et al. (2010) by calculating 95% confidence intervals of known reference specimens via the approach described in Littrell et al. (2007).

Genetic Differentiation Analyses

Two types of genetic differentiation tests were conducted for each pair of collections. First, pairwise F_{ST} estimates were produced via the estimator developed by Weir and Cockerham (1984) using the program GENETIX (Belkhir et al. 1996–2004). Statistical significance was determined via permutation tests that evaluated the probability of finding equal or higher F_{ST} values due to chance. Rejection of the null hypothesis indicated that the two groups were significantly differentiated (i.e., data were not consistent with the two collections having been collected from a single, panmictic population). The second type of differentiation test was the genic differentiation test, conducted using GENEPOP software (Rousset 2008). This analysis determined whether two collections were found to have significantly different allele frequency distributions at each locus (1,000 Markov chain permutations).

Effective Population Size Estimation

Effective population size was estimated for each collection using the single-sample linkage disequilibrium estimator from the program LDNe (Waples and Do 2008). Confidence intervals were determined using the parametric procedure and the inclusion of all alleles that were present at 2% frequency or greater.

Results

Collections

Three hundred twenty-eight fish from the Escambia River system and 124 fish from the Yellow River system were successfully genotyped. Salinity varied from 0.06 to 1.61 ppt during fall collections and 0.02–0.14 ppt during spring collections on the Escambia River and from 0.02 to 0.40 ppt during fall collections on the Yellow River. There were no consistent patterns of genetic disequilibrium across the collections. A single locus was found to be out of HWE in the 2012 collections from the Escambia River system (river habitat collection, *Msaf1*, $P < 0.0001$). For the 2013 collections, one locus was found to be out of HWE among the fish collected from the Escambia salt marsh habitat (*Msaf6*, $P = 0.0004$) and two from the Yellow river habitat (*Msaf14*, $P < 0.0001$; *Msaf27*, $P < 0.0001$). One locus was out of HWE from the Escambia 2015 collection (*Msaf6*, $P = 0.0002$), and none were out of HWE from the Escambia 2016 sample. No linked loci were isolated among the fish collected from the Escambia system in any of the collection years. For the Yellow River, four pairs of loci showed evidence of linkage disequilibrium among the fish from the salt marsh habitat (*Msaf1* and *Msaf9*, $P = 0.0001$; *Msaf5* and *Msaf6*, $P =$

0.0001; *Msaf5* and *Msaf9*, $P = 0.0001$; *Msaf6* and *Msaf9*, $P = 0.0001$); two were detected from the river habitat (*Msaf5* and *Msaf24*, $P = 0.0003$; *Msaf6* and *Msaf24*, $P < 0.0001$). These results suggest there generally was random mating and genetic equilibrium within the collections (i.e., habitat types). However, Largemouth Bass production and catch-per-unit-effort (CPUE) was less in Yellow River than the Escambia River (Bass et al. 1979) and may have led to disequilibrium between a greater number of markers than in the Escambia system.

Genetic Composition

The STRUCTURE analysis indicated that fish captured in salt marsh habitats had a greater proportion of LMB alleles than did fish captured in river habitats (Table 1). This pattern was found in both systems and consistent over several years of sampling in the Escambia salt marsh and river habitats. Over four years, proportions of FLB alleles from the fish captured in the Escambia salt marsh ranged from 0.030 to 0.056, while the two years of samples from the river habitat were assigned as having 0.091 and 0.094 FLB alleles. In the Yellow River, fish collected from the river habitat had more than twice the proportion of FLB alleles as the fish captured in the salt marsh (0.340 versus 0.146; Table 1). A greater proportion of fish collected from the Yellow River drainage were hybrids than were collected from the Escambia River drainage (Table 1). Nearly all of the hybrids identified in the Escambia River system had genetic assignments that suggested they were produced by varying degrees of backcrossing with Largemouth Bass (Figure 2). Most of the hybrids from the Yellow River system also appeared to result from backcrossing with Largemouth Bass, however, there was a greater proportion of Florida Bass alleles overall and a minority of the individuals were backcrosses with Florida Bass (Figure 2).

Genetic Differentiation

More than three-quarters (22 of 28) of the pairwise F_{ST} estimates represented statistically significant genetic differentiation (Table 2). This included all the estimates of differentiation between the two rivers (i.e., F_{ST} estimates for Escambia versus Yellow drainage collections). While the largest differentiation estimates were between rivers (Escambia versus Yellow), all estimates for inter-habitat (river versus salt marsh) comparisons were also statistically significant (Table 2). Half of these estimates of inter-habitat differentiation were of a magnitude similar to that identified for the Escambia River versus Yellow River comparisons (Escambia salt marsh versus river F_{ST} range = 0.024–0.036), which is notable since there are no physical barriers to fish moving between the river and salt marsh habitats. The F_{ST} estimate among collections from the salt marsh and river habitats in the Yellow River drainage was also significant, but the F_{ST} value was considerably smaller than that within the Escambia drainage. There was no evidence of differentiation between the two temporal samples for river habitat from the Escambia river habitat (2012 versus 2013 river habitat collections). Five of the six estimates between groups of fish collected from the salt marsh habitat in the Escambia river (e.g., Escambia 2012 salt marsh versus 2013 salt marsh) were not significant. Only one temporal F_{ST} estimate from the Escambia salt marsh was statistically significant (2013 versus 2015), which was consistent with extremely low levels of genetic differentiation between groups of fish collected in salt marsh habitat several years apart.

There was considerable variation in the magnitude of allele frequency differentiation found between pairs of collections. The number of genically differentiated loci ranged from 0 to 12 (Table 2). The greatest differentiated was found between collections from the Escambia and Yellow rivers (ranged

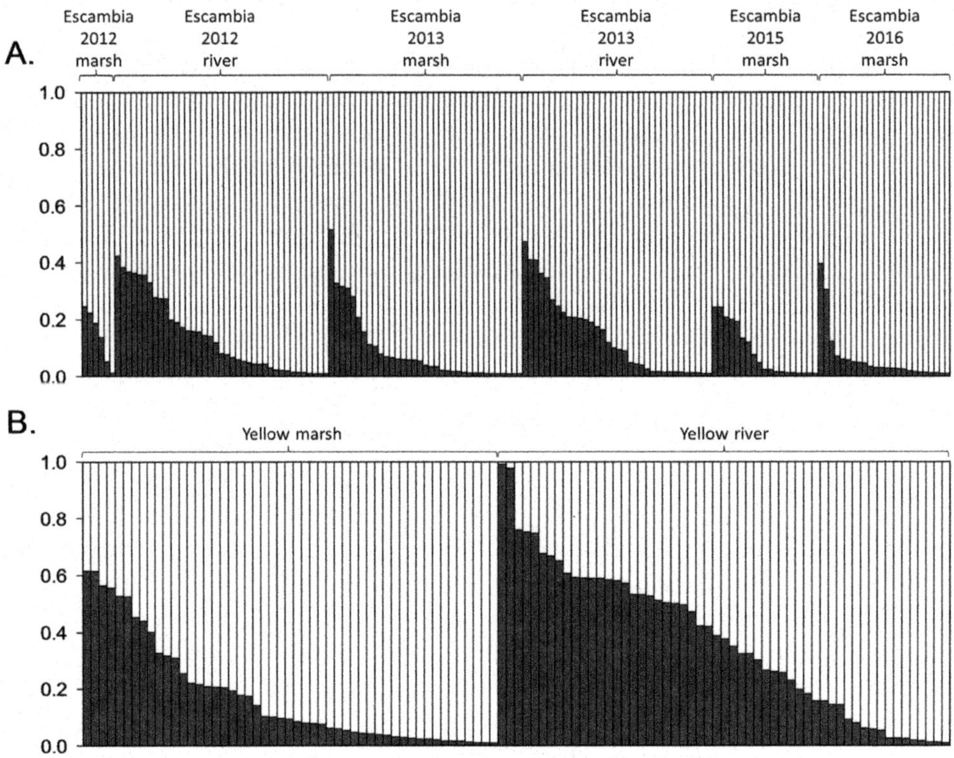

Figure 2. Genetic composition of individual hybrid Largemouth Bass × Florida Bass from the (A) Escambia and (B) Yellow River, Florida. The white portion of each bar represents the q-values for Largemouth Bass ancestry and the dark gray bar colors are q-values for Florida Bass ancestry according to STRUCTURE analyses.

from 7 to 12 loci). But, similar to the F_{ST} results, half of the comparisons between fish from salt marsh and river habitats within the Escambia collections had similar levels of differentiation (4–9 genically differentiated loci) as was detected between Escambia and Yellow River collections. There is a spatial component that would be expected to limit the amount of gene flow between the two drainages, whereas, the saltmarsh and river habitats are contiguous within the Escambia drainage. Because there is no physical separation, genetic differences would result from something more than just isolation by distance; yet, the differences we detected are equivalent to inter-drainage comparisons in many cases.

Allele frequencies were not significantly different between the 2012 and 2013 samples from the Escambia River habitat (*P*-values ranged from 0.173 to 0.943 across the tests for the 15 loci). There also were no differentiated loci in 4 of the 6 comparisons between salt marsh habitats within the Escambia system. The two exceptions were comparisons that differed at a single locus, and both involved the 2013 collection from the salt marsh.

Some loci were found to be consistently differentiated between fish collected from the salt marsh and river habitats in the Escambia collections. Markers *Msaf1* and *Msaf28* were significantly differentiated in all 8 comparisons of fish between the salt marsh and river collections (Table 3). Loci *Msaf14*, *Msaf17*,

Table 2. Genetic differentiation estimates for Largemouth Bass and intergrade Largemouth Bass × Florida Bass collected from the salt marsh and freshwater river habitats of the Escambia River (2012, 2013, 2015, and 2016) and Yellow River (2013 only), Florida. The Escambia collections are identified with 'Esc', then the year the collection was made, and a single letter that denotes whether the sample came from the salt marsh (M) or river (R) habitat. Similarly, the Yellow River collections are identified with the same habitat designation. Statistically significant F_{ST} estimates (above diagonal) are identified with an asterisk. The numbers of genically differentiated loci are presented below the diagonal. Genic differentiation tests determined whether pairs of populations had statistically significant different allele frequencies at each locus.

	Esc2012M	Esc2012R	Esc2013M	Esc2013R	Esc2015M	Esc2016M	YellowM	YellowR
Esc2012M		0.026*	0.006	0.024*	−0.002	0.002	0.035*	0.042*
Esc2012R	7		0.035*	0.001	0.036*	0.026*	0.029*	0.029*
Esc2013M	0	7		0.036*	0.006*	0.003	0.043*	0.045*
Esc2013R	4	0	8		0.034*	0.025*	0.028*	0.028*
Esc2015M	0	9	1	5		0.001	0.036*	0.046*
Esc2016M	0	6	1	5	0		0.034*	0.043*
YellowM	9	10	10	9	10	12		0.014*
YellowR	10	12	11	7	11	10	5	

Table 3. Microsatellite loci found to be genically differentiated between pairs of Largemouth Bass and intergrade Largemouth Bass × Florida Bass collections from the Escambia River (2012, 2013, 2015, and 2016) and Yellow River (2013 only) drainages, Florida.

System	Comparison	Differentiated loci											
Escambia	2012 salt marsh vs. 2012 river	*Msaf1*			*Msaf10*	*Msaf14*	*Msaf17*	*Msaf22*				*Msaf28*	*Msaf29*
	2012 salt marsh vs. 2013 river	*Msaf1*				*Msaf14*						*Msaf28*	*Msaf29*
	2012 river vs. 2013 salt marsh	*Msaf1*		*Msaf6*	*Msaf10*	*Msaf14*	*Msaf17*	*Msaf22*				*Msaf28*	*Msaf29*
	2012 river vs. 2015 salt marsh	*Msaf1*			*Msaf10*	*Msaf14*	*Msaf17*	*Msaf22*	*Msaf24*		*Msaf27*	*Msaf28*	*Msaf29*
	2012 river vs. 2016 salt marsh	*Msaf1*			*Msaf10*	*Msaf14*	*Msaf17*	*Msaf22*				*Msaf28*	
	2013 salt marsh vs. 2013 river	*Msaf1*		*Msaf6*		*Msaf14*	*Msaf17*	*Msaf22*		*Msaf25*		*Msaf28*	*Msaf29*
	2013 river vs. 2015 Salt marsh	*Msaf1*				*Msaf14*	*Msaf17*					*Msaf28*	*Msaf29*
	2013 river vs. 2016 salt marsh	*Msaf1*					*Msaf17*	*Msaf22*				*Msaf28*	*Msaf29*
	2013 salt marsh vs. 2015 salt marsh			*Msaf6*									
	2013 salt marsh vs. 2016 salt marsh			*Msaf6*									
	2012 salt marsh vs. 2013 salt marsh												
	2012 salt marsh vs. 2016 salt marsh												
	2012 salt marsh vs. 2015 salt marsh												
	2015 salt marsh vs. 2016 salt marsh												
Yellow	2012 river vs. 2013 river												
	2013 salt marsh vs. 2013 river		*Msaf5*		*Msaf10*	*Msaf14*		*Msaf22*	*Msaf24*				

and *Msaf29* were differentiated for 7 of the 8 comparisons. Conversely, *Msaf5*, *Msaf31*, and *Msaf32* were not differentiated in any of the comparisons. These patterns may be attributable to low rates of genetic drift or, alternatively, one or more of the consistently differentiated loci may be physically linked (i.e., located in close proximity on a particular chromosome) to a gene that is under divergent selection between fish in the salt marsh and river habitats.

Effective Population Size

Estimated effective sizes of the populations ranged from small to inestimably large, with variation among years at a given site. The effective population size estimates (N_e) were particularly low in the Escambia salt marshes in 2012 (N_e = 89.2) and the salt marsh (N_e = 205.6) and river (N_e = 173.2) habitats of the Yellow River collections in 2013 (Table 4). All the other estimates suggested that populations were relatively large and robust (N_e estimates ranged from 471.2 to infinity). The variation among estimates from the Escambia salt marshes (N_e estimates ranged from 89.2 to infinity) suggests the number of fish contributing offspring to year classes fluctuated between years.

Discussion

Largemouth Bass and intergrade individuals did not exist as single, panmictic populations between salt marsh and riverine habitats in two coastal Florida Rivers. Low to moderate levels of genetic differentiation were detected between fish in the river and salt marsh habitats of each system, indicating that gene flow does occur between fish in the two habitats, but that the rate is lower than would be expected if mating had been random. There are many possible explanations as to why gene flow might be restricted between groups of fish in these habitats. Restriction of gene flow could occur if fish from the salt marsh and those from riverine habitats mostly remained in their natal habitat (i.e., fish did not move between habitat types as frequently as would be expected under random dispersal) or if they assortatively mated with fish from the same habitat from which they had originated. Previous studies have reported limited movement out of the river for adult and juvenile individuals in the Mobile-Tensaw system (Norris et al. 2005; Lowe et al. 2009). If coastal bass have adapted to the mesohaline environment in ways fish in the upstream habitats have not (DeVries et al. 2015), divergent selection may reduce the likelihood that

Table 4. Estimated effective population sizes for Largemouth Bass and Largemouth Bass × Florida Bass collections from the Escambia River (2012, 2013, 2015, and 2016) and Yellow River drainages (2013 only), Florida.

River	Year	Habitat	N_e	95% CI
Escambia	2012	salt marsh	89.2	53.5–223.5
Escambia	2012	riverine	1707.9	338.3–infinity
Escambia	2013	salt marsh	infinity	339.4–infinity
Escambia	2013	riverine	infinity	512.5–infinity
Escambia	2015	salt marsh	471.2	214.1–infinity
Escambia	2016	salt marsh	669.9	249.7–infinity
Yellow	2013	salt marsh	205.6	137.0–385.4
Yellow	2013	riverine	173.2	122.9–280.8

fish can successfully spawn outside their natal habitat, or, when such matings do happen, the offspring may have reduced fitness. Purcell et al. (2012) reported similar genetic differentiation patterns between Western Mosquitofish *Gambusia affinis* from brackish and freshwater habitats of two estuaries in Louisiana. In that study, the fish from salt marsh habitat were known to have adaptations for increased salinity tolerance, and the magnitude of genetic differentiation between the groups of fish in freshwater versus salt marsh habitats (pairwise FST values were 0.009 and 0.011; Purcell et al. 2012) was smaller than the estimates from the present study. Further work will be required to determine what factors are responsible for restricting gene flow between groups of fish in the river and salt marsh habitats in these rivers.

The fish in both systems and all habitats were found to be predominately of Largemouth Bass ancestry. Bailey and Hubbs (1949) used meristic and morphometric evaluations to determine that 10 specimens collected from Pensacola Bay drainages were Largemouth Bass. More recently, a statewide genetic study (Barthel et al. 2010) found that, depending on the type of markers and sample analyzed, 80–90% of the alleles from fish in the Escambia River were of Largemouth Bass descent (microsatellite loci identified greater frequencies of Largemouth Bass alleles than allozyme loci). The present study represents the first attempt to compare the composition of groups of fish captured in salt marsh versus riverine habitats. Our genetic analyses consistently showed that fish from riverine habitats had a low to moderate but greater proportion of Florida Bass ancestry than fish in the salt marsh. The lower frequencies of Florida Bass alleles in the salt marsh collections may indicate that fish with Florida Bass genes have decreased fitness in mesohaline habitat. DeVries et al. (2015) reported that all 240 fish collected and analyzed from the Mobile-Tensaw Delta were genetically similar to Largemouth Bass and distinct from Florida Bass. They considered this noteworthy, as several hundred thousand Florida Bass had been stocked in the Delta from 1988 to 2000 (Armstrong et al. 2000; Ranier 2010), and Florida Bass alleles were detected in collections from the Delta at very low frequencies by an earlier allozyme study (Hallerman et al. 1986). Additionally, 2,000 Florida Bass fingerlings were marked with microtags and stocked in the Escambia River salt marsh in 2000 (Krause 2002). However, few individuals were recovered within 6 months of stocking and internal examinations revealed abnormalities that were likely to preclude long-term survival (R. DeMauro, FWC, personal communication). While there are many possible explanations for stocked fish failing to survive and contribute to recruitment, it has been suggested that no contribution was detected, at least in part, because of strong selection against Florida Bass phenotypes in the salt marsh habitats (DeVries et al. 2015). The greater proportion of Florida Bass alleles and hybrids in the Yellow River system may have resulted from the fact that it has much less salt marsh habitat than the Escambia system, possibly leading to less of a selective cost to Florida Bass ancestry within that system. An alternative explanation for the greater abundance is escapement of stocked fish from Karick Lake, a state-managed reservoir that has an outflow into the Yellow River. Florida Bass were stocked into Karick Lake in 1997, and it is possible that some individuals escaped from Karick Lake into the Yellow River after being stocked. A third, natural explanation for the different compositions is that the Yellow River system has more Florida Bass alleles because it is located closer to the range of pure Florida Bass in the Florida peninsula. The closer geographical proximity may have made it easier for Florida Bass to disperse into the Yellow River drainage than the Escambia system when the taxa came into secondary contact in this area during the Pleistocene epoch.

While the present study used markers that are expected to be selectively neutral (i.e., not under direct selection), it is possible that some might be linked (i.e., close on the chromosomes) to genes that were under divergent selection between fish in the salt marsh and river habitats. If so, selection acting upon genetically determined phenotypic differences in the groups inhabiting salt marsh versus river habitats would also lead to differentiation at some of the study markers. A growing number of studies have linked specific genes to phenotypic variation between sympatric groups (e.g., dwarf and normal Lake Whitefish *Coregonus clupeaformis*, Renaut et al. 2012; anadromous and freshwater Rainbow and Steelhead trout *Oncorhynchus mykiss*, Hecht et al. 2013; anadromous and freshwater resident Atlantic Salmon *Salmo salar*, Perrier et al. 2013; ecologically and morphologically divergent crater-lake cichlids *Amphilophus zaliosus* and *A. astorquii*, Franchini et al. 2014; snail ecotypes *Littorina saxatilis*, Westram et al. 2014; piscivorous and planktivorous ecotypes of Lake Trout *Salvelinus namaycush*, Bernatchez et al. 2016), including cases where the groups occupy brackish versus freshwater segments in a coastal river system (e.g., Threespine Stickleback, Jones et al. 2012; European Whitefish, Papakostas et al. 2012; killifish *Fundulus majalis* and *F. heteroclitus*, Whitehead et al. 2013; American Eel, Pavey et al. 2015; Prickly Sculpin, Dennenmoser et al. 2017). Dennenmoser et al. (2017) used whole-genome sequencing to determine that numerous candidate genes were linked to genetic differentiation between groups of Prickly Sculpin that spawned in brackish-water environments versus freshwater environments. Differentiated genes were related to physiological functions such as ion transport, structural permeability of gills, and other traits related to the differences in the osmotic environments experienced by the two groups. Dennenmoser et al. (2017) concluded that their findings were consistent with a scenario in which a number of genes were responsible for groups of fish being adapted to different habitat types while still sharing gene flow. While the present study was not designed to identify genes underlying phenotypic differences, observation of similar differentiation patterns in the Escambia and Yellow rivers would be expected if similar forces have been driving genetic divergence in the two river systems. For example, if adaptation has occurred via parallel pathways in the two stream systems, markers linked to genes under divergent selection would be expected show similar patterns of differentiation in both the Escambia and Yellow river collections. Several markers (e.g., *Msaf14* and *Msaf22*) were genetically differentiated in the majority of the comparisons between groups collected from salt marsh and river habitats in both drainages. These findings alone cannot be considered as evidence that these markers are linked to genes under selection, as it is also possible that certain markers may be more likely to become divergent due to greater polymorphism levels or by the fish in the two systems having a shared genetic background (i.e., they have been founded from the same source population). However, a number of studies have shown that parallel adaptations can also develop from natural selection acting on different sets of genes in different groups (e.g., Perrier et al. 2013; Westram et al. 2014; Bernatchez et al. 2016; Dennenmoser et al. 2017). Future genomics work may be able to determine whether the differences identified in this study are linked to genes under selection.

Estimates of effective population sizes suggested that in most years considerable numbers of fish successfully reproduce in both the river and the salt marsh habitats of the Escambia River. The only exception is the low effective size estimate for the 2012 collection from the salt marsh habitat. No environmental or meteorological anomalies were observed during this time and thus, differences in effective population size cannot be directly contributed to rainfall or drought events that could have caused abnormal salinity levels. A low effective size for one of the four temporal collections may reflect fluctuations in the number of breeding adults in the salt marsh over time or lower reproductive success and recruitment due to environmental conditions in some years, or it could be an artifact of the smaller sample size and greater size range (older age composition) of that collection. Moderate to large effective size estimates in the other three sampling years indicates successful spawning occurred in the salt marsh habitat in most years. This suggests fish in salt marsh habitat are part of an abundant breeding population, not simply a situation in which juveniles are frequently a product of a small number of breeding pairs, despite what is generally considered atypical habitat for Largemouth Bass. Further, collection locations of juveniles were randomly distributed throughout the salt marsh habitat, not concentrated in clusters close to riverine habitat or other freshwater inputs (i.e., small tidal creeks). Low visibility due to water clarity made it difficult to observe spawning nests in salt marsh habitat. However, FWC biologists have observed schools of Largemouth Bass fry shortly after hatching in this environment (Richard Krause, FWC, Personal Communication). Therefore, spawning likely occurred in many places in the salt marsh habitat and juveniles were able to survive after hatching. The Ne estimates for both the salt marsh and river habitats in the Yellow drainage were relatively low. Although we have evaluated only one set of samples from the habitats in the Yellow River system, the breeding population in the Yellow River may be relatively small. Additional samples would need to be evaluated to determine whether the results from this study are representative of the Yellow River Largemouth Bass population in other years.

Management Implications

Future management strategies for coastal bass should consider the genetic composition of fish inhabiting salt marsh and riverine habitats. If divergent selection has led to different sets of genetic adaptations among fish in the two habitat types, then stocking individuals into habitats to which they are not well adapted may result in low fitness and poststocking survival. Therefore, strategies such as stocking Florida Bass into mesohaline habitats would not be likely to be effective and could even threaten the genetic composition of these populations. The Florida Fish and Wildlife Conservation Commission has created four genetic management units for Largemouth Bass (Panhandle, North Florida, West Coast, and St. Johns/Everglades; Porak et al. 2015), whereby individuals or progeny from one regional unit are not stocked into another unit. This study indicates that there are cases where groups of fish differ genetically at small spatial scales. If future work determines that genetic differences are related to adaptations for survival in mesohaline habitats, it may be prudent to evaluate how management strategies conserve populations exhibiting the coastal life history. The information collected to date suggests that sufficient spawning occurs annually in these rivers so stocking is not needed.

Acknowledgments

We thank Andy Strickland, Katie Woodside, Chris Paxton, Matt Szatkowski, Wes Porak, William Kitrell Jr., Chris Hickey, Roger Bitz, Tim Churchill, Joseph Love, David Philipp, Steve Crawford, and Gina DelPizzo for assistance with specimen collection. Nathan Van Bibber performed many of the analyses in the genetics laboratory.

References

Armstrong, D. L., W. H. Tucker, and R. G. Spray. 2000. Mobile Delta management report 1999. Alabama Game and Fish Division, Montgomery.

Bailey, R. M., and C. L. Hubbs. 1949. The black basses (*Micropterus*) of Florida, with description of a new species. University of Michigan Museum of Zoology, Occasional Papers 516, Ann Arbor.

Barthel, B. L., D. J. Lutz-Carrillo, K. E. Norberg, W. F. Porak, M. D. Tringali, T. W. Kassler, W. E. Johnson, A. M. Readel, R. A. Krause, and D. P. Philipp. 2010. Genetic relationships among populations of Florida Largemouth Bass. Transactions of the American Fisheries Society 139:1615–1641.

Bass, D. G., Jr., D. Yeager, and V. G. Hitt 1979. Ecology of the Yellow River system, Florida. Florida Game and Fresh Water Fish Commission, De Funiak Springs, Florida.

Belkhir K., P. Borsa, L. Chikhi, N. Raufaste, and F. Bonhomme. 1996–2004. GENETIX 4.05, logiciel sous Windows TM pour la génétique des populations. Laboratoire Génome, Populations, Interactions, CNRS UMR 5171, Université de Montpellier II, Montpellier (France).

Berg, P. R., S. Jentoft, B. Star, K. H. Ring, H. Knutsen, S. Lien, K. S. Jakobsen, and C. André. 2015. Adaptation to low salinity promotes genomic divergence in Atlantic Cod (*Gadus morhua* L.). Genome Biology and Evolution 7:1644–1663.

Bernatchez, S., M. Laporte, C. Perrier, P. Sirois, and L. Bernatchez. 2016. Investigating genomic and phenotypic parallelism between piscivorous and planktivorous Lake Trout (*Salvelinus namaycush*) ecotypes by means of RADseq and morphometrics analyses. Molecular Ecology 25:4773–4792.

Dennenmoser, S., S. M. Vamosi, A. W. Nolte, and S. M. Rogers. 2017. Adaptive genomic divergence under high gene flow between freshwater and brackish-water ecotypes of Prickly Sculpin (*Cottus asper*) revealed by pool-seq. Molecular Ecology 26:25–42.

DeVries, D. R., and R. A. Wright. 2010. Largemouth Bass in the Mobile River Delta: information needed for managing this unique resource. Final Report, Alabama Department of Conservation and Natural Resources, Division of Wildlife and Freshwater Fisheries, Montgomery.

DeVries, D. R., R. A. Wright, D. C. Glover, T. M. Farmer, M. R. Lowe, A. J. Norris, and A. C. Peer. 2015. Largemouth Bass in coastal estuaries: a comprehensive study from the Mobile-Tensaw River Delta, Alabama. Pages 297–310 *in* M. D. Tringali, J. M. Long, T. W. Birdsong, and M. S. Allen, editors. Black bass diversity: multidisciplinary science for conservation. American Fisheries Society, Symposium 82, Bethesda, Maryland.

Excoffier, L., and H. E. L. Lischer. 2010. Arlequin suite ver 3.5: a new series of programs to perform population genetics analyses under Linux and Windows. Molecular Ecology Resources 10:564–567.

Excoffier, L., and M. Slatkin. 1998. Incorporating genotypes of relatives into a test of linkage disequilibrium. American Journal of Human Genetics 62:171–180.

Ferchaud, A-L., S. H. Pedersen, D. Bekkevold, J. Jian, Y. Niu, and M. M. Hansen. 2014. A low-density SNP array for analyzing differential selection in freshwater and marine populations of Threespine Stickleback (*Gasterosteus aculeatus*). BMC Genomics 15:867–877.

Franchini, P., C. Fruciano, M. L. Spreitzer, J. C. Jones, K. R. Elmer, F. Henning, and A. Meyer. 2014. Genomic architecture of ecologically divergent body shape in a pair of sympatric crater lake cichlid fishes. Molecular Ecology 23:1828–1845.

FWC (Florida Fish and Wildlife Conservation Commission). 2012. Fisheries independent monitoring (sport fish sampling). Annual report for federal aid. Wallop-Breaux Annual Report F-175-R-1, Tallahassee, Florida.

Guo, S., and E. Thompson. 1992. Performing the exact test of Hardy-Weinberg proportion for multiple alleles. Biometrics 48:361–372.

Haffner, J. B., J. J. McHugh, and N. Nichols. 2003. Bass anglers information team: 2003 annual report. Alabama Wildlife and Freshwater Fisheries Division, Montgomery.

Hallerman, E. M., R. O. Smitherman, R. B. Reed, W. H. Tucker, and R. A. Dunham. 1986. Biochemical genetics of largemouth bass in mesosaline and freshwater areas of the Alabama River system. Transactions of the American Fisheries Society 115:15–20.

Hecht, B. C., N. R. Campbell, D. E. Holecek, and S. R. Narum. 2013. Genome-wide association reveals genetic basis for the propensity to migrate in wild populations of Rainbow and Steelhead trout. Molecular Ecology 22:3061–3076.

Jones, F. C., Y. F. Chan, J. Schmutz, J. Grimwood, SD. Brady, A. M. Southwick, D. M. Absher, R. M. Myers, T. E. Reimchen, B. E. Deagle, D. Schluter, and D. M. Kingsley. 2012. A genome-wide SNP genotyping array reveals patterns of global and repeated species-pair divergence in sticklebacks. Current Biology 22:83–90.

Krause, R. A. 2002. Exploitation of an estuarine Largemouth Bass population in northwest Florida. Pages 553–558 *in* D. P. Philipp and M. S. Ridgway, editors. Black bass: ecology, conservation, and management. American Fisheries Society, Symposium 31, Bethesda, Maryland.

Littrell, B. M., D. J. Lutz-Carrillo, T. H. Bonner, and L. T. Fries. 2007. Status of an introgressed Guadalupe Bass population in a Central Texas Stream. North American Journal of Fisheries Management 27:785–791.

Lowe, M. R., D. R. DeVries, R. A. Wright, S. A. Ludsin, and B. J. Fryer. 2009. Coastal Largemouth Bass (*Micropterus salmoides*) movement in response to changing salinity. Canadian Journal of Fisheries and Aquatic Sciences 66:2174–2188.

Markham, J. L., D. T. Cosden, and R. K. Schaffer. 2002. Use of GIS generated habitat maps for sampling Largemouth Bass in Maryland's tidal rivers. Pages 593–601 *in* D. P. Philipp and M. S. Ridgway, editors. Black bass: ecology, conservation, and management. American Fisheries Society, Symposium 31, Bethesda, Maryland.

Meador, M. R., and W. E. Kelso. 1989. Behavior and movements of Largemouth Bass in response to salinity. Transactions of the American Fisheries Society 118:409–415.

Meador, M. R., and W. E. Kelso. 1990a. Growth of Largemouth Bass in low-salinity environments. Transactions of the American Fisheries Society 119:545–552.

Meador, M. R., and W. E. Kelso. 1990b. Physiological responses of Largemouth Bass, *Micropterus salmoides*, exposed to salinity. Canadian Journal of Fisheries and Aquatic Sciences 47:2358–2363.

Nack, S. B., D. Bunnell, D. M. Green, and J. L. Forney. 1993. Spawning and nursery habitats of Largemouth Bass in the tidal Hudson River. Transactions of the American Fisheries Society 122:208–216.

Nagylaki, T. 1975. Conditions for the existence of clines. Genetics 80:595–615.

Near, T., T. W. Kassler, J. B. Koppelman, C. B. Dillman, and D. P. Philipp. 2003. Speciation in North American black basses, *Micropterus* (Actinopterygii: Centrarchidae). Evolution 57:1610–1621.

Norris, A. J., D. R. DeVries, and R. A. Wright. 2010. Coastal estuaries as habitat for a freshwater fish species: exploring population-level effects of salinity on Largemouth Bass. Transactions of the American Fisheries Society 139: 610–625.

Norris, A. J., R. A. Wright, D. R. DeVries, D. L. Armstrong, Jr., and J. Zolczynski. 2005. Movement patterns of coastal Largemouth Bass in the Mobile-Tensaw River Delta, Alabama: a multi-approach study. Proceedings of the Annual Conference of the Southeastern Association of Fish and Wildlife Agencies 59:200–216.

Olinger, L.W., R. G. Rogers, P. L. Fore, R. L. Todd, B. L. Mullins, F. T. Bisterfeld, and L. A. Wise. 1975. Environmental and recovery studies of Escambia Bay and the Pensacola Bay system, Florida. EPA 904/9–76–016. U.S. Environmental Protection Agency, Region IV, Atlanta.

Papakostas, S., A. Vagemägi, J-P. Vähä, M. Himberg, L. Peil, and C. R. Primmer. 2012. A proteomics approach reveals divergent molecular responses to salinity in populations of European Whitefish (*Coregonus lavaretus*). Molecular Ecology 21:3516–3530.

Pavey, S. A., J. Gaudin, E. Normandeau, M. Dionne, M. Castonguay, C. Audet, and L. Bernatchez. 2015. RAD sequencing highlights polygenic discrimination of habitat ecotypes in the panmictic American Eel. Current Biology 25:1666–1671.

Peer, A. C., D. R. DeVries, and R. A. Wright. 2006. First-year growth and recruitment of coastal Largemouth Bass (*Micropterus salmoides*): spatial patterns unresolved by critical periods along a salinity gradient. Canadian Journal of Fisheries and Aquatic Sciences 63:1911–1924.

Perrier, C., V. Bourret, M. P. Kent, and L. Bernatchez. 2013. Parallel and nonparallel genome-wide divergence among replicate population pairs of freshwater and anadromous Atlantic Salmon. Molecular Ecology 22:5577–5593.

Peterson, M. S. 1988. Comparative physiological ecology of centrarchids in hyposaline environments. Canadian Journal of Fisheries and Aquatic Sciences 45:827–833.

Peterson, M. S. 1991. Differential length-weight relationships among centrarchids (Pisces: Centrarchidae) from tidal freshwater and oligohaline wetland habitats. Wetlands 11:325–332.

Porak, W. F., B. L. Barthel, R. B. Stout, M. D. Tringali, M. Matthews, W. F. Pouder, and D. P. Philipp. 2015. The development of a genetic conservation program for the Florida Bass in its native range. Pages 585–600 *in* M. D. Tringali, J. M. Long, T. W. Birdsong, and M. S. Allen, editors. Black bass diversity: multidisciplinary science for conservation. American Fisheries Society, Symposium 82, Bethesda, Maryland.

Pritchard, J. K., M. Stephens, and P. Donnelly. 2000. Inference of population structure using multilocus genotype data. Genetics 155:945–959.

Purcell, K. M., A. Hitch, S. Martin, P. L. Klerks, and P. L. Leberg. 2012. The role of genetic structure in the adaptive divergence of populations experiencing saltwater intrusion due to relative sea-level rise. Journal of Evolutionary Biology 25:2623–2632.

Ranier, D. 2010. Delta bass: fisheries biologists are developing a new strain of bass brood-stock to release in the Mobile-Tensaw Delta. Outdoor Alabama (July):12–14.

Renaut, S., N. Maillet, E. Normandeau, C. Sauvage, N. Derome, S. M. Rogers, and L. Bernatchez. 2012. Genome-wide patterns of divergence during speciation: the Lake Whitefish case study. Philosophical Transactions of the Royal Society B 367:354–363.

Rousset, F. 2008. Genepop'007: a complete re-implementation of the Genepop software for Windows and Linux. Molecular Ecology Resources 8:103–106.

Seyoum, S., M. D. Tringali, B. L. Barthel, M. C. Davis, S. L. Schmitt, P. S. Bellotti, and W. F. Porak. 2013. Isolation and characterization of eighteen microsatellite loci for the Largemouth Bass, *Micropterus salmoides*, and cross amplification in congeneric species. Conservation Genetics Resources 5:697–701.

Swingle, H. A., and D. G. Bland. 1974. A study of the fishes of the coastal watercourses of Alabama. Alabama Marine Resources Bulletin 10:22–69.

Tebo, L. B., Jr., and E. G. McCoy. 1964. Effect of seawater concentration on the reproduction and survival of Largemouth Bass and Bluegills. Progressive Fish-Culturist 26:99–106.

Tringali, M., W. Porak, P. Bellotti, S. Seyoum, N. Trippel, A Strickland, J. Knight, C. Paxton, C. Puchulutegui, and M. Davis. 2010. Population genetics of freshwater fishes for fish management and stock enhancement. Completion report for the Federal Aid in Sport Fish Restoration Act Grant F-128 to the U.S. Department of the Interior, Fish and Wildlife Service from the Florida Fish and Wildlife Conservation Commission, St. Petersburg.

Waples, R. S., and C. Do. 2008. LDNE: a program for estimating effective population size from data on linkage disequilibrium. Molecular Ecology Resources 8:753–756.

Weir, B. S., and C. C. Cockerham. 1984. Estimating *F*-statistics for the analysis of population structure. Evolution 38:1358–1370.

Westram, A. M., J. Galindo, M. Alm Rosenblad, J. W. Grahame, M. Panova, and R. K. Butlin. 2014. Do the same genes underlie parallel phenotypic divergence in different *Littorina saxatilis* populations? Molecular Ecology 23:4603–4616.

Whitehead, A., J. L. Roach, S. Zhang, and F. Galvez. 2011. Genomic mechanisms of evolved physiological plasticity in killifish distributed along an environmental salinity gradient. Proceedings of the National Academy of Sciences U.S.A. 108:6193–6198.

Whitehead, A., S. Zhang, J. L. Roach, and F. Galvez. 2013. Common functional targets of adaptive micro- and macro-evolutionary divergence in killifish. Molecular Ecology 22:3780–3796.

American Fisheries Society Symposium 87:167–192, 2019

Population and Life-History Characteristics of Two *Micropterus* Species in the Flint River System, Georgia

Steven M. Sammons*
*Department of Fisheries, 203 Swingle Hall
Auburn University, AL 38849, USA*

Travis R. Ingram and John M. Kilpatrick
*Georgia Department of Natural Resources
2024 Newton Road, Albany, Georgia 31701, USA*

Abstract.—Black bass *Micropterus* spp. are important components of river fisheries in the southeastern U.S., but little is known about their basic life history and population characteristics in these systems. Largemouth Bass *M. salmoides* and Shoal Bass *M. cataractae* were collected from three areas in the Flint River on multiple occasions from 2005 through 2013. Collections were taken from the undammed section of the Flint River above Lake Blackshear, within Lake Blackshear, and the regulated reach of river below Lake Blackshear. Shoal Bass do not occur in Lake Blackshear, so only Largemouth Bass were collected. Population and life history data such as body condition, growth, mortality, and recruitment variability were examined for these species in all three areas. Shoal Bass and Largemouth Bass appeared to have similar scopes of growth in this system, reaching similar maximum sizes. However, Largemouth Bass grew slower than Shoal Bass in both river sections. Growth of both species was slower in the upper versus lower area of the river; growth of Largemouth Bass was similar between the lower Flint River and Lake Blackshear. Female Largemouth Bass grew faster than males in both sections of the river; whereas, growth of Shoal Bass was similar between sexes. Longevity of Largemouth Bass appeared to be slightly higher than Shoal Bass, with correspondingly lower annual mortality. Age frequencies and annual mortality was similar between the sexes for both species in all areas. Recruitment of Largemouth Bass was negatively affected by high flows in the spring in the upper Flint River and Lake Blackshear; whereas, Shoal Bass recruitment in the upper Flint River was

*Corresponding author: sammosm@auburn.edu

not correlated to any flow measure. The opposite pattern was found in the lower Flint River, with recruitment of Largemouth Bass unaffected by flow but Shoal Bass recruitment negatively affected by high flows. Shoal Bass exhibited numerous differences from Largemouth Bass in the Flint River, including growth, mortality, and recruitment, demonstrating the danger in applying rate function data or basic biology concepts from one species to another closely related one. Results from this study will help biologists better understand the potential and limitations of black bass fisheries in rivers, and thereby enable more effective management.

Introduction

Collecting data to describe population characteristics of fishes was the genesis of fisheries management and remains an integral part of this field today. Population characteristics of fishes are vital data that allow biologists to more completely understand the animals of interest. Furthermore, rate data such as growth, mortality, and recruitment can be used by fisheries managers to make informed decisions about fisheries under their purview. For instance, tradeoffs between growth and mortality have been shown to be major determinants in the ultimate effectiveness of harvest restriction in a variety of fishes (Allen and Miranda 1995; Slipke et al. 1998; Sammons et al. 2006; Churchill and Black 2007).

Black bass *Micropterus* spp. have been studied extensively over the past 50 years; however, comparatively little of this work has occurred in rivers and streams. Also, most of these studies have been conducted on the more widespread species such as Largemouth Bass *M. salmoides* and Smallmouth Bass *M. dolomeiu*. In contrast, little work has been done on Spotted Bass *M. punctulatus* and population characteristics and rate function data are virtually nonexistent for species endemic to the southeastern U.S., such as Shoal Bass *M. cataractae*, Suwannee Bass *M. notius*, and Redeye Bass *M. coosae* (Leitner and Earley 2015; Nagid et al. 2015; Sammons et al. 2015). Although some of these endemic species occur in reservoirs and rivers, others are obligate riverine species, and many river fisheries throughout the southeastern U.S. are based on these species (Bonvechio and Allen 2005; Sammons 2016).

Collectively, black bass constitute one of the most popular and economically important freshwater fisheries in North America. In 2011, an estimated 10.6 million anglers spent 171 million days fishing for black bass in the U.S., representing approximately 39% of all anglers and angling effort in freshwater systems other than the Great Lakes (USFWS and USBOC 2014). In Georgia, an estimated 763,000 people fished freshwaters an average of 11 d and spent over $870 million in 2011 (USFWS and USBOC 2014). Forty-five percent of these anglers targeted black bass, and 35% fished in rivers and streams. Thus, recreational fishing on the rivers and streams in the state have great economic importance; yet relatively little is known about the population characteristics of the sportfish in these systems.

Shoal Bass is a species endemic to the Apalachicola Basin, USA, including the Chattahoochee and Flint river systems in Alabama, Florida, and Georgia. The species is a fluvial specialist, occupying shallow, rocky riffles and shoals in medium- to large-sized streams and rivers and is intolerant of reservoir conditions (Sammons et al. 2015). In Georgia, significant fisheries have developed for Shoal Bass on the Flint, Ocmulgee, and upper Chattahoochee rivers, with a growing segment of the angler population seeking large, trophy (≥2.25 kg) fish (Sammons 2016). In terms of abundance and size structure, the highest quality Shoal Bass fisheries are found in Georgia, which also contains most of the remaining habitat for this species (Sammons et al. 2015). Shoal Bass occur in sympatry with native stocks of Largemouth Bass throughout their native range, but the two species are rarely found in the same habitat in streams. While Shoal Bass are commonly found in shallow riffle areas and fast current, Largemouth Bass more typically occur in pools and slower runs (Wheeler and Allen 2003; Goclowski et al. 2013).

Understanding the population dynamics of recreational fishes is important to sustaining and managing these resources. Shoal Bass rate functions have been little studied in its natural range (Williams and Burgess 1999; Sammons et al. 2015), and biologists lack the basic life history data that are vital for successful management of any fishery. Furthermore, little is known about population dynamics of lotic populations of Largemouth Bass throughout their range. Therefore, this study was undertaken to obtain basic population data from populations of Shoal Bass and Largemouth Bass in the Flint River system, Georgia. Two river reaches and one reservoir were selected for this study. Specifically, the objectives of this study were to compare growth, mortality, recruitment, and condition among areas, species, and gender.

Methods

Study Area

A major tributary (mean annual discharge 114.12 m^3/s) of the Apalachicola River drainage, the Flint River flows 565 km from its headwaters near Atlanta, Georgia, USA, to its confluence with the Chattahoochee River at Lake Seminole (Figure 1). The upper reaches of the Flint River flow through the piedmont physiographic region of Georgia and are characterized by a series of wide, granite shoal areas with shallow water and fast current interspersed with narrower run and pool areas with deeper water and less current (Marcinek 2003). At the Fall Line (i.e., the transition from the piedmont to coastal plain physiographic region), the river drops approximately 125 m over 80 km. Below the Fall Line the river becomes a typical coastal plain stream, characterized by sandy substrate with some limestone outcroppings, greater amounts of woody debris present in the channel, and greater base flows due to the addition of groundwater from aquifers (Opsahl et al. 2007). The upper Flint River (UFR) flows over 320 km before being impacted by the first of three mainstem impoundments, making it one of only 42 rivers in the USA with >200 km of unimpeded flow (Benke 1990).

Lake Blackshear is the upper-most impoundment on the Flint River (Figure 1). Completed in 1930, the 3,440-ha reservoir is a typical mainstem, lowland impoundment, with mean depth of 3.2 m and a mean annual retention time of 16 d (U.S. EPA 1975). Much of the middle and lower reservoir is heavily vegetated with Bald Cypress *Taxodium distichum* and a variety of emergent and submergent aquatic vegetation. Downstream of Lake Blackshear is the second reservoir on the Flint River, Lake Chehaw, which is a 567-ha, shallow, riverine impoundment (Figure 1). Between these impoundments is a 15-km reach of the Flint River regulated by hydropeaking discharges from Warwick Dam, which forms Lake Blackshear (hereafter, LAKE).

Geomorphology of the Flint River below Warwick Dam downstream to the headwaters of Lake Seminole (hereafter, the lower Flint River [LFR]) is characterized by limestone outcroppings, a deep thalweg, and a hydraulic connection to the Upper Floridan aquifer resulting in large groundwater inputs (Opsahl et al. 2007). Land use within this reach is

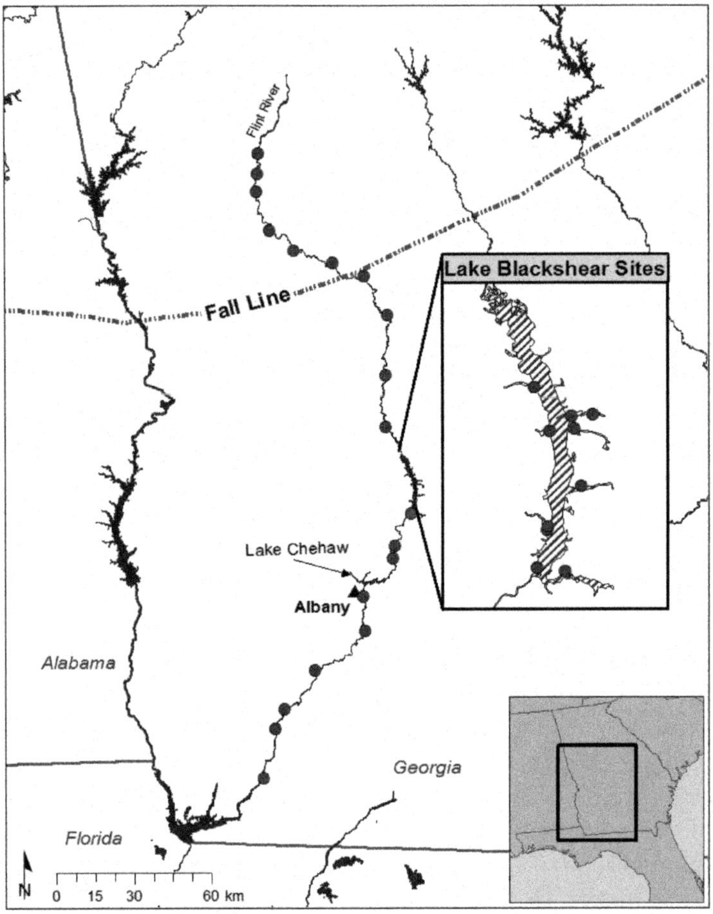

Figure 1. Sampling sites for black bass on the Flint River system, Georgia, from 2005 to 2013. Sites above Lake Blackshear were considered the upper Flint River and those below were considered the lower Flint River.

dominated by agriculture; however, the river corridor remains mostly forested and other than the Albany, Georgia (2013 population ~ 76,200) area, there is little urban development within the watershed. Numerous springs are found throughout this reach that add to stream flow and provide summer thermal refugia for species such as Gulf-strain Striped Bass *Morone saxatilis* (Van Den Avyle and Evans 1990; Opsahl et al. 2007; Ingram and Kilpatrick 2015). In the late 1800s, the U.S. Army Corps of Engineers opened and maintained a 1-m low-water navigation channel from Albany, Georgia, to Bainbridge, Georgia, to facilitate steamboat and barge navigation (Thurston 1973). This was accomplished by dredging and blasting shoals, which resulted in man-made rock islands that are still extant today. Although the effects of these activities on aquatic biota are unknown, they likely have resulted in deeper channels, swifter currents, and possibly exacerbating the impacts of hydropeaking flows from upstream hydropower dams. Due to concerns about low and inconsistent recruitment observed in the LFR by Georgia Department of Natural Resources (GADNR) biologists, Shoal Bass have been stocked annually, and have contributed significantly to overall year-class strength in some years (T. Ingram, unpublished data).

Fish Collection and Processing

In each study area (i.e., UFR, LAKE, and LFR), Largemouth Bass and Shoal Bass were collected from ten randomly-chosen, fixed sites using a boom-mounted boat electrofishing unit with pulsed DC; transects at each site were 1-h long. All of these sites were those used by GADNR biologists for standardized sampling, and samples were collected during this program using established protocols. These sites were distributed across a 200-km stretch of the UFR, throughout LAKE, and across a 109-km reach of the LFR (Figure 1). All collections from the UFR were made in the late summer and fall (i.e., August to October) of 2005, 2009, and 2011; whereas, all collections in LAKE were made in the spring (March–April) of 2007, 2009, and 2011. In the LFR, Largemouth Bass were collected in the fall of 2008 and in spring and fall of 2013; whereas, Shoal Bass were collected in the spring of 2008, 2011, and 2013. Shoal Bass do not occur in LAKE, and only Shoal Bass were collected from the UFR in 2005. All bass collected were measured for total length (TL, mm) and weighed (g). In LAKE and in some years on the LFR, a subsample of fish (approximately 10 fish per 25-mm group) was returned to the laboratory for aging. All bass collected on the UFR were returned to the laboratory for aging. In the laboratory, fish were sexed by visual examination of gonads, and saggital otoliths were extracted and aged according to methods described in Maceina (1988) and Buckmeier and Howells (2003).

Data Analyses

Because Shoal Bass do not currently have a standard weight equation available to calculate Relative Weight, condition of all species were assessed using the Relative Fulton Condition Factor (Kn; Neumann et al. 2012) using the statewide standard length-weight relations developed by GADNR. Only fish caught in the fall were used for this analysis, to eliminate seasonal bias in the data, meaning that LAKE data were not used for Largemouth Bass. Differences in Kn among species and areas were examined within each area using two-way analysis of variances (SAS Institute 2004).

Ages were assigned to all unaged fish using an age-length key (Chih 2009). Growth of each species in each area was described using a von Bertalanffy (1938) model using data pooled across years; ages of fish collected in the fall were corrected by adding 0.5 to each age to account for growth past the last annulus (Sammons and Maceina 2008). Time for each species to reach 356 mm TL and 457 mm TL were also estimated using this model. Growth of age 1–6 fish was compared directly between species within each area and among areas for each species using analysis of covariance (ANCOVA) to compare the slopes of TL to $\log_{10}$ age regressions (Bartlett et al. 1984; SAS Institute 2004). For both species, annuli were assumed to be fully visible by June 1, thus ages of all fish were age-corrected by adding 0.083 (1/12) for each month the fish were collected past May to account for growth past the last annulus and smooth out the regression lines. Growth of each species in each area was compared among years using an ANCOVA as described above and no differences were found; thus, data were pooled and used to test for differences among species and areas. Differences in growth between male and female fish of each species in each area were assessed using von Bertalanffy (1938) models and ANCOVA similar to other growth analyses. However, these analyses were only conducted for samples where few or no fish were subsampled to avoid biases described by Bettoli and Miranda (2001), as sex was unable to be determined for unaged fish.

Age frequencies were tabulated for each species and area; data were pooled across years to dampen the effects of strong and weak year classes that might be observed for a single sampling year (Miranda and Bettoli 2007; Maceina and Sammons 2016). We expected this technique to also dampen any possible effects of various stocking rates of contributions of stocked Shoal Bass in LFR on mortality rates. Instantaneous rate of total annual mortality (Z) was estimated using weighted catch-curve regressions on the descending limb (age ≥3) using the Fisheries Analysis and model Simulator (FAMS) software developed by Slipke and Maceina (2014). Annual survival (S) of each species in each area was calculated using the equation $S = e^{-Z}$, and total annual mortality (A) was calculated from $A = 1–S$ (Ricker 1975). Because age ranges for mortality estimates varied among species and rivers, mortality was standardized for each species and river combination using numbers of only age 3–10 fish in $\log_{10}$ number to age regressions. These standardized mortality estimates were compared between Largemouth Bass and Shoal Bass within each area using ANCOVA to compare the slopes of the regressions (SAS Institute 2004; Miranda and Bettoli 2007; Sammons and Maceina 2009a). Furthermore, mortality of age 3–10 fish was compared among areas for each species, pooling across years, using ANCOVA. Age frequencies were also tabulated for males and females of Largemouth Bass and Shoal Bass in each area pooled across years; differences in age frequencies of fish age 2 and older between sexes were examined in each area using Kolmogorov-Smirnov tests. Additionally, differences in A estimates between sexes of age 2–10 fish were tested in each area using ANCOVA as described above (SAS Institute 2004). As for growth, these analyses were not conducted for samples where the majority of fish had ages assigned using the age-length key.

Recruitment variation of both species in each area was examined using residual analysis from catch curves (Maceina 1997). Residuals were pooled across years and compared to annual discharge data in each river using regression analysis (Maceina 1997; Sammons and Maceina 2009b). Recruitment variation of each species in each river was also compared to river discharge in four seasons (prespawn [January–March], spawn [March–April], summer [May–August], and fall [September–December]) in an attempt to identify critical periods for recruitment (Sammons and Maceina 2009b). River discharge data were obtained from gauging stations maintained by the U.S. Geological Survey (www.usgs.gov/water) from gauges located approximately midway between the longitudinal distribution of sampling sites in each Flint River area (UFR: Carsonville, 02347500; LFR: Newton, 02353000). Flow through Lake Blackshear was assumed to be approximated by discharge data from the gauge located on the Flint River about 42 km above the reservoir (Montezuma, 02349605). Mean discharge was calculated for each river section for each growth year, and for each quarter of the growth year (Sammons and Maceina 2009b).

Multiple regression analysis was used to assess relationships among river flow data and catch-curve residuals in each area (Maceina 1997), using a generalized regression model:

$$\text{LOGNUM} = b_0 - b_1(\text{AGE}) \pm b_i(\text{DISCHARGE})$$

where b_0, b_1, and b_i were the regression coefficients for the intercept and slope coefficients, LOGNUM was $\log_e$ (Number of fish at each age) and DISCHARGE was either a single or multiple measures of annular or seasonal flow volume or variation in flow volume. In each case, the candidate models were chosen based on Akaike's Information Criteria (AIC) and Mallows' (1973) Cp statistic (Burnham and Anderson 2002; SAS Institute 2004). Semipartial correlation coefficients (SCORR1; SAS Institute 2004) were calculated to determine the relative contribution of each independent variable to the overall coefficient of determination of the model in sequential order after age was entered. Squared partial correlation coefficients (PCORR2; SAS Institute 2004) were calculated for each independent variable to define the amount of variation in number explained by flow variables after removing the effects of the other variables. Variance Inflation and Condition indices were used to choose the best model from among the candidate models, defined as the model that explained the greatest amount of variation in number with the fewest number of independent variables. Significance level for all statistical tests was set at $P \leq 0.05$; except for ANCOVA analyses comparing mortality rates, where the significance level was set at $P \leq 0.10$, due to smaller sample sizes affecting statistical power.

Results

Field Collection and Condition

Pooled electrofishing samples comprised 277, 1,062, and 618 Largemouth Bass in the UFR, LAKE, and LFR, respectively (Figure 2). Samples comprised 516 and 909 Shoal Bass in the UFR and LFR, respectively. Length-frequency distributions for Largemouth Bass were generally unimodal in LAKE and LFR, but fish were more uniformly distributed over a range of lengths in the UFR (Figure 2). Shoal Bass length frequency was skewed towards fish <300 mm TL in the UFR; whereas, it was skewed towards fish >350 mm TL in the LFR. Condition of Largemouth Bass was higher in the UFR than the LFR ($F_{1, 646} = 5.11$; $P = 0.0241$); whereas, condition of Shoal Bass was similar among areas ($F_{1, 479} = 0.76$; $P = 0.3840$; Figure 3). Condition of Shoal Bass was higher than Largemouth Bass in both areas ($F \geq 26.29$; $P < 0.0001$).

Growth

Growth of Largemouth Bass differed among areas ($F_{2, 1438} = 6.82$; $P = 0.011$), with slowest growth in the UFR. Growth was similar between the LFR and LAKE ($P = 0.1256$). von Bertalanffy models predicted that growth was initially similar among areas, but fish in LAKE diverged away from those in the LFR and UFR by age 2 (Figure 4). Largemouth Bass growth was similar between the two Flint River areas up to about age 4, whereupon fish in the LFR maintained fast growth; whereas, growth in the UFR began leveling off. By age 9, predicted Largemouth Bass growth in LAKE and LFR was virtually identical (Figure 4). Growth models predicted that Largemouth Bass reached 356 mm TL in 4.26, 3.27, and 3.87 years in the UFR, LAKE, and LFR, respectively. In contrast, fish reached 457 mm TL in 9.62 years in the UFR, but only in 5.89 and 6.53 years in the LAKE and LFR, respectively. Similar to Largemouth Bass, growth of Shoal Bass was higher in the LFR than in the UFR ($F_{1, 1202} = 539.05$; $P < 0.0001$). Growth rapidly diverged between the areas after age 1, but unlike Largemouth Bass, maintained a similar trajectory between areas throughout the lifespan of the fish (Figure 4). Shoal Bass were predicted to reach 356 mm TL in 4.03 and 3.09 years in the UFR and LFR, respectively, and 457 mm TL in 6.86 and 5.43 years in the UFR and LFR, respectively. Shoal Bass grew faster than Largemouth Bass in both the UFR ($F_{1, 596} = 7.25$; $P = 0.0073$) and the LFR ($F_{1, 1342} = 1.97$; $P = 0.0465$).

Differences in growth between sexes was unable to be examined in LAKE, as all samples were aged using an age-length key. Further, in the LFR only the 2013 sample was able to be used, as a majority of fish in the other two years had ages assigned by an age-length key. Growth of female Largemouth Bass was faster than males in UFR and LFR; von Bertalanffy models predicted that growth

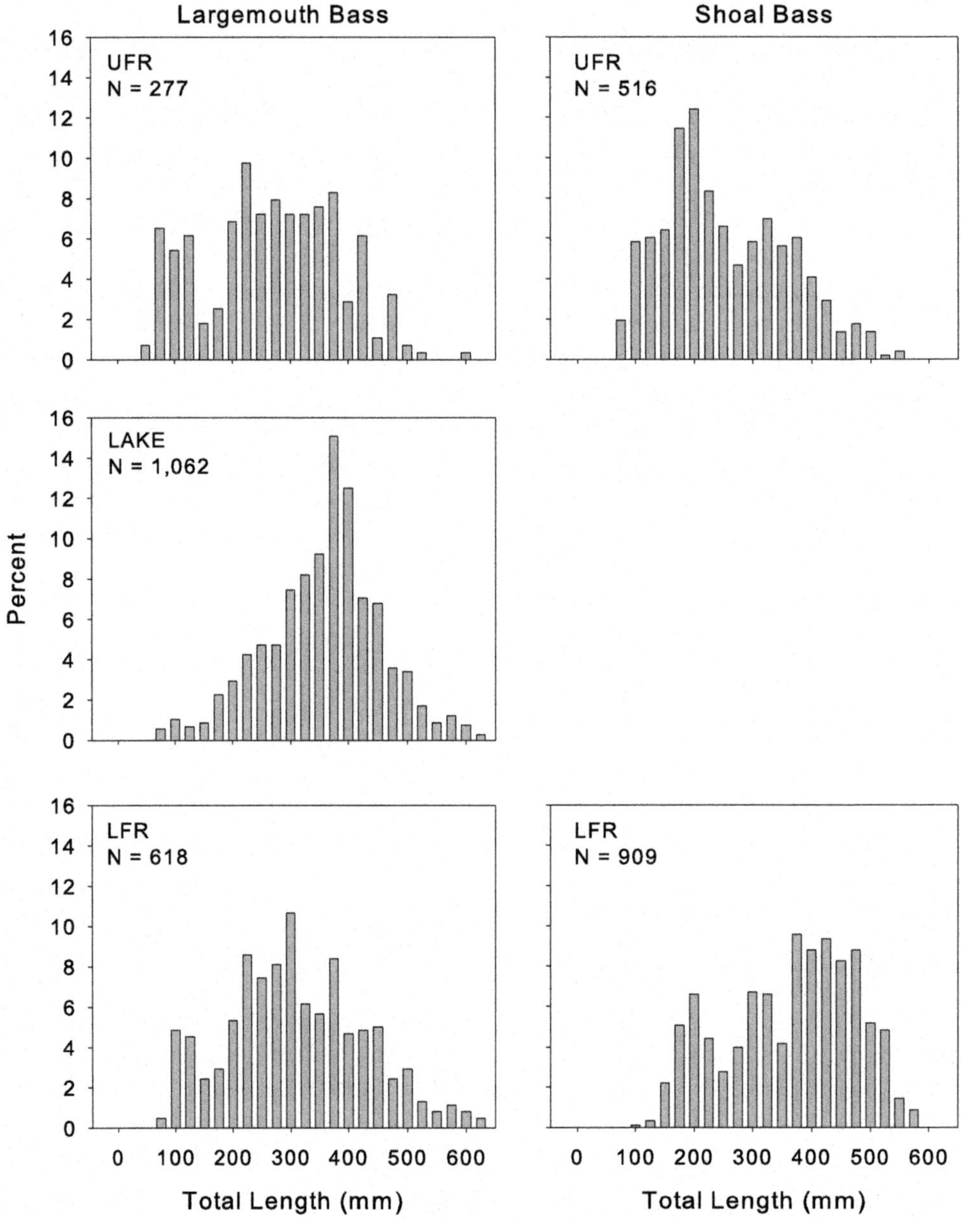

Figure 2. Length-frequencies (25-mm length-groups) of two black bass species collected from the upper Flint River (UFR), Lake Blackshear (LAKE), and the lower Flint River (LFR), Georgia.

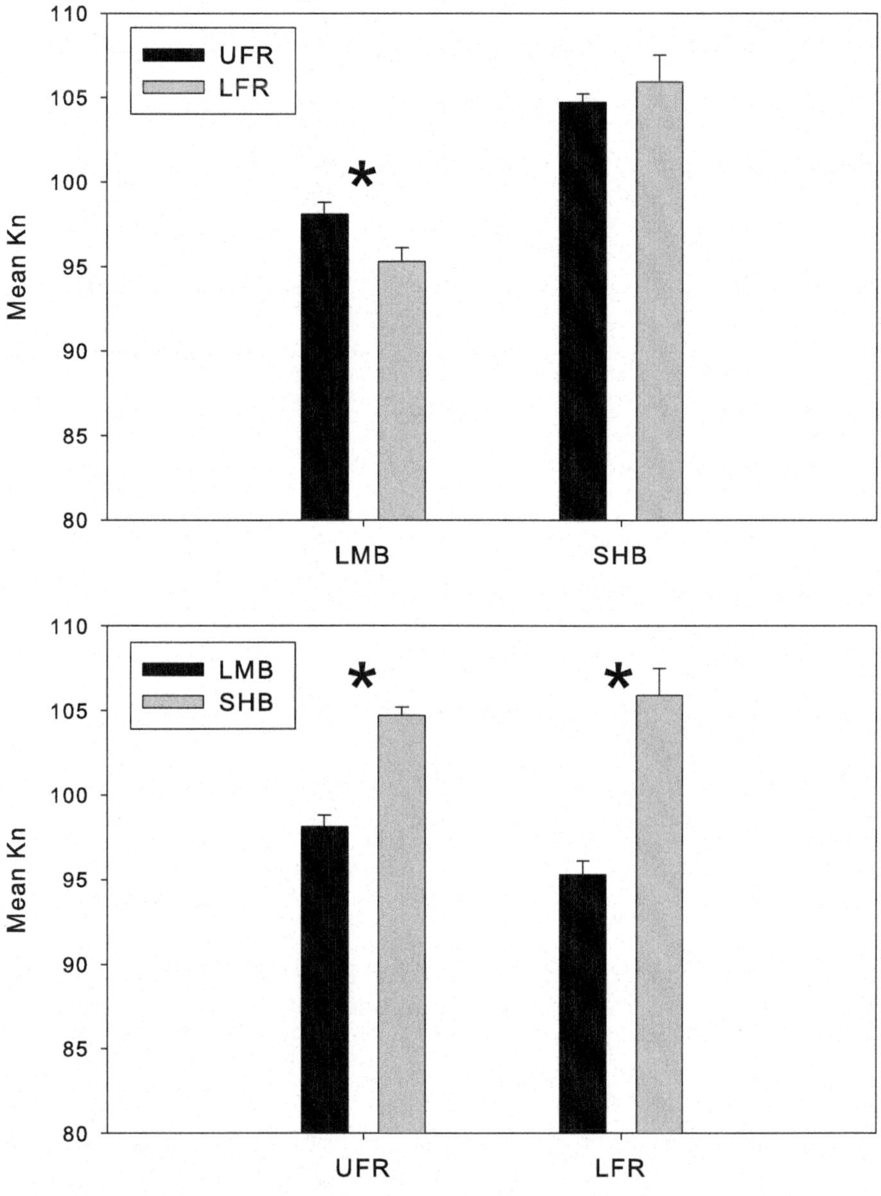

Figure 3. Mean Fulton Relative Condition (Kn) factors of two species of black bass from the upper Flint River (UFR) and the lower Flint River (LFR), Georgia. Asterisks denote significant differences in Kn (Tukey test, $P < 0.05$); error bars represent standard errors.

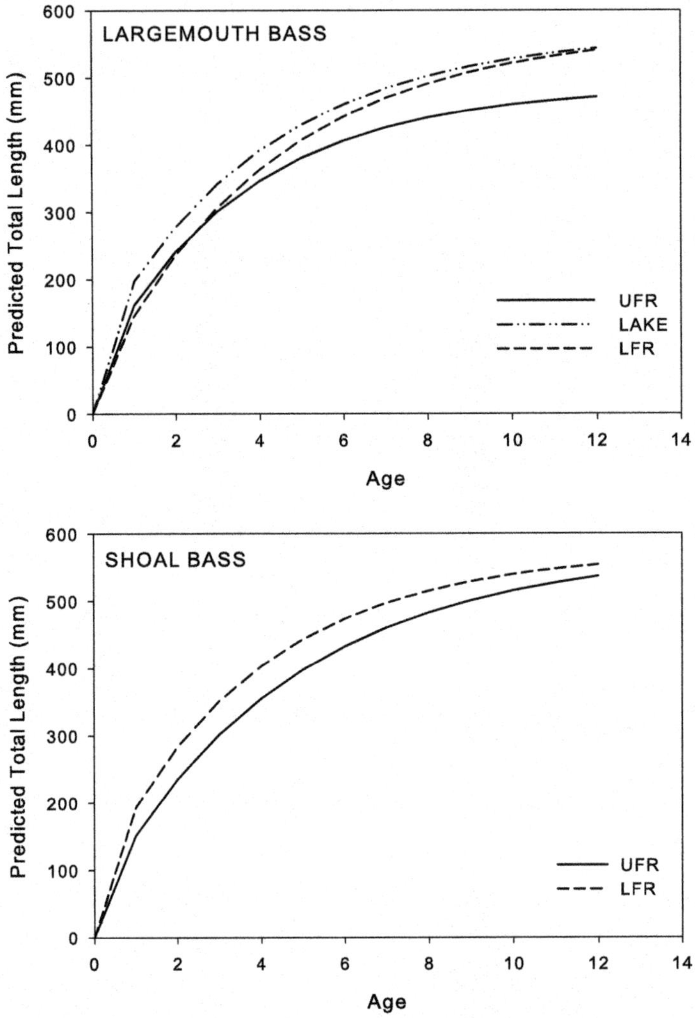

Figure 4. Predicted mean total length at age from von Bertalanffy growth models for Largemouth Bass and Shoal Bass collected from the upper Flint River (UFR), Lake Blackshear (LAKE), and the lower Flint River (LFR), Georgia. No Shoal Bass were collected in LAKE.

diverged sharply after about age 3 (Figure 5). In contrast, Shoal Bass growth was similar between sexes in both areas, and von Bertalanffy models predicted that growth was similar throughout their lifespan.

Mortality

Aged Largemouth Bass ranged from 1 to 13 years in the UFR and LAKE, and from 1 to 12 years in the LFR (Table 1). In contrast, aged Shoal Bass ranged from 1 to 11 years in both areas of the Flint River. Age structure of both species was skewed younger than age 6 in the UFR; whereas, older fish were more common in the LFR, especially for Largemouth Bass (Table 1). Largemouth Bass in LAKE were commonly collected up to age 10. Total annual mortality of age-3–10 Largemouth Bass ranged from 0.33 to 0.40 across the three study areas, and did not vary among areas (Table 1; $F_{2, 18} = 0.85$; $P = 0.4450$). Similarly, Shoal Bass A was 0.49 in the UFR and 0.44 in the LFR and was simi-

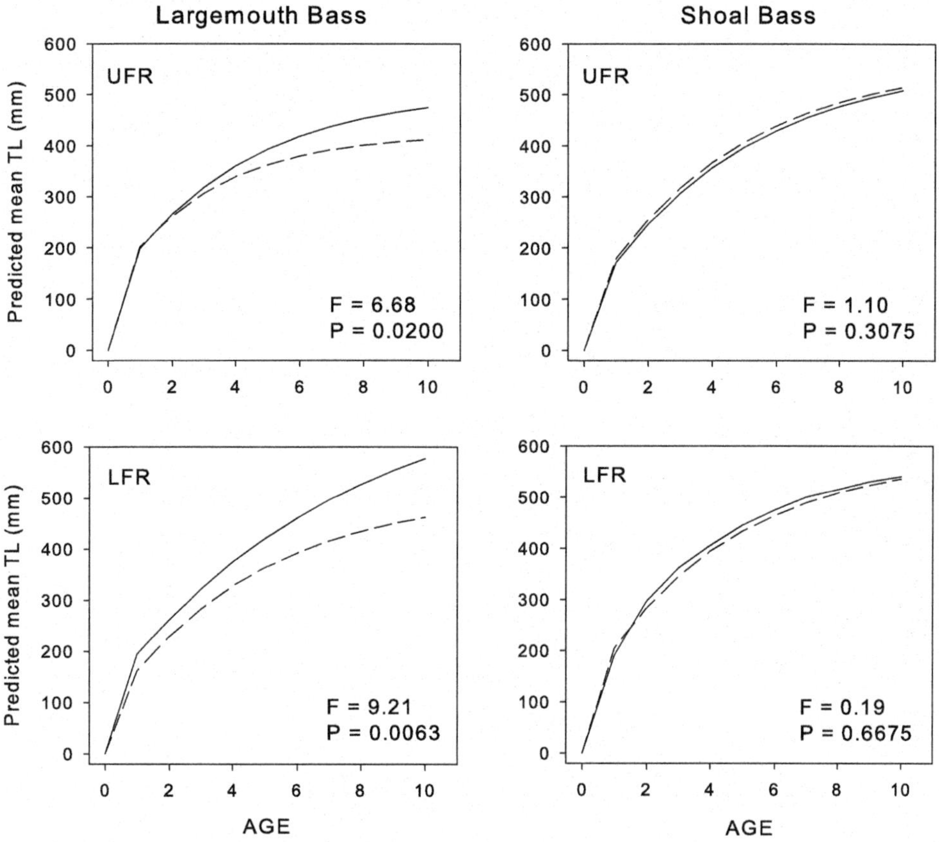

Figure 5. Predicted mean total length (TL) at age from von Bertalanffy growth models for male (dotted line) and female (solid line) Largemouth Bass and Shoal Bass collected from the upper Flint River (UFR) and the lower Flint River (LFR), Georgia. Results of ANCOVA analyses for assessing growth differences between sexes are given in each panel.

Table 1. Number of fish per age of Largemouth Bass (LMB) and Shoal Bass (SHB) collected from the upper Flint River (UFR), the lower Flint River (LFR), and Lake Blackshear (LAKE), Georgia. Numbers of each species in each area are pooled across three samples collected from 2005–2013. Results of weighted catch curves are presented for fish age 3–10 for each species in each area; parameters are total instantaneous mortality (Z), total annual mortality (A), and theoretical maximum age (T_{max}).

		Ages							Catch Curve		
Area	Species	1	2	3	4	5	6	7	Z	A	T_{max}
UFR	LMB	57	51	49	29	15	3	4	0.403	0.33	11.5
	SHB	189	83	93	35	13	15	4	0.678	0.49	9.3
LAKE	LMB	186	219	332	87	84	50	51	0.518	0.40	13.2
LFR	LMB	71	175	145	54	56	26	19	0.463	0.37	13.4
	SHB	105	127	100	175	134	52	43	0.586	0.44	12.3

		Ages						Catch Curve		
Area	Species	8	9	10	11	12	13	Z	A	T_{max}
UFR	LMB	3	3	2	3	1	1	0.403	0.33	11.5
	SHB	1	2	1	1	-	-	0.678	0.49	9.3
LAKE	LMB	13	23	12	1	2	2	0.518	0.40	13.2
LFR	LMB	10	18	6	0	1	-	0.463	0.37	13.4
	SHB	6	5	0	3	-	-	0.586	0.44	12.3

lar between areas ($F_{1,11}$ = 0.99; P = 0.3416). However, A of Shoal Bass was greater than Largemouth Bass in both the UFR ($F_{1,12}$ = 4.10; P = 0.0658) and the LFR (F = 6.09; df = 1, 12; P = 0.0296).

Age frequencies of male and female Largemouth Bass in the UFR were similar (KSa = 0.81; P = 0.5293), with fish of both sexes collected up to age 11 (Figure 6); total annual mortality was similar likewise between sexes ($F_{1,14}$ = 0.55; P = 0.4694). In contrast, Shoal Bass in the UFR older than age 6 were dominated by females (KSa = 1.43; P = 0.0332), although one age 10 male was collected (Figure 6). However, A was still similar between sexes ($F_{1,14}$ = 0.11; P = 0.7400). Similar to the UFR, age frequencies of male and female Largemouth Bass in the LFR were similar (KSa = 01.96; P = 0.3126), with females and males collected up to age 9 and age 10, respectively (Figure 7). Total annual mortality was similar between sexes ($F_{1,14}$ = 1.00; P = 0.3349), despite a 0.11-unit difference, likely due to low sample size. Age

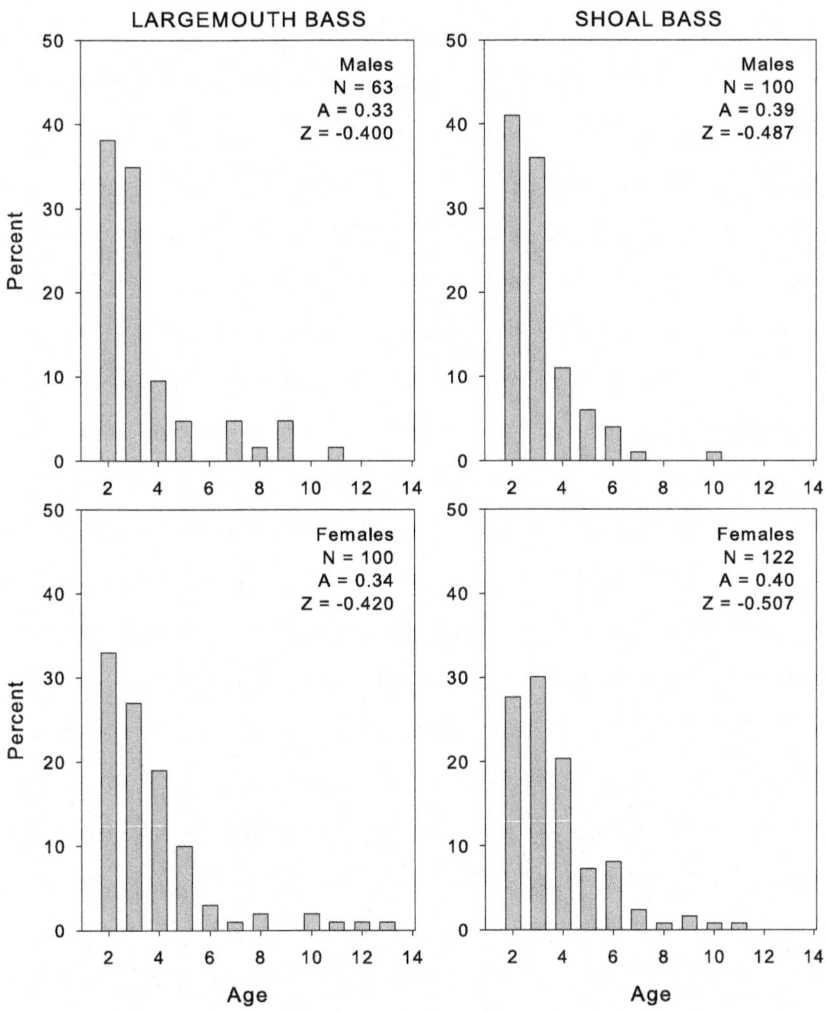

Figure 6. Age structure (age 2 and older) of male and female Largemouth Bass and Shoal Bass collected from the upper Flint River, Georgia. Sample size (N) and total annual mortality (A) and instantaneous rate of mortality (Z) derived from catch-curve analyses are presented in each panel.

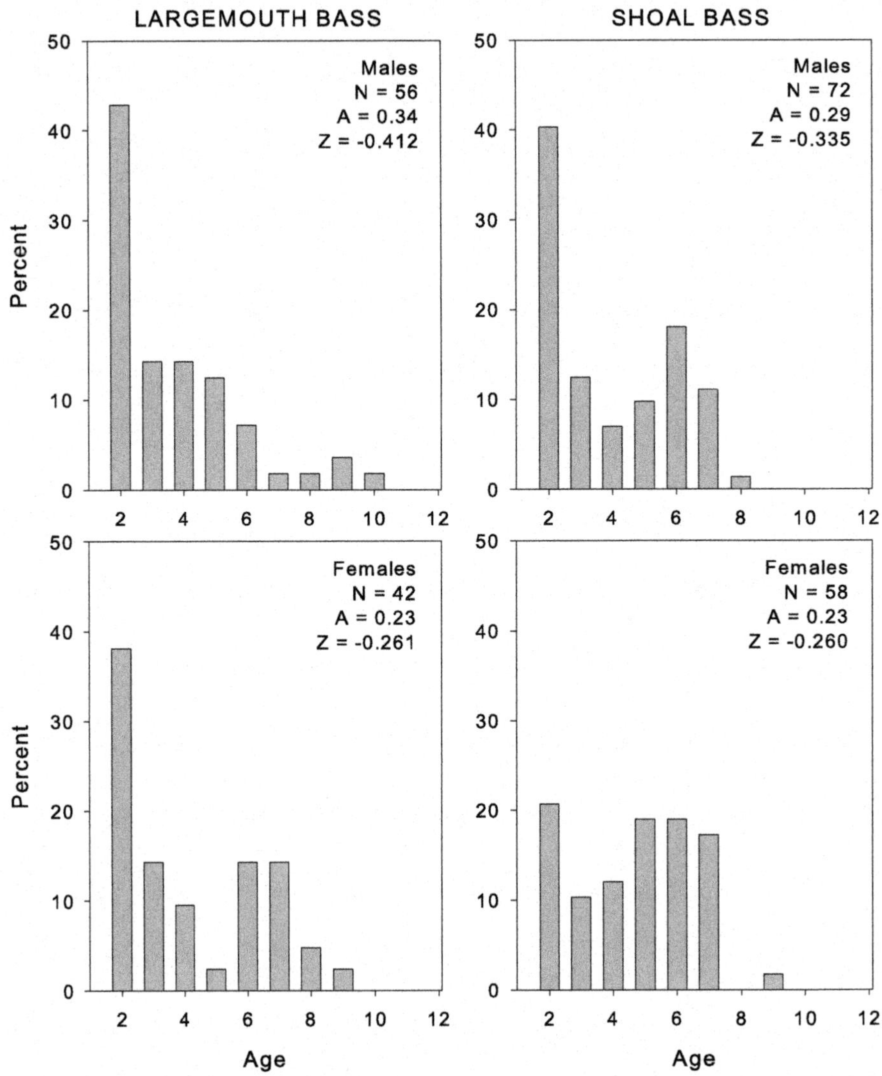

Figure 7. Age structure (age 2 and older) of male and female Largemouth Bass and Shoal Bass collected from the lower Flint River, Georgia. Sample size (N) and total annual mortality (A) and instantaneous rate of mortality (Z) derived from catch-curve analyses are presented in each panel.

frequency of male Shoal Bass in the LFR was more skewed towards age-2 fish than the female age frequency (Figure 7; KSa = 1.23; $P = 0.0959$), but ANCOVA results indicated that A was similar between sexes ($F_{1, 12} = 0.25$; $P = 0.6251$).

Recruitment

Mean monthly discharge over the course of the study ranged from 2.19 to 408.63 m^3/s in the UFR, 13.51–679.24 m^3/s in the LAKE, and 21.49–778.35 m^3/s in the LFR. Residuals of the catch-curve regressions for Largemouth Bass in the UFR were negatively correlated with annual mean discharge and mean discharge during the summer and fall (Table 2). Similarly, catch-curve residuals for Largemouth Bass in LAKE were negatively correlated with annual mean discharge and mean discharge during the summer. In contrast, catch-curve residuals for Largemouth Bass in the LFR were not correlated with mean discharge during any time period (Table 2). The opposite pattern was observed for Shoal Bass, where catch-curve residuals were not correlated with mean discharge in any time period in the UFR, but were negatively correlated with annual mean discharge and discharge in the LFR during the spawning season, summer, and in fall.

Based on AIC scores and Variance Inflation and Condition indices, the best model explaining variation in year-class strength of Largemouth Bass in the UFR was:

LOGe (NUMBER) = 4.4358–0.3452 (AGE) – 0.0004 (ANNUAL DISCHARGE)

where, NUMBER = number of fish collected in each year-class, AGE = age of fish in that year-class when collected, and ANNUAL DISCHARGE = mean annual discharge during the year fish were spawned. This model explained 78% of the variation and was highly significant ($P < 0.0001$). As expected, age explained the greatest amount of variation in number collected, but mean annual discharge explained an additional 7% of the variation after accounting for age, and when holding for the effects of age, mean annual discharge explained 25% of the variation in number collected.

Similarly, the best model explaining variation in year-class strength of Largemouth Bass in LAKE was:

LOGe (NUMBER) = 6.1906–0.4101 (AGE) – 0.0003 (SUMMER DISCHARGE)

where, NUMBER = number of fish collected in each year-class, AGE = age of fish in that year-class when collected, and SUMMER DISCHARGE = mean discharge during the summer that fish were spawned. This model explained 72% of the variation and was highly significant ($P < 0.0001$). Annual summer discharge explained an additional 14% of the variation after accounting for age, and when holding for the effects of age, explained 33% of the variation in number collected.

The best model explaining variation in year-class strength of Shoal Bass in the LFR was:

LOGe (NUMBER) = 6.2675–0.4710 (AGE) – 0.0003 (ANNUAL DISCHARGE)

where, NUMBER = number of fish collected in each year-class, AGE = age of fish in that year-class when collected, and ANNUAL DISCHARGE = mean annual discharge during the year that fish were spawned. This model explained 69% of the variation and was highly significant ($P < 0.0001$). Annual summer discharge explained an additional 10% of the variation after accounting for age, and when holding for the effects of age, explained 24% of the variation in number collected. No significant models relating year-class strength to river discharge were found for Shoal Bass in the UFR or Largemouth Bass in the LFR.

Table 2. Pearson correlations (P-value) between catch-curve residuals of Largemouth Bass (LMB), and Shoal Bass (SHB) and mean annual discharge and in four seasons during the spawning year of each year class. Fish were collected in the upper Flint River UFR), Lake Blackshear (LAKE), and in the lower Flint River (LFR).

Area	Species	Annual	Prespawn	Spawn	Summer	Fall
UFR	LMB	–0.56 (0.0069)	–0.28 (0.2135)	–0.38 (0.0850)	–0.44 (0.0404)	–0.49 (0.0209)
	SHB	–0.12 (0.5357)	0.03 (0.8699)	–0.25 (0.2050)	–0.25 (0.1925)	0.05 (0.7899)
LAKE	LMB	–0.41 (0.0390)	–0.12 (0.5400)	–0.25 (0.2139)	–0.59 (0.0015)	–0.22 (0.2859)
LFR	LMB	–0.17 (0.4858)	–0.33 (0.1661)	0.08 (0.7558)	–0.10 (0.6751)	–0.11 (0.6543)
	SHB	–0.69 (0.0002)	–0.12 (0.5861)	–0.51 (0.0117)	–0.64 (0.0008)	–0.65 (0.0005)

Discussion

Other than Largemouth Bass and Smallmouth Bass, little work has been done on most species of *Micropterus*, and relatively little of the studies conducted on Largemouth Bass have occurred in rivers. Thus, studies such as this one can provide basic data on population and life history characteristics of these fishes that can be useful for biologists to better manage river fisheries for these species. Much of the current understanding of the genus *Micropterus* has been drawn from studies of Largemouth Bass and Smallmouth Bass, and the unstated assumption has always been that other species in the genus have similar life history characteristics. However, failure by biologists to recognize and understand differences in life history characteristics among species in other genera has resulted in poor management decisions, some of which have taken decades to rectify (e.g., Sowa and Rabeni 1995; Greene and Maceina 2000; Hilderbrand and Kershner 2000; Huckins et al. 2000; Wares et al. 2004; Schoenebeck and Hansen 2005).

The Flint River system supports robust fisheries for Largemouth Bass and Shoal Bass, as demonstrated by the wide range of sizes and ages of these species collected during this study. Both species are native to this system, but only Largemouth Bass are found in LAKE, because Shoal Bass are fluvial specialists that do not withstand impoundment (Sammons et al. 2015). Shoal Bass and Largemouth Bass appeared to have similar scopes of growth throughout the system, reaching similar maximum sizes. However, Largemouth Bass grew slower than Shoal Bass in each river area. Typically, Largemouth Bass are commonly the fastest growing *Micropterus* in systems occupied by multiple species (Beamesderfer and North 1995; Novinger 1987; Sammons and Bettoli 1998; Long and Fisher 2001; Churchill and Black 2007). Growth of fishes is mediated by a variety of factors, including climate, productivity of the system, food supply, and environmental factors such as river discharge or reservoir water levels (Zolczynski and Davies 1976; Keast and Eadie 1985; McCauley and Kilgour 1990; Beamesderfer and North 1995; Diana 1995; DiCenzo et al. 1995; Crawford et al. 1998; Sammons and Maceina 2009b). While found in rivers and streams, Largemouth Bass are more commonly found in lentic waters and oxbows and backwaters of larger rivers (Boschung and Mayden 2004). Habitat choice and distribution across habitats likely reflects evolutionary adaptation over thousands of years, thus most rivers probably represent less than ideal environments for this species. In contrast, Shoal Bass are lotic specialists and are adapted for rocky, fast-flowing riverine habitats such as the Flint River (Williams and Burgess 1999; Sammons et al. 2015). This divergence in evolutionary adaptation may be reflected in the disparity in growth rates between these two species observed in this study.

Diet and food supply can also influence growth of fishes (Adams et al. 1982; Ellison 1984; Jones et al. 1994; Lott et al. 1996; Margenau et al. 1998). In particular, occurrence of fish prey in the diet of piscivores is often correlated with faster growth rates (Adams et al. 1982; Margenau et al. 1998; Olson and Young 2003). Growth of Largemouth Bass was similar among areas up to age 1, but at older ages fish in LAKE grew faster than in either river area. Growth of Largemouth Bass in the UFR began to level off after age 4, but in the LFR growth of older fish continued at a high rate up to the maximum age collected. These observations suggest that LAKE and the LFR may have a better forage base for larger Largemouth Bass than the UFR, likely consisting of large-bodied forage fishes such as shad or suckers. Previous work has documented that Largemouth Bass diet in the UFR was dominated by sunfishes *Lepomis* spp. (Sammons 2012), which are not as energetically efficient as soft-rayed fishes such as suckers and shad (Miranda and Muncy 1990). Anglers routinely catch both Largemouth Bass and Shoal Bass >2.25 kg from the Ocmulgee River, Georgia, but catches of fish of this size appear to be much rarer in the Flint River. Thus, trophy Largemouth Bass production in the UFR may be constrained by appropriate forage availability for large fish, which has been documented by researchers working on piscivores in other systems (Ellison 1984; Eberts et al. 1998; Margenau et al. 1998). In contrast, the scope of growth for Shoal Bass appeared to be more similar than Largemouth Bass between the UFR and LFR. Unlike Largemouth Bass, Shoal Bass diet often consist of a mixture of fish, crayfish, and a variety of smaller macroinvertebrates (Wheeler and Allen 2003; Sammons 2012), which may be less variable across the system. Regardless, growth of both species was faster in the LFR than in the UFR. River productivity tends to increase over an upstream to downstream gradient due to increased tributary inputs and higher allocthanous material (Vannote et al. 1980). This leads to higher food densities which may be reflected in faster growth of predacious fishes such as black bass. Also, the study was conducted over a north-south gradient of over 400 km, with corresponding variations in temperatures and growing season that may have contributed to the observed faster growth in the downstream-most area (Conover 1990; Beamesderfer and North 1995). Thus responses to management actions, such as harvest restrictions may vary across the entire length of larger rivers, similar to what has been observed across large spatial scales (Allen and Miranda 1995; Isermann et al. 2002).

Largemouth Bass longevity appeared to be greater than Shoal Bass throughout the system, with fish collected up to age 13 in the UFR and LAKE and up to age 12 in the LFR. This is comparable to what Woodside et al. (2015) found for Largemouth Bass in the Chipola River, Florida. Largemouth Bass are known to exhibit greater longevity than most congenerics across their range. Crawford et al. (1998) found Largemouth Bass as old as 16.5 years in Florida; whereas, the current longevity record for this species is 24 years old, collected in upstate New York (Green and Heidinger 1994). Coble (1975) stated that the maximum age of Smallmouth Bass was likely around 15 years, but this species has been collected up to age 19 from the Hudson River, New York (Maceina and Sammons 2006). Thus, longevity of these two species is likely similar. Bonvechio et al. (2005) found that maximum age of Largemouth Bass and Suwannee Bass *M. notius* was relatively similar (age 11–12) in three Florida rivers. However, other species of *Micropterus* appear to have shorter lifespans. Longevity of Spotted Bass has been reported to be age 7–8 by a variety of authors (Vogele 1975; Novinger 1987; Sammons and Bettoli 1998), although Spotted Bass up to age 11 have been collected in Virginia and Kentucky (Long and Fisher 2001).

Longevity of Shoal Bass appears to be similar to that reported for Spotted Bass and Alabama Bass. Fish up to age 11 were collected from both areas of the Flint River in this study and were collected up to age 10 in the Ocmulgee River by Sammons and Goclowski (2012). Similarly, Woodside et al. (2015) collected Shoal Bass up to age 10 in the Chipola River, Florida. DiCenzo et al. (1995) reported a maximum age of age 11 for Alabama Bass collected from 10 Alabama reservoirs; whereas, Alabama Bass up to age 14 were collected from Allatoona Reservoir, Georgia (Hakala and Sammons 2015). Based on the results of this study, it appears that Shoal Bass longevity is relatively long, maybe only a year or two shorter than Largemouth Bass. This was reflected by the estimates of A of each species. Largemouth Bass consistently had lower A than Shoal Bass, but Shoal Bass mortality fell within the range commonly reported for other black bass species (Beamesderfer and North 1995; Bonvechio et al. 2005; Churchill and Black 2007; Shepherd and Maceina 2009). Also, Shoal Bass in the UFR were found to be exploited at moderately high rates (22–25%) in a concurrent study (Sammons 2016), which may have increased A; however, exploitation of Largemouth Bass in the UFR, and of both species in the LFR, is unknown. Woodside et al. (2015) reported ranges of annual mortality for Largemouth Bass and Shoal Bass in the Chipola River, Florida, that were similar to what was found in this study, with Largemouth Bass having significantly lower mortality than Shoal Bass (32% versus 57%).

Sexual dimorphism has been reported in age and growth for a variety of centrarchid species. In sunfish, the males commonly exhibit faster growth than the females, presumably as an adaptive advantage for enhanced ability to defend nests against potential egg and larvae predators (Ehlinger 1997; Sammons et al. 2006; Sammons and Maceina 2009a). The prevailing paradigm for *Micropterus* species is that females commonly grow faster than males and attain larger sizes, but this has been based mostly on studies of Largemouth Bass (Hubert 1976; Porak et al. 1987; Schramm and Smith 1988; Crawford et al. 1998; Bonvechio et al. 2005). However, Heidinger (1975) pointed out that the literature regarding the relative growth of male and female fish is contradictory even for Largemouth Bass, with many studies reporting no difference in growth between sexes. Coble (1975) and Olmsted and Kilambi (1978) reported similar growth between sexes of Smallmouth Bass and Spotted Bass, respectively. Suwannee Bass in Florida rivers exhibited sexual dimorphic growth patterns similar to Largemouth Bass, with females growing larger than males (Bonvechio et al. 2005). Growth between sexes of Largemouth Bass, Smallmouth Bass, and Spotted Bass was not consistent among years and year classes in Pickwick Lake, Alabama; however, the general trend over time was for the females of all three species to be larger than males at each age-class (Hubert 1976). In the current study, female Largemouth Bass grew faster than males; however, longevity and A was similar between the sexes. In contrast, growth and A of Shoal Bass was similar between males and females in both areas, but females consistently lived longer than males. Thus, while there was a trend for larger Shoal Bass to be female, it appeared that the larger sizes were attained more through increased longevity rather than any growth advantage between the sexes.

Recruitment of Largemouth Bass in the UFR and LAKE was negatively affected by high flows, which has been observed in other Largemouth Bass and Smallmouth Bass populations subject to the effects of flows. Possible reasons advanced for this relationship are nest disruption and physical advection and mortality of larvae during flood events (Maceina and Bettoli 1998; Bonvechio and Allen 2005; Smith et al. 2005). Discharge was not related to the recruitment of Shoal Bass in the UFR, and it is not surprising that recruitment of an obligate river species such as Shoal Bass would be less affected by floods than a lentic specialist like Largemouth Bass. However, the opposite pattern was observed in the LFR, where recruitment of Shoal Bass, but not Largemouth Bass, was negatively affected by high flows. Reasons behind this disparity are not immediately obvious; however, the LFR has been heavily altered by man compared to the relatively intact UFR system. Two hydropower dams near the top of the LFR area have altered the flow regime to the point that it has affected Shoal Bass recruitment (Sammons et al. 2015). Further, the main channel of the LFR below Albany, Georgia, was dredged by the U.S. Army Corps of Engineers in the late 1800s to create and maintain navigability between Bainbridge, Georgia, and Albany (Thurston 1973). It seems likely that these alterations have affected the ability of Shoal Bass to recruit successfully in years of high water. Although variable stocking rates and contribution of stocked Shoal Bass in this reach may have confounded our results somewhat, pooling residuals across years likely minimized this bias (Maceina and Sammons 2016). It appears that recruitment of both stocked and wild fish are negatively affected by high flows in this system, but future research is needed in this reach to better define these relations for stocked and wild fish.

Research on the Chipola River, Florida, found that Shoal Bass spawned in small protected areas in the middle of high-flow areas (Bitz et al. 2015). This matches the observation by Johnston and Kennon (2007) that Shoal Bass larvae in a small Alabama creek were found in current refugia behind large boulders in shoals. Thus, access to flow refugia during the spawning and early juvenile periods are likely critical for successful Shoal Bass recruitment during years of higher flows. Hydraulic refugia are critical for organisms to withstand the variations of physical variables (Valentin et al. 1996). High relief caused by the large shoals found in the UFR likely provide ample flow refugia at almost any discharge level. However, shoal dredging in the LFR has likely increased water velocities and eliminated many main channel flow refugia areas, and this has likely been exacerbated by hydropeaking flows for the upstream dams (De Jalon et al. 1994; Freeman et al. 2001; Bunn and Arthington 2002). Strong year classes of Shoal Bass in the Chipola River, Florida, were associated with low flows and low flow variation (Woodside et al. 2015); however, shoals in the Chipola River are low relief compared to those in the UFR, and thus likely offer less refugia to spawning and juvenile Shoal Bass. Certainly, more work remains to be done to better characterize the effects of annual and seasonal discharge patterns on Shoal Bass recruitment.

Management Implications

Growth and mortality of black bass in the Flint River approximated those seen for congeners in reservoirs across the southeastern U.S. (Olmsted and Kilambi 1978; Novinger 1987; DiCenzo et al. 1995; Churchill and Black 2007; Shepherd and Maceina 2009). This demonstrates that these populations are capable of supporting quality fisheries with potential for anglers to catch trophy-sized individuals. Although currently small directed fisheries do occur on this river, they likely do not compare in terms of visitation rates and subsequent economic value compared to comparable-quality reservoir fisheries. A recent study on Lewis Smith Reservoir, Alabama, found that the Striped Bass fishery alone in this reservoir was worth approximately $700,000 annually through direct expenditures by anglers, despite Striped Bass anglers representing only 21% of the total annual effort (Lothrop 2012). It is likely that river fisheries in Georgia are underutilized by anglers, but development and promotion of these fisheries can result in direct benefits to the surrounding communities and the state in general.

Shoal Bass exhibited numerous differences from Largemouth Bass in the Flint River, including growth, mortality, and recruitment. Previous research has demonstrated comparable differences in behavior, movement, and habitat use between these species (Goclowski et al. 2013; Sammons 2015). While this may not be surprising given that they are fluvial specialists, it further demonstrates the importance of studying rate functions of all major sportfish species in a system. The common practice of "species borrowing" (sensu Ney 1993), whether unconsciously or by design, can result in erroneous conclusions about population resilience in the face of anthropogenic changes such as increased fishing pressure, urbanization, or climate change. Quantifying differences in growth and survival of fishes across large spatial scales allows fisheries biologists to better understand and manage fisheries (Isermann et al. 2002; Paukert et al. 2002; Porath and Hurley 2005; Tomcko and Pierce 2005). Results from this study will help biologists better understand the potential and limitations of black bass fisheries in rivers, and thereby enable more effective management.

References

Adams, S. M., R. B. McLean, and M. M. Huffman. 1982. Structuring of a predator population through temperature-mediated effects on prey availability. Canadian Journal of Fisheries and Aquatic Sciences 39:1175–1184.

Allen, M. S., and L. E. Miranda. 1995. An evaluation of the value of harvest restrictions in managing crappie fisheries. North American Journal of Fisheries Management 15:766–772.

Bartlett, J. R., P. F. Randerson, R. Williams, and D. M. Ellis. 1984. The use of analysis of covariance in the back-calculation of growth in fish. Journal of Fish Biology 24:201–213.

Beamesderfer, R. C. P., and J. A. North. 1995. Growth, natural mortality, and predicted response to fishing for Largemouth Bass and Smallmouth Bass populations in North America. North American Journal of Fisheries Management 15:688–704.

Benke, A. C. 1990. America's vanishing streams. Journal of the North American Benthological Society 9:77–88.

Bettoli, P. W., and L. E. Miranda. 2001. Cautionary note about estimating mean length at age with subsampled data. North American Journal of Fisheries Management 21:425–428.

Bitz, R., A. Strickland, T. Alferman, and J. Bock. 2015. Associated habitat of nesting Shoal Bass *Micropterus cataractae* in the Chipola River, Florida. Pages 237–248 *in* M. D. Tringali, J. M. Long, T. W. Birdsong, and M. S. Allen, editors. Black bass diversity: multidisciplinary science for conservation. American Fisheries Society, Symposium 82, Bethesda, Maryland.

Bonvechio, T. F., and M. S. Allen. 2005. Relations between hydrological variables and year-class strength of sportfish in eight Florida waterbodies. Hydrobiologia 532:193–207.

Bonvechio, T. F., M. S. Allen, and R. L. Cailteux. 2005. Relative abundance, growth, and mortality of Suwannee Bass in four Florida rivers. North American Journal of Fisheries Management 25:275–283.

Boschung, H. T., and R. L. Mayden. 2004. Fishes of Alabama. Smithsonian Books, Washington, D.C.

Buckmeier, D. L., and R. G. Howells. 2003. Validation of otoliths for estimating ages of Largemouth Bass to 16 years. North American Journal of Fisheries Management 23:590–593.

Bunn, S. E., and A. H. Arthington. 2002. Basic principles and ecological consequences of altered flow regimes for aquatic biodiversity. Environmental Management 30:492–507.

Burnham, K. P., and D. R. Anderson. 2002. Model selection and multimodel inference, a practical information-theoretical approach, 2nd edition. Springer, New York.

Chih, C. P. 2009. The effects of otolith sampling methods on the precision of growth curves. North American Journal of Fisheries Management 29:1519–1528.

Churchill, T. N., and W. P. Black. 2007. Use of mortality caps to determine effectiveness of black bass harvest restrictions in Tennessee reservoirs. Proceedings of the Annual Conference of the Southeastern Association of Fish and Wildlife Agencies 59(2005):217–226.

Coble, D. W. 1975. Smallmouth Bass. Pages 21–33 *in* R. H. Stroud and H. Clepper, editors. Black bass biology and management. Sport Fishing Institute, Washington D.C.

Conover, D. O. 1990. The relation between capacity for growth and length of growing season: evidence for and implications of countergradient variation. Transactions of the American Fisheries Society 119:416–430.

Crawford, S., W. F. Porak, D. J. Renfo, and R. L. Cailteux. 1998. Age and growth of trophy Largemouth Bass in Florida. Proceedings of the Annual Conference of the Southeastern Association of Fish and Wildlife Agencies 50(1996):212–220.

De Jalon, D. G., P. Sanchez, and J. A. Camargo. 1994. Downstream effects of a new hydropower impoundment on macrophyte, macroinvertebrates and fish communities. Regulated Rivers: Research & Management 9:253–261.

Diana, J. S. 1995. Biology and ecology of fishes. Cooper Publishing Group, LLC, Carmel, Indiana.

DiCenzo, V. J., M. J. Maceina, and W. C. Reeves. 1995. Factors related to growth and condition of the Alabama subspecies of Spotted Bass in reservoirs. North American Journal of Fisheries Management 15:794–798.

Eberts, R. C., Jr., V. J. Santucci, Jr., and D. H. Wahl. 1998. Suitability of the Lake Chubsucker as prey for Largemouth Bass in small impoundments. North American Journal of Fisheries Management 18: 295–307.

Ehlinger, T. J. 1997. Male reproductive competition and sex-specific growth patterns in Bluegill. North American Journal of Fisheries Management 17:508–515.

Ellison, D. G. 1984. Trophic dynamics of a Nebraska Black Crappie and White Crappie population. North American Journal of Fisheries Management 4:355–364.

Freeman, M. C., Z. H. Bowen, K. D. Bovee, and E. R. Irwin. 2001. Flow and habitat effects on juvenile fish abundance in natural and altered flow regimes. Ecological Applications 11:631–631.

Goclowski, M. R., A. J. Kaeser, and S. M. Sammons. 2013. Movement and habitat differentiation among adult Shoal Bass, Largemouth Bass, and Spotted Bass in the upper Flint River, Georgia. North American Journal of Fisheries Management 33:56–70.

Green, D. M., and Heidinger, R. C. 1994. Longevity record for Largemouth Bass. North American Journal of Fisheries Management 14:464–465.

Greene, J. C., and M. J. Maceina. 2000. Influence of trophic state on Spotted Bass and Largemouth Bass spawning time and age-0 population characteristics in Alabama reservoirs. North American Journal of Fisheries Management 20:100–108.

Hakala, J. P., and S. M. Sammons. 2015. Exploitation and length-limit evaluation of Alabama Bass in a south-eastern USA reservoir. Fisheries Management and Ecology 22:518–529.

Heidinger, R. C. 1975. Life history and biology of the Largemouth Bass. Pages 11–20 *in* R. H. Stroud and H. Clepper, editors. Black Bass Biology and Management. Sport Fishing Institute, Washington, D.C.

Hilderbrand, R. H., and J. L. Kershner. 2000. Conserving inland Cutthroat Trout in small streams: how much stream is enough? North American Journal of Fisheries Management 20:513–520.

Hubert, W. A. 1976. Age and growth of three black bass species in Pickwick Reservoir. Proceedings of the Southeastern Association of Game and Fish Commissioners 29(1975):126–134.

Huckins, C. J. F., C. W. Osenberg, G. G. Mittelbach. 2000. Species introductions and their ecological consequences: an example with congeneric sunfish. Ecological Applications 10:612–625.

Ingram, T. R., and J. Kilpatrick. 2015. Assessment of the Shoal Bass population in the Lower Flint River, Georgia. Pages 157–168 *in* M. D. Tringali, J. M. Long, T. W. Birdsong, and M. S. Allen, editors. Black bass diversity: multidisciplinary science for conservation. American Fisheries Society, Symposium 82, Bethesda, Maryland.

Isermann, D. A., S. M. Sammons, P. W. Bettoli, and T. N. Churchill. 2002. Predictive evaluations of size restrictions as management strategies for Tennessee reservoir crappie fisheries. North American Journal of Fisheries Management 22:649–657.

Johnston, C. E., and R. A. Kennon. 2007. Habitat use of the Shoal Bass, *Micropterus cataractae*, in an Alabama stream. Journal of Freshwater Ecology 22:493–498.

Jones, M. S., J. P. Goettl, Jr., and S. A. Flickinger. 1994. Changes in Walleye food habits and growth following a Rainbow Smelt introduction. North American Journal of Fisheries Management 14:409–414.

Keast, A., and J. M. Eadie. 1985. Growth depensation in year-0 Largemouth Bass—the influence of diet. Transactions of the American Fisheries Society 114:204–213.

Leitner, J., and L. A. Earley. 2015. Redeye Bass *Micropterus coosae*. Pages 61–66 *in* M. D. Tringali, J. M. Long, T. W. Birdsong, and M. S. Allen, editors. Black bass diversity: multidisciplinary science for conservation. American Fisheries Society, Symposium 82, Bethesda, Maryland.

Long, J. M., and W. L. Fisher. 2001. Precision and bias of Largemouth, Smallmouth, and Spotted Bass ages estimated from scales, whole otoliths, and sectioned otoliths. North American Journal of Fisheries Management 21:636–645.

Lothrop, R. L. 2012. Economic impact of Striped Bass angler visitation to Lewis Smith Lake, Alabama. Master's thesis, Auburn University, Auburn, Alabama.

Lott, J. P., D. W. Willis, D. O. Lucchesi. 1996. Relationship of food habits to Yellow Perch growth and population structure in South Dakota lakes. Journal of Freshwater Ecology 11:27–37.

Maceina, M. J. 1988. A simple grinding procedure for sectioning otoliths. North American Journal of Fisheries Management 8:141–143.

Maceina, M. J. 1997. Simple application of using residuals from catch-curve regressions to assess year-class strength in fish. Fisheries Research 32:115–121.

Maceina, M. J., and P. W. Bettoli. 1998. Variation in Largemouth Bass recruitment in four mainstream impoundments of the Tennessee River. North American Journal of Fisheries Management 18:998–1003.

Maceina, M. J., and S. M. Sammons. 2006. An evaluation of different structures to age freshwater fish from a northeastern U.S. river. Fisheries Management and Ecology 13:237–242.

Maceina, M. J., and S. M. Sammons. 2016. Assessing the accuracy of published natural mortality estimators using rates determined from five unexploited freshwater fish populations. North American Journal of Fisheries Management 36:433–446.

Mallows, C. L. 1973. Some comments on Cp. Technometrics 15:661–675.

Marcinek, P. A. 2003. Variation of fish assemblages and species abundances in the upper Flint River shoals, Georgia. Master's thesis, University of Georgia, Athens.

Margenau, T. L., P. W. Rasmussen, and J. M. Kampa. 1998. Factors affecting growth of Northern Pike in small northern Wisconsin lakes. North American Journal of Fisheries Management 18:625–639.

McCauley, R. W., and D. M. Kilgour. 1990. Effect of air-temperature on growth of Largemouth Bass in North America. Transactions of the American Fisheries Society 119:276–281.

Miranda, L. E., and P. W. Bettoli. 2007. Mortality. Pages 229–277 *in* C. S. Guy and M. L. Brown, editors. Analysis and interpretation of freshwater fisheries data. American Fisheries Society, Bethesda, Maryland.

Miranda, L. E., and R. J. Muncy. 1990. Bioenergetic values of shads and sunfishes as prey for Largemouth Bass. Proceedings of the Annual Conference of Southeastern Associated Fish and Wildlife Agencies 43(1989):153–163.

Nagid, E. J., T. F. Bonvechio, K. I. Bonvechio, and W. F. Porak. 2015. Suwannee Bass *Micropterus notius* Bailey & Hubbs, 1949. Pages 67–74 *in* M. D. Tringali, J. M. Long, T. W. Birdsong, and M. S. Allen, editors. Black bass diversity: multidisciplinary science for conservation. American Fisheries Society, Symposium 82, Bethesda, Maryland.

Neumann R. M., C. S. Guy, and D. W. Willis. 2012. Length, weight, and associated indices. Pages 637–676 *in* A.V. Zale, D. L. Parrish, and T. M. Sutton, editors. Fisheries Techniques, 3rd edition. American Fisheries Society, Bethesda, Maryland.

Ney, J. J. 1993. Bioenergetics modeling today—growing pains on the cutting edge. Transactions of the American Fisheries Society 122:736–748.

Novinger, G. D. 1987. Evaluation of a 15.0-inch minimum length limit on Largemouth Bass and Spotted Bass catches at Table Rock Lake, Missouri. North American Journal of Fisheries Management 7:260–272.

Olmsted, L. L., and R. V. Kilambi. 1978. Age and growth of Spotted Bass (*Micropterus punctulatus*) in Lake Fort Smith, Arkansas. Transactions of the American Fisheries Society 107:21–25.

Olson, M. H., and B. P. Young. 2003. Patterns of diet and growth in co-occurring populations of Largemouth Bass and Smallmouth Bass. Transactions of the American Fisheries Society 132:1207–1213.

Opsahl, S. P., S. E. Chapal, D. W. Hicks, and C. K. Wheeler. 2007. Evaluation of groundwater and surface-water exchanges using streamflow difference analyses. Journal of the American Water Resources Association 43:1132–1141.

Paukert, C. P., D. W. Willis, and D. W. Gablehouse, Jr. 2002. Effect and acceptance of Bluegill length limits in Nebraska natural lakes. North American Journal of Fisheries Management 22: 1306–1313.

Porak, W., W. S. Coleman, and S. Crawford. 1987. Age, growth, and mortality of Florida Largemouth Bass utilizing otoliths. Proceedings of the Annual Conference of Southeastern Associated Fish and Wildlife Agencies 40:206–215.

Porath, M. T., and K. L. Hurley. 2005. Effects of waterbody type and management actions on Bluegill growth rates. North American Journal of Fisheries Management 25:1041–1050.

Ricker, W. E. 1975. Computation and interpretation of biological statistics of fish populations. Fisheries Research Board of Canada Bulletin 191.

Sammons, S. M. 2012. Diets of juvenile and subadult size classes of three *Micropterus* spp. in the Flint River, Georgia: potential for trophic competition. Southeastern Naturalist 11:387–404.

Sammons, S. M. 2015. First evidence of potadromy and partial migration in black basses: Shoal Bass *Micropterus cataractae* (Atinopterygii, Centrarchidae) in the Upper Flint River, USA. Hydrobiologia 751:135–146.

Sammons, S. M. 2016. Catch and exploitation of Shoal Bass in the Flint River, Georgia, USA: implications for harvest restrictions. North American Journal of Fisheries Management 36:606–620.

Sammons, S. M., and P. W. Bettoli. 1998. Influence of water levels and habitat manipulations on fish recruitment in Normandy Reservoir. Final Report submitted to Tennessee Wildlife Resources Agency, Nashville, and Tennessee Valley Authority, Chattanooga.

Sammons, S. M., and M. R. Goclowski. 2012. Relations between Shoal Bass and congeneric black bass species in Georgia rivers with emphasis on movement patterns, habitat use and recruitment. Final Report to Georgia Department of Natural Resources, Social Circle, Georgia.

Sammons, S. M., and M. J. Maceina. 2008. Evaluating the potential effectiveness of harvest restrictions on riverine sunfish populations in Georgia, USA. Fisheries Management and Ecology 15:167–178.

Sammons, S. M., and M. J. Maceina. 2009a. Effects of river flows on growth of Redbreast Sunfish (Centrarchidae) in Georgia rivers. Journal of Fish Biology 74:1580–1593.

Sammons, S. M., and M. J. Maceina. 2009b. Variation in growth and survival of Bluegill and Redbreast Sunfish in Georgia rivers. North American Journal of Fisheries Management 29:101–108.

Sammons, S. M., D. G. Partridge, and M. J. Maceina. 2006. Differences in population metrics between Bluegill and Redear Sunfish: implications for the effectiveness of harvest restrictions. North American Journal of Fisheries Management 26:777–787.

Sammons, S. M., K. L. Woodside, and C. J. Paxton. 2015. Shoal Bass *Micropterus cataractae* Williams and Burgess, 1999. Pages 75–81 *in* M. D. Tringali, J. M. Long, T. W. Birdsong, and M. S. Allen, editors. Black bass diversity: multidisciplinary science for conservation. American Fisheries Society, Symposium 82, Bethesda, Maryland.

SAS Institute, Inc. 2004. SAS system for linear models. Release 9.1. Cary, North Carolina.

Schoenebeck, C. W., and M. J. Hansen. 2005. Electrofishing catchability of Walleyes, Largemouth Bass, Smallmouth Bass, Northern Pike, and Muskellunge in Wisconsin lakes. North American Journal of Fisheries Management 25:1341–1352.

Schramm, H. L., and D. C. Smith. 1988. Differences in growth rate between sexes of Florida Largemouth Bass. Proceedings of the Annual Conference of the Southeastern Association of Fish and Wildlife Agencies 41(1987):76–84.

Shepherd, M. D., and M. J. Maceina. 2009. Effects of Striped Bass stocking on Largemouth Bass and Spotted Bass in Lewis Smith Lake, Alabama. North American Journal of Fisheries Management 29:1232–1241.

Slipke, J. W., and M. J. Maceina. 2014. Fishery Analysis and Model Simulator (FAMS), Version 1.64. American Fisheries Society, Bethesda, Maryland.

Slipke, J. W., M. J. Maceina, V. H. Travnichek, and K. C. Weathers. 1998. Effects of a 356-mm minimum length limit on the population characteristics and sport fishery of Smallmouth Bass in the shoals reach of the Tennessee River, Alabama. North American Journal of Fisheries Management 18:76–84.

Smith, S. M., J. S. Odenkirk, and S. J. Reeser. 2005. Smallmouth Bass recruitment variability and its relation to stream discharge in three Virginia rivers. North American Journal of Fisheries Management 25:1112–1121.

Sowa, S. P., and C. F. Rabeni. 1995. Regional evaluation of the relation of habitat to distribution and abundance of Smallmouth Bass and Largemouth Bass in Missouri streams. Transactions of the American Fisheries Society 124:240–251.

Thurston, W. N. 1973. The Apalachicola-Chattahoochee-Flint River transportation route in the Nineteenth Century. The Georgia Historical Quarterly 57:200–212.

Tomcko, C. M., and R. B. Pierce. 2005. Bluegill recruitment, growth, population size structure, and associated factors in Minnesota lakes. North American Journal of Fisheries Management 25:171–179.

U.S. Environmental Protection Agency (U.S. EPA) 1975. Report on Lake Blackshear, Georgia. EPA Region IV, Working Paper No. 283.

U.S. Fish and Wildlife Service and U.S. Bureau of the Census (USFWS and USBOC). 2014. 2012 national survey of fishing, hunting, and wildlife-associated recreation. U.S. Government Printing Office, Washington, D.C.

Valentin, S., F. Lauters, C. Sabaton, P. Breil, and Y. Souchon. 1996. Modeling temporal variations of physical habitat for Brown Trout (*Salmo trutta*) in hydropeaking conditions. Regulated Rivers: Research & Management 12:317–330.

Van Den Avyle, M. J., and J. W. Evans. 1990. Temperature selection by Striped Bass in a Gulf of Mexico coastal river system. North American Journal of Fisheries Management 10:58–66.

Vannote, R. L., G. W. Minshall, K. W. Cummins, J. R. Sedell, and C. E. Cushing. 1980. The River Continuum Concept. Canadian Journal of Fisheries and Aquatic Sciences 37:130–137.

Vogele, L. E. 1975. The Spotted Bass. Pages 34–45 *in* R. H. Stroud and H. Clepper, editors. Black bass biology and management. Sport Fishing Institute, Washington D.C.

von Bertalanffy, L. 1938. A quantitative theory of organic growth. Human Biology 10:181–213.

Wares, J. P., D. Alo, and T. F Turner. 2004. A genetic perspective on management and recovery of federally endangered trout (*Oncorhynchus gilae*) in the American Southwest. Canadian Journal of Fisheries and Aquatic Sciences 61:1890–1899.

Wheeler, A. P., and M. S. Allen. 2003. Habitat and diet partitioning between Shoal Bass and Largemouth Bass in the Chipola River, Florida. Transactions of the American Fisheries Society 132:438–449.

Williams, J. D., and G. H. Burgess. 1999. A new species of bass, *Micropterus cataractae* (Teleostei: Centrarchidae), from the Apalachicola River Basin in Alabama, Florida, and Georgia. Bulletin of the Florida Museum of Natural History 42(2):81–114.

Woodside, K. L., C. J. Paxton, and N. C. Kierl. 2015. Stock assessment of Shoal Bass in the Chipola River, Florida. Pages 139–156 Pages 237–248 *in* M. D. Tringali, J. M. Long, T. W. Birdsong, and M. S. Allen, editors. Black bass diversity: multidisciplinary science for conservation. American Fisheries Society, Symposium 82, Bethesda, Maryland.

Zolczynski, S. J., Jr., and W. D. Davies. 1976. Growth characteristics of the northern and Florida subspecies of Largemouth Bass and their hybrid, and a comparison of catchability between the subspecies. Transactions of the American Fisheries Society 105:240–243.

American Fisheries Society Symposium 87:193–211, 2019

Spatial Ecology of Shoal Bass in Ichawaynochaway Creek, Georgia

Travis R. Ingram*
Georgia Department of Natural Resources, Albany, Georgia 31701, USA

Steven M. Sammons
School of Fisheries, Aquaculture, and Aquatic Sciences
Auburn University, 203 Swingle Hall, Auburn, Alabama, 36849, USA

Adam J. Kaeser
U.S. Fish and Wildlife Service, Panama City Fish and Wildlife Conservation Office, 1601 Balboa Avenue, Panama City, Florida, 32405, USA

Rachel A. Katz
U.S. Fish and Wildlife Service, 300 Westgate Center Drive Hadley, Massachusetts, 01035, USA

Sean C. Sterrett
Massachusetts Cooperative Fish and Wildlife Research Unit
Department of Environmental Conservation, University of Massachusetts
160 Holdsworth Way, Amherst, Massachusetts 01003, USA

Abstract.—Shoal Bass *Micropterus cataractae* are fluvial specialists endemic to the Apalachicola-Chattahoochee-Flint River Basin that are considered to be in decline throughout their native range. Effective conservation requires a comprehensive understanding of the migratory behavior and multi-scale habitat associations of Shoal Bass with riverine shoals, the critical mesohabitat upon which the species depends. We assessed movement patterns and habitat use of Shoal Bass using radio telemetry in the lower 24 km of Ichawaynochaway Creek, a 6th-order tributary of the Flint River and one of the few relatively undisturbed streams inhabited by this species. In general, Shoal Bass exhibited relatively low movement rates with increased movement in the spring, and

*Corresponding author: travis.ingram@gadnr.org

no tagged Shoal Bass migrated from the creek into the Flint River during the study period. Most study fish preferred moderate depths (<2 m) and swift velocities during the year, and higher velocities in the winter, potentially reflecting seasonal changes in flow. These conditions were routinely satisfied through occupation of a 9-km reach with a network of large shoal complexes. Shoal Bass exhibited a distinct preference for close proximity to large shoals, and an affinity for greater depth variability associated with edge and boundary conditions within discrete shoal complexes. Despite previous studies that have documented high movement of this species in other systems, these findings suggest that the Ichawaynochaway Creek Shoal Bass population may be relatively sedentary and associate to specific areas that provide suitable habitat. This may have implications for assessing the integrity, distribution, and abundance of suitable Shoal Bass habitat in small karst limestone streams, designing projects for restoration or enhancement of existing habitat, and gauging the species vulnerability to threats such as habitat loss, introgression, and hybridization.

Introduction

Shoal Bass *Micropterus cataractae* has been recognized as a unique riverine black bass for many years (Bailey and Hubbs 1949), but has only been formally described since 1999 (Williams and Burgess 1999). Shoal Bass are among a suite of aquatic species endemic to the Apalachicola-Chattahoochee-Flint (ACF) basin in Alabama, Florida, and Georgia (Couch et al. 1996; Brim Box and Williams 2000). Shoal Bass populations appear to be declining across parts of their range, and are nearly extirpated in the Chattahoochee River below Atlanta due to dam construction, habitat destruction, and introgressive hybridization with nonnative congeners (Sammons et al. 2015). Thus, with a growing emphasis on native black bass management and the rising popularity of Shoal Bass among anglers, attention has turned to species conservation. Effective Shoal Bass conservation requires knowledge of species movements and habitat use patterns to help identify priority management actions such as habitat restoration or enhancement and harvest regulations (Sammons et al. 2015).

Movements and habitat use of black bass *Micropterus* spp. can vary greatly depending on species and environmental conditions, supporting the need for inter- and intraspecific knowledge to inform management across the genus. For example, Smallmouth Bass *M. dolomieu* are known to migrate substantial distances in large rivers and reservoirs at northern latitudes, but home ranges can be smaller in streams and more productive lakes (Langhurst and Schoenike 1990; Ridgway and Shuter 1996). Largemouth Bass *M. salmoides* are usually more sedentary, with home ranges often <5 ha (Winter 1970; Mesing and Wicker 1986). Overall, Shoal Bass were thought to be relatively sedentary until recent studies revealed relatively large-scale movements (>5 km) occurred throughout the year (Goclowski et al. 2013; Ingram et al. 2013; Sammons and Earley 2015). Furthermore, the species exhibits potamodromy, or migration for reproduction within freshwater systems, in the upper Flint River in Georgia, moving up to 200 km to access spawning habitat (Sammons 2015). Compared to mainstem rivers, Shoal Bass were observed to have limited movement within smaller tributaries (Stormer and Maceina 2009), but at times fish from mainstem rivers have been found to move up into tributaries during the spawning season (Sammons and Goclowski 2012; Sammons and Earley 2015). Given the observed range of movement behavior within mainstem rivers and between tributaries and mainstem rivers, understanding potential drivers of movements is imperative to informing management at the proper scale (Northcote 1997).

Shoal Bass are considered fluvial habitat specialists based on their common association with rocky substrates in swift flowing environments (Wheeler and Allen 2003; Goclowski et al. 2013). Individuals typically occur in larger rivers and streams that contain appropriate "shoal" habitat (i.e., shallow channel segments characterized by rocky substrates, swift current, and high gradient), but are rarely found in smaller streams (<4th order; Ramsey 1975; Katechis 2015). As habitat specialists, the entire life history of Shoal Bass is closely linked to shoal habitat. Shoals provide substrate for spawning, refuge from predators, and invertebrate (insect and crustacean) and vertebrate food items for all developmental stages (Wheeler and Allen 2003; Sammons 2012; Sammons et al. 2015). Shoal Bass are generally intolerant of reservoir conditions but have been documented moving through them to reach lotic habitat when translocated (Ingram et al. 2013), but not on their own initiative (Sammons and Earley 2015). Throughout their life history, use of shoal habitat may also be a mechanism for reducing interspecific competition and hybridization with native or introduced congeners (Wheeler and Allen 2003; Goclowski et al. 2013). Despite the importance of rocky, swift velocity habitats to Shoal Bass, little is known about differences between male and female habitat use, or the relative importance of shoal size and distribution within occupied streams. This knowledge is needed to assess the current state of Shoal Bass habitat at a scale relevant to its life history, and to assess population vulnerability and potential for restoration.

The objectives of this study were to describe movement and habitat use of adult Shoal Bass in Ichawaynochaway Creek, a relatively undisturbed 6th-order tributary of the lower Flint River. Specifically, we set out to (1) quantify and assess variation in movement patterns among individuals, (2) to compare individual-level habitat associations across seasons and sexes, and (3) to assess population-level habitat preferences for local and neighboring shoal habitat characteristics within a habitat use versus availability framework.

Methods

Study Area

With a watershed area of 2,750 km^2, Ichawaynochaway Creek is one of the largest tributaries to the Flint River and is mainly fed by groundwater from the Floridan Aquifer during lower flow conditions (Golladay et al. 2007). Because of this karst geology, lower portions of Ichawaynochaway Creek contain numerous rocky shoals and limestone outcrops, unlike most coastal plain streams. This study was conducted on a 24-km reach in Baker County, where the creek flows through Ichauway, a 12,000-ha private reserve that is the research site of the Joseph W. Jones Ecological Research Center (Figure 1). This reach was bounded by a breached, 1920s-era power dam (RKM 23.6) and the confluence with the Flint River (RKM 0; Figure 1).

Fish Tagging and Tracking

Adult Shoal Bass were collected during the prespawning season (March–April 2015) by electrofishing (Figure 1). Equal effort was

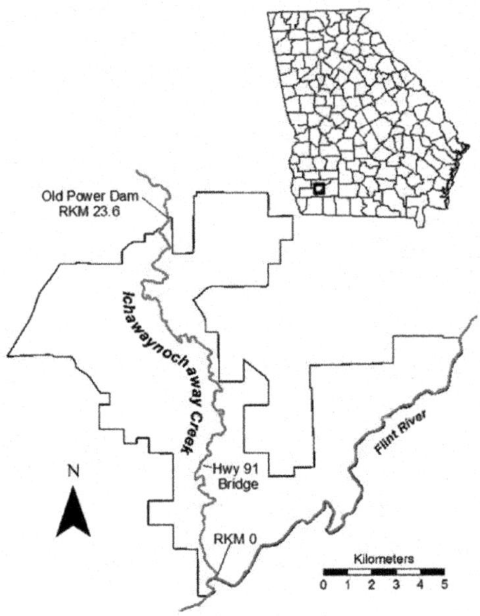

Figure 1. Ichauway reserve, located in Baker County, Georgia. The entire study reach of Ichawaynochaway Creek (RKM 0–23.6) flows through the center of the reserve and the Flint River forms the eastern border.

expended throughout the study area to minimize location bias, however few fish were captured below the Hwy 91 Bridge (Figure 1). Upon capture, the first 30 Shoal Bass that met the minimum 185-g weight requirements for the tags were measured (total length TL; mm) and weighed (g). Sex was visually determined during transmitter insertion. Fish were anesthetized using tricane methanesulfonate (150 mg/L) and surgically implanted with a 3.6-g transmitter with a battery life of 365 d (model F1580, Advanced Telemetry Systems) and an internal anchor tag (Floy Tag FM-84, Seattle, Washington, USA; Maceina et al. 1999). Shoal Bass were released at their capture location after regaining equilibrium. Although tricane methanesulfonate is regulated by a 21-d withdrawal period for use in public waters, public access is restricted at Ichauway and angling is not allowed in the study reach of Ichawaynochaway Creek.

Tracking began after a 14-d, postsurgery recovery period by either motorboat, kayak, or wading, and was conducted approximately every 14 d between April 2015 and April 2016. Tracking surveys of the entire study area were conducted using an ATS model R2000 receiver (Isanti, MN) and a directional yagi antenna. The signal range of transmitters was 0.8 km under normal conditions and may have been reduced in deeper water (>10 m; Sammons and Earley 2015); however, depths of the stream channel rarely exceeded 8 m. Daily streamflow ranged from 4.8 to 99.1 m^3/s during tracking (U.S. Geological Survey stream gauge site 02355350). When telemetered fish were not located in the study area, tracking was conducted in a 7-km reach above the dam, and within a reach of the Flint River ± 10 km from the Ichawaynochaway Creek confluence. In addition, a passive ATS model R4500 data-logger was deployed at river kilometer (RKM) 0 to detect any Shoal Bass leaving Ichawaynochaway Creek.

Fish locations were obtained as described by Sammons and Earley (2015) and marked using a Global Positioning System unit (Garmin GPSmap 78 s; Olathe, KS). Tracking events were conducted downstream by boat until a signal was detected. A fish location was marked when the signal was strongest with the antenna pointed at the water's surface. At each fish location we recorded time and transmitter number, and measured water depth and midwater column velocity (via portable flowmeter model 2100, Swoffer Instruments, Seattle, Washington, USA). Fish locations obtained via telemetry were later used to associate each fish location with the substrate present at the location of the fish in a previously developed habitat map.

Investigating Habitat Associations

The instream habitat of Ichawaynochaway Creek was mapped using side scan sonar in 2008 (Kaeser and Litts 2008, 2010). The resulting map included classified substrates, mid-channel water depth, and large woody debris (defined as any piece of wood ≥10 cm diameter and ≥1.5 m in length; Figure 2). Substrate class descriptions, and details of field-assessed map accuracy are provided by Kaeser and Litts (2008, 2010). Because much of the substrate in Ichawaynochaway Creek

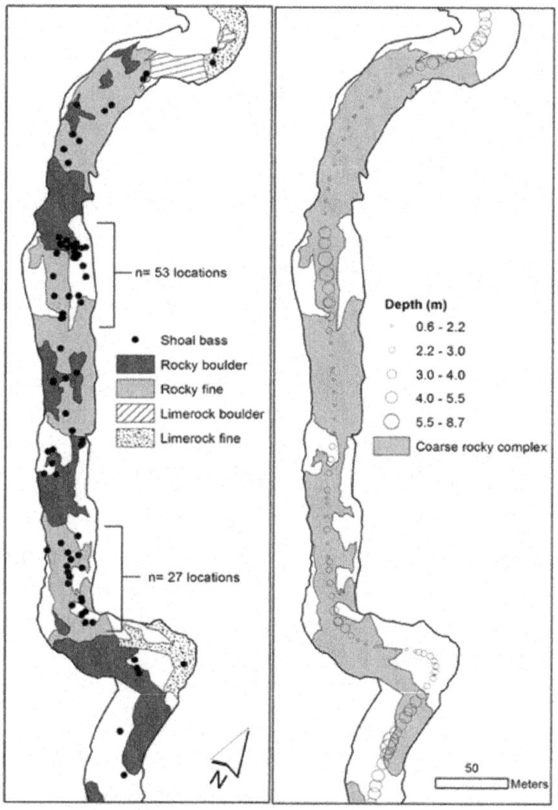

Figure 2. Map of a single coarse rocky complex in Ichawaynochaway Creek, Georgia spanning river kilometer 16.5–17.1, where the largest number of telemetered Shoal Bass locations (N = 121) occurred. Figure denotes Shoal Bass telemetry locations (left panel) and center channel depth (right panel). The complex (14,700 m^2) was composed of alternating patches of rocky boulder and rocky fine substrates; fish were frequently observed in proximity to the margins of these substrate patches, in areas of transition from shallow to deeper depths.

was large and coarse, and the channel was entrenched and highly stable, we did not expect any major changes in substrate distribution to have occurred between the habitat mapping and this study.

Fish locations obtained via telemetry were used to characterize population-level habitat use, as opposed to individual microhabitat use, within a 17.5-km focal reach of the study area. Among the 30 study fish, only 1 fish was ever located downstream of RKM 7.0; we chose to remove this individual ($N = 9$ locations) from the habitat-use analysis, and restricted the area for analysis to a focal reach spanning RKM 6.0–23.5. We used the distance-based approach, described in detail by Sterrett et al. (2015), to associate fish locations with map data (Conner and Plowman 2001; Conner et al. 2003). A 15-m buffer was generated around each fish location to represent the area in "use." To evaluate habitat use versus availability, we generated random locations ($N = 475$), one for each observed fish location (Northrup et al. 2013) that represented habitat available within the focal reach using the Generalized Random Tessellated Stratification (GRTS) sampling algorithm (Kincaid et al. 2013). We assumed all random availability locations were independent. We defined available habitat as habitat within a fish's potential home range (Johnson 1980) and assumed that habitat measured at the 18-km scale was proportionally similar to habitat available at the home-range scale (i.e., 1 km). To prevent overlap of used and random locations, buffered areas around used locations were excluded from the area available for random location assignment.

We identified a set of predictor variables representing local habitat and neighboring shoal habitat from the map that might explain patterns of habitat use by adult Shoal Bass in Ichawaynochaway Creek. To examine the influence of local features we used the NEAR tool in Arc GIS 10.2 and calculated the linear distances from all locations to the edge of the nearest representative of each of the following substrate features: (A) boulder (i.e., the nearest distance to either rocky boulder or limerock boulder substrates), (B) boulder-bedrock (i.e., nearest distance to either rocky boulder, limerock boulder, or limerock fine—a class of limestone bedrock substrate as defined in Kaeser and Litts (2010)), and (C) coarse rocky complex (CRC), a composite feature developed to represent potential shoal areas. We defined "coarse rocky complex" as an area spanning the creek channel composed of contiguous patches of rocky boulder, limerock boulder, and rocky fine substrates. Of note, our CRCs did not incorporate elements of depth and current velocity typically recognized as "shoal" habitat features. Patches of limerock fine substrate, narrow bands of rocky substrate along the river margins, and any complexes less than 200 m^2 in total area were excluded from the CRC class. Total counts of LWD were summarized within the 15-m buffer of all locations. To extract relevant water depth data, all used and random locations were snapped to the mid-channel line that represented the sonar survey track (Sterrett et al. 2015). At the snapped position, a 15 m buffer was generated around each location, and all depth measurements within the buffer ($n \geq 4$) were used to yield a measurement of average local depth and depth variability (standard deviation of depths) for each location (Sterrett et al. 2015).

To assess the influence of neighborhood habitat features, we calculated three variables representing near shoal-complex characteristics: total area (m^2) of the nearest CRC, distance (shortest channel length, m) to nearest neighboring CRC, and total area (m^2) of nearest neighboring CRC. If neighborhood habitats were found to be important, we inferred that features of the nearest shoal and instream, neighborhood context (i.e., shoal arrangement and sizes) influenced the likelihood of Shoal Bass occurring at any particular location in the stream channel.

Data Analyses

Mean daily movement rates of Shoal Bass were calculated as the distance moved between locations divided by elapsed time (Sammons et al. 2003). Mean daily movement (log-transformed), and stream velocity and depth at observed locations were compared among seasons (spring: March–May, summer: June–August, fall: September–November, and winter: December–February) and sexes using a mixed-model analysis of variance (ANOVA) (SAS Institute 2000). A Bonferroni correction ($P < 0.10$/number of comparisons) was used for multiple comparisons. Substrate use by male and female fish was compared among seasons (as percent use) using Z-tests (Sammons and Earley 2015). Significance for these tests was set $P = 0.10$, unless a Bonferroni correction was used as described above.

Following the habitat use-availability analyses of Sterrett et al. (2015), we used logistic regression (Hosmer and Lemeshow 2000) and an information theoretic approach (Burnham and Anderson 2002) to evaluate Shoal Bass habitat use at the population level given the habitat available measured by random locations (Gillies et al. 2006; Beyer et al. 2010). We developed a set of 11 a priori candidate models to explore hypotheses related to local habitat features and six a priori candidate models to explore additional hypotheses related to neighborhood habitat features (Table 1). Local habitat models examined combinations of distance to boulder, boulder-bedrock, and CRC, mid-channel depth, and mid-channel depth variability. Although LWD was measured during the habitat survey, we did not include LWD in any candidate model because Shoal Bass are not known to use LWD in rocky habitats (Goclowski et al. 2013) and variation in LWD among fish and random locations was low. Correlated variables ($r^2 > 0.50$) were never included in the same candidate model. We also included a global model that contained all local and neighborhood habitat variables to evaluate model fit. We included an individual random effect in all models to account for nonindependence among repeat observations of habitat used by the same fish and unequal number of observations per fish. The random effect was centered on zero with a large variance in each candidate model, which allowed for robust population-level inference in habitat use (Gillies et al. 2006). We used all random locations ($N = 475$) to represent available habitat to each individual and assumed that actual habitat availability (i.e., within each fish's home range) was proportional to habitat available in the study area.

We evaluated the relative support of each local and neighborhood habitat use-availability model using Akaike's Information Criterion (AIC) with a small-sample bias adjustment (AICc) (Burnham and Anderson 2002). The number of parameters (K) for each candidate model included the number of variables in the model and an additional three parameters representing the intercept, error term, and variance of the random effect. We calculated Akaike weights (w) for each model, which ranged from zero to one with the best approximating model having the highest weight; models with delta AICc of less than two were considered equally plausible (Burnham and Anderson 2002). We first identified a plausible candidate set of local habitat models, then evaluated relative support for the addition of neighborhood shoal habitat variables in a second candidate model set. We based inferences on variables included in the top approximating model(s) to understand the importance of local habitat and neighborhood features. The precision of each parameter was estimated using 95% credibility intervals. We calculated scaled odds ratios for each predictor in top model(s) to facilitate interpretation of habitat and neighborhood variables (Hosmer and Lemeshow 2000).

Table 1. Total length (TL, mm), weight (WT, g), number of times relocated (LOC), maximum range (Max range, km), days at large during the study (Days), and the fate of each male (M) and female (F) Shoal Bass tracked from April 2015–April 2016 in Ichawaynochaway Creek, Georgia. Maximum range was defined as the distance between the farthest upstream and farthest downstream location a fish was relocated.

Transmitter	Sex	TL	WT	LOC	Max range	Days	Fate
8010	M	321	409	4	2.30	331	Tag Died
8028	F	360	640	21	4.63	364	Alive
8052	M	356	667	18	8.49	364	Alive
8071	F	351	563	18	0.46	316	Tag Died
8090	F	538	620	24	14.70	345	Tag Died
8110	M	335	552	9	9.32	100	Fish Died
8131	M	390	820	19	0.63	331	Tag Died
8151	M	431	1110	19	8.11	345	Tag Died
8170	F	378	800	17	0.30	364	Alive
8191	F	397	932	14	0.56	330	Tag Died
9001	M	319	361	20	0.50	345	Tag Died
9014	M	315	389	17	5.66	304	Tag Died
9023	F	322	373	17	7.93	148	Unknown
9031	F	356	456	14	0.32	304	Tag Died
9044	F	375	803	18	0.15	345	Tag Died
9052	M	331	501	17	0.27	345	Tag Died
9063	M	374	612	21	0.30	364	Alive
9073	M	466	1367	10	0.08	321	Tag Died
9082	F	312	352	24	11.12	364	Alive
9091	F	353	631	12	12.55	203	Unknown
9104	M	350	606	17	1.47	364	Alive
9112	M	340	602	18	0.11	364	Alive
9120	M	338	560	9	5.12	330	Tag Died
9130	M	360	675	17	0.21	330	Tag Died
9142	M	328	487	19	0.74	364	Alive
9151	F	315	398	22	4.90	345	Tag Died
9159	F	348	617	10	0.23	345	Tag Died
9171	F	340	575	18	1.41	364	Alive
9183	M	382	710	12	2.45	286	Unknown
9192	F	362	631	16	0.47	364	Alive

All models used Markov chain Monte Carlo (MCMC) sampling implemented in JAGS (version 3.4.0) using the Rjags package in R (version 3.3.1), and models were fit using 5,000 iterations, a 500 burn-in, a thinning rate of three to reduce autocorrelation among samples, three chains, and diffuse priors. Model convergence was assessed using visual inspection of chains and Gelman-Rubin's convergence diagnostic (R-hat <1.1; Gelman and Rubin 1992). We evaluated model fit of the global model using a posterior predictive check and a Bayesian *P*-value based on the sums of squares discrepancy metric, where a fit model has a *P*-value near 0.5 (Gelman and Hill 2007).

Results

We captured and implanted transmitters in 16 male and 14 female adult Shoal Bass ranging from 312 to 466 mm TL and 352–1,367 g (Table 2). Twenty-one tracking events were conducted after initial capture from April 14, 2015 to April 12, 2016, with individual fish at large for 100–364 d and relocated 4–24 times, resulting in a total of 484 fish locations (Table 2). Of the 30 telemetered fish, one fish was relocated downstream past the Hwy 91bridge after initial capture, and remained at RKM 1.5–1.6 for the duration of the study. Two fish emigrated upstream of the breached power dam during the study (one in May and one in March). The fish located in May later returned to the study area and remained for the entirety of the study. The fish located in March remained upstream of the study area until the study ended. No fish were located in a 5.3-km section below the Hwy 91 bridge (i.e., between RKM 1.6 and 7.0). This section was deeper with slower average velocity than other sections of the study area. No fish were detected by the passive data logger located near the confluence of the Flint River, indicating that all study fish remained in Ichawaynochaway Creek for the duration of the study.

Most fish locations (446 of 484; 92%) occurred in a 9-km reach spanning RKM 14–22. A total of 24 CRCs existed within this reach, representing 54% of the total channel area. The ratio of rocky boulder to rocky fine substrates in this reach was nearly 50:50 (i.e., 82,300:83,800 m2). Eleven study fish were consistently located in a single, large (14,700 m2) shoal complex spanning RKM 16.5–17.1 (Figure 2).

Table 2. Seasonal mean and standard error (SE) daily movement rates (m/d), depth (m), and velocity (m/s) at capture locations of male and female Shoal Bass in Ichawaynochaway Creek, Georgia (April 2015–April 2016). Means of each variable followed by the same letter were similar among seasons ($P \geq 0.1$).

Sex	Parameter	Spring	Summer	Fall	Winter
Male	Movement	134.0y (64.8)	9.1z (4.0)	5.6z (4.0)	3.2z (0.7)
	Depth	2.4y (0.2)	1.5z (0.1)	2.1y (0.2)	3.2yz (0.2)
	Velocity	0.3z (0.0)	0.3z (0.0)	0.3z (0.0)	0.8y (0.1)
Female	Movement	131.1y (56.3)	104.9yz (71.3)	5.25z (1.3)	24.7yz (20.5)
	Depth	2.3y (0.2)	1.5z (0.1)	1.8z (0.2)	2.9yz (0.2)
	Velocity	0.3z (0.0)	0.3z (0.0)	0.3z (0.0)	0.8y (0.1)

Movement Patterns

All locations for each of nearly half of the study fish (16 of 30) fell within a reach of creek spanning less than 1 km per individual. Thirteen of these limited-range fish appeared to remain in a reach <500 m long for the duration of the study (Figure 4). Shoal Bass daily movement was higher for females (mean = 73.8 m/d) compared to males (mean = 35.9 m/d; F = 3.86; df = 1, 454; P = 0.04; Table 1) and females, on average, had greater maximum movement distances (mean = 7.23 km; SE = 2.7; maximum = 15.00 km) compared to males (mean = 4.07 km; SE = 1.4; maximum = 9.00 km; Table 2). In general, average daily movement was less than 150 m for both sexes throughout the year (Table 1; Figure 3). Daily movements of both sexes were higher in the spring compared to the fall and winter (F = 2.62; df = 3, 98; P < 0.01; Table 1; Figure 3). Daily movement of males was 14–44 times greater in spring than movement in the other seasons. However, female movement was similar between spring and summer, and 4–24 times greater in spring and summer compared to fall and winter. Mean movement was similar between the sexes across seasons (F = 1.26; df = 4, 20; P = 0.32; Figure 3).

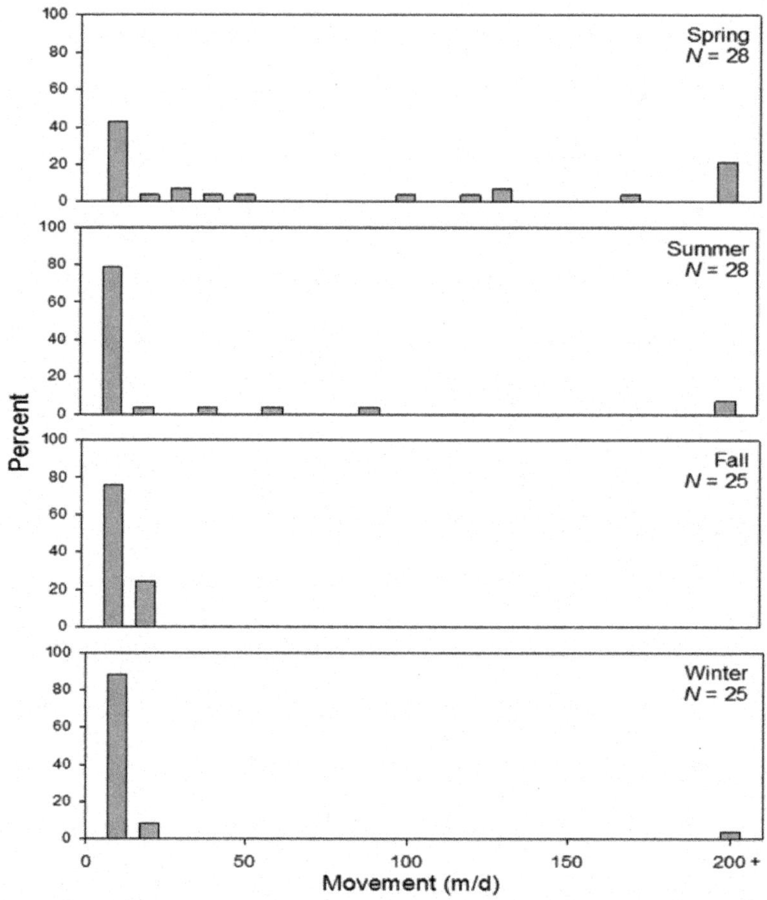

Figure 3. Seasonal movement distribution of telemetered Shoal Bass in the study reach of Ichawaynochaway Creek, Georgia.

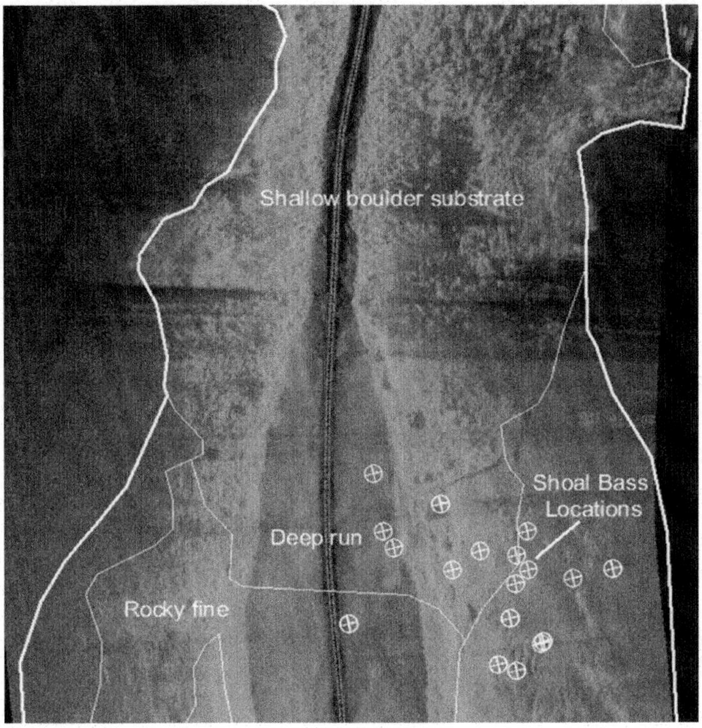

Figure 4. A close-up look at a high use area in Ichawaynochaway Creek, Georgia, within the coarse rocky complex at river kilometer 16.9, as it appeared in sonar imagery. A total of 33 Shoal Bass locations (white circle with crosshairs; many overlapping) were observed in the area shown. Here, fish occupied a 6-m deep run immediately downstream of a shallow, boulder outcrop that spanned the river channel. Water flow over the boulders and into the run created much turbulence, a phenomenon that is manifest in the sonar imagery as a fuzzy, whitish tone to the otherwise dark, water column region in the center of the image.

Individual-level Habitat Associations

Both male and female Shoal Bass occurred in areas with similar, moderate stream velocity (i.e., 0.5 m/s; $F = 3.86$; df = 1, 484; $P = 0.95$), except during the winter, when both sexes occurred in higher velocities (i.e., 0.8 m/s; $F = 2.47$; df = 4, 92; $P < 0.01$; Table 1). Mean depth was also similar among sexes ($F = 3.86$; df = 1, 484; $P = 0.12$), but it differed across seasons ($F = 2.47$; df = 4, 92; $P < 0.01$) with fish of both sexes located in deeper depths in the winter (Table 1).

Shoal Bass were located in all substrate classes over the duration of the study (Table 3). Percent limerock substrates (limerock fine and limerock boulder) and rocky fine substrates were similar among sexes ($Z = -0.68$; $P > 0.05$, $Z = 0.45$; $P > 0.05$, $Z = 0.78$; $P \geq 0.05$, respectively; Table 3). Male and female substrate use slightly differed in the fall and winter months with females utilizing a higher percentage of rocky fine substrate ($Z = -6.51$; $P < 0.05$ and $Z = -1.73$; $P < 0.05$, respectively; Table 3). Both sexes were found in similar substrates during the summer and spring spawning season (Z range –0.40 to –0.93; $P > 0.05$, respectively). During the summer, Shoal Bass were located in areas with lower percentages (23–37%) of rocky fine substrates compared to the winter and fall and in areas with higher percentages

Table 3. Percentage of fish locations associated with each mapped substrate class for male and female Shoal Bass during each season of the study in Ichawaynochaway Creek. Substrate class acronyms represent: sand (S), rocky fine (Rf), rocky boulder (Rb), rocky limerock fine (Lf), and limerock boulder (Lb).

		Substrate				
Sex	Season	S	Rf	Rb	Lf	Lb
Male	Spring	37	29	25	4	6
	Summer	27	24	39	5	5
	Fall	22	48	18	10	2
	Winter	29	36	22	12	2
Female	Spring	31	46	20	3	0
	Summer	19	37	37	5	2
	Fall	15	55	28	2	0
	Winter	13	67	20	0	0

(37–39%) of rocky boulder (Table 3). In the spring, fish were found in areas with higher percentages (31–37%) of sandy substrates compared to other substrates, and this was similar between sexes ($Z = -1.54$; $P > 0.05$). Use of sandy substrates was low during other seasons, and again, this was similar between sexes ($Z = 1.65$; $P > 0.05$; Table 3).

Population-level Habitat Use

Local and neighborhood habitat variables examined for use-availability analysis varied among fish and random locations (Table 4). Frequently relocated individuals had similar ranges in habitat-use variables compared to individuals observed less often, suggesting a limited influence of unequal location on habitat use.

The best-approximating local habitat model contained all of the model support (wi), included the variables distance to CRC and depth variability, and it was 5.4 times more likely to be the best approximating model than the second best model (ΔAICc = 3.4), which included the interaction between CRC and depth variability (Table 5). Neighborhood habitat variables improved the best-approximating local habitat model and included the area of the nearest and nearest neighboring CRC (Table 5). This model was 5.7 times more likely the best neighborhood model than the next model (ΔAICc = 3.5) which additionally included distance to the nearest neighboring CRC. Parameter estimates (scaled odds ratio) from the top neighborhood model indicated that Shoal Bass were, on average, 68 times less likely to occur with every 20-m increase in distance from the nearest CRC, and 15 times more likely to occur with every 5,000-m^2 increase in the area of the nearest CRC (Table 6). Additionally, Shoal Bass were 43 times more likely to occur with every 0.5-m increase in depth variability (standard deviation in mid-channel depths) and less likely to occur as the size of the nearest neighboring shoal increased (Table 6). The intra-class correlation coefficient (a measure of consistency among individuals) was high (0.71), suggesting similar habitat use-availability responses within and among individuals. Bayesian P-values indicated that the global and best-approximating models adequately fit the observed data ($P = 0.51$ and 0.51, respectively).

Table 4. Mean and standard error (SE) of local and neighborhood habitat variables quantified at capture (*N* = 475) location of Shoal Bass and random locations (*N* = 475) throughout the 24-km Ichawaynochaway Creek study area. The variables large woody debris (LWD), mid-channel depth, and mid-channel depth variability represent conditions within a 15-m buffer located mid-channel and adjacent to the position of the fish and proximity metrics represent distance to nearest substrate type. CRC = coarse rocky complex.

	Random Locations		Capture Locations	
Local Habitat Variables	Mean	SE	Mean	SE
Depth (m)	3.0	0.04	3.1	0.06
Depth variability (SD)	0.35	0.02	0.52	0.02
Proximity to bedrock outcrop substrate (m)	14.8	1.3	8.7	0.70
Proximity to boulder substrate (m)	41.7	2.70	12.0	0.95
Proximity to nearest CRC (m)	24.7	2.12	2.9	0.37
Neighborhood Habitat Variables				
Area of nearest CRC (m^2)	7093	340	11708	336
Proximity to nearest neighbor CRC (m)	52.5	2.1	38.3	1.1
Area of nearest neighbor CRC (m^2)	4194	204	3520	172

Discussion

Shoal Bass in Ichawaynochaway Creek were sedentary, with few large movement events occurring during the study period. When long-range movements did occur, Shoal Bass moved during the spring, most likely in relation to spawning activity (Bitz et al. 2015). Shoal Bass in a Chattahoochee River tributary were observed to have extremely low movement rates (<7 m/week), but these fish were confined to a 600-m reach of shoal habitat between two natural barriers (Stormer and Maceina 2009). Similarly, Shoal Bass in a 2-km reach of the Chattahoochee River bounded by an upstream dam and a downstream large reservoir also exhibited low movement (Sammons and Earley 2015). In contrast, Shoal Bass in the Flint River have been documented to move longer distances (>10 km), especially during the spring spawning period, when fish have been documented to move up to 200 km (Goclowski et al. 2013; Sammons 2015; T. Ingram, unpublished data). Similar to our study, Bitz et al. (2015) found low movements for Shoal Bass in the Chipola River, Florida, which, like Ichawaynochaway Creek, is a karst limestone stream with no movement barriers near the southern limit of the species' range. Movement of Shoal Bass may be at least in part dictated by geomorphology and hydrology of streams, combined with the quality and distribution of shoal habitats (Goclowski et al. 2013; Ingram et al. 2013; Sammons 2015). Thus, it is possible that behavior of this species in streams with abundant deep and complex shoal habitat (i.e., karst limestone features) and high groundwater influence may differ from those living in more typical lotic systems, similar to what has been found for Smallmouth Bass (Ettinger-Dietzel et al. 2016; Westhoff et al. 2016). Consistent and strong effects of rocky substrate and shoal area, along with movement patterns, suggest that a relatively small reach containing large

Table 5. Candidate habitat use-availability models with mean deviance (Dev), effective number of parameters (K), AIC, AICc, the difference between the best model (bold type) and each subsequent model (delta-AICc), model weight (wi), and the standard deviation of the individual random effects (RE SD) for habitat selection of Shoal Bass in Ichawaynochaway Creek, Georgia, from April 14 2015 to April 12 2015. CRC = coarse rocky complex.

Local Habitat Model	Dev	K	AICc	ΔAICc	wi	RE SD
Proximity to CRC, depth variability	3828.9	5.0	3839.0	0.0	1.00	0.07
Proximity to CRC x depth variability	3830.2	6.0	3842.4	3.4	0.18	0.06
Proximity to CRC x depth	3868.3	6.0	3880.5	41.5	0.00	0.07
Proximity to CRC, depth	3872.8	5.0	3883.0	44.0	0.00	0.07
Proximity to CRC	3887.3	4.0	3895.4	56.4	0.00	0.06
Proximity to boulder, depth variability	3904.2	5.0	3914.3	75.3	0.00	0.07
Proximity to boulder	3943.3	4.0	3951.4	112.4	0.00	0.07
Proximity to boulder, depth	3941.9	5.0	3952.1	113.1	0.00	0.08
Proximity to boulder-bedrock, depth variability	4060.5	5.0	4070.6	231.6	0.00	0.06
Proximity to boulder-bedrock	4130.8	4.0	4138.9	299.9	0.00	0.07
Proximity to boulder-bedrock	4130.4	5.0	4140.5	301.5	0.00	0.07

Neighborhood Habitat Candidate Models	Dev	K	AICc	ΔAICc	wi	RE SD
Proximity to CRC, depth variability, area of the nearest CRC, area of the nearest neighboring CRC	3783.3	7.0	3793.5	0.0	1.00	0.07
Proximity to CRC, depth variability, area of the nearest CRC, area of the nearest neighboring CRC, proximity of the nearest CRC	3784.8	8.0	3796.9	3.5	0.17	0.06
Proximity to CRC, depth variability, area of the nearest CRC	3802.7	6.0	3810.8	17.4	0.00	0.06
Proximity to CRC, depth variability, area of the nearest neighboring CRC	3807.6	6.0	3815.6	22.2	0.00	0.07
Proximity to CRC, depth variability, depth variability	3828.9	5.0	3834.9	41.5	0.00	0.07
Proximity to CRC, depth variability, proximity of the nearest neighboring CRC	3830.4	6.0	3838.5	45.0	0.00	0.07

Table 6. Parameter estimates and scaled-odds ratios (mean and 95% credible intervals, CI) from top supported neighborhood habitat use-availability model for proximity to CRC, depth variability (standard deviation of mid-channel depth), area of the nearest CRC, area of the nearest neighboring CRC). Scaled odds ratios less than 1 represent decreased odds of habitat use and values greater than 1 represent increased odds of habitat use with every 1 increase in the corresponding unit scalar. CRC = coarse rocky complex.

Parameter	Estimate	SD	Unit Scaler	Scaled Odds Ratio	Lower 95% CI	Upper 95% CI
Intercept	–3.7096	0.1576	–	–	–	–
Proximity to CRC	–2.0143	0.4133	20 m	0.3158	0.2426	0.6852
Depth variability	0.2991	0.0450	0.50 SD	1.4348	1.2894	1.5837
Area of nearest CRC	0.2159	0.0427	5000 m^2	1.1503	1.0916	1.2174
Area of nearest neighboring CRC	–0.1930	0.0483	5000 m^2	0.7911	0.7050	0.8889

continuous shoals provided suitable spawning and year-round habitat for Shoal Bass in Ichawaynochaway Creek, Georgia.

Potamodromy is well documented in salmonids (Northcote 1997) but has only recently been described in Shoal Bass (Sammons 2015). Shoal Bass in the Chattahoochee and Flint rivers have been found to migrate into tributary streams during the spawning season, presumably to access spawning habitat (Sammons and Goclowski 2012; Sammons and Earley 2015). Furthermore, several studies have demonstrated spawning potamodromy behavior of this species within the mainstem Flint River (Goclowski et al. 2013; Ingram et al. 2013; Sammons 2015), and to a lesser extent in the Chipola River, Florida (Bitz et al. 2015). Both male and female Shoal Bass in Ichawaynochaway Creek exhibited an increase in daily movement rates in the spring that may have been linked to spawning migrations. Females exhibited ten times greater movement than males in the summer that may be attributed to postspawning migration. This behavior has been documented in salmonids with males making short-distance migrations and females undertaking longer migrations (Jonsson and Jonsson 1993; Northcote 1997). However, Sammons (2015) suggested both sexes of Shoal Bass exhibited similar migratory behavior in the Flint River.

Habitat use-availability analyses indicated that not all rocky habitat in the creek was utilized equally by Shoal Bass. Previous studies used proximity to various coarse rocky substrate classes, such as rocky boulder and bedrock, as indicators of shoal association (Goclowski et al. 2013; Sterrett et al. 2015). However, we found that the best substrate-based predictor of Shoal Bass habitat use in Ichawaynochaway Creek was proximity to CRCs—a variable we synthesized that combined contiguous, coarse rocky substrates into a single, composite shoal unit. Proximity to limerock fine did not improve our ability to predict habitat use, unlike other rocky substrates. Although Shoal Bass have been associated with bedrock substrate in other studies (Stormer and Maceina 2009; Goclowski et al. 2013; Bitz et al. 2015; Sammons and Earley 2015), these results generally indicate that bedrock associations likely differ according to type of bedrock, stream context, and hydrogeomorphic setting. In Ichawaynochaway Creek, limerock fine bedrock outcroppings typically occur in runs, and areas of low bathymetric relief (i.e., uniformly shallow or deep reaches). The lack of surface complexity of this bedrock type may limit its attractiveness for foraging or cover.

Habitat use-availability analyses further indicated that not all shoal habitat in the creek was utilized equally by Shoal Bass. Coarse rocky complexes, particularly the larger complexes, were used in higher proportion than their availability. High site fidelity of Shoal Bass observed between RKM 14 and 22 suggests an underlying benefit to occupation of reaches with numerous, large shoals arranged in series. Moreover, we found no evidence that reaches composed of smaller, scattered shoals with similar total area were used by Shoal Bass as often as larger continuous shoals. Levels of benthic organic matter and macroinvertebrates are higher in shoals than in pools (O'Connell and Campbell 1953; Bogan et al. 1995; Grubaugh et al. 1997; Marcinek et al. 2005). As islands of biological productivity, large shoals in series may trap and retain allochthonous material being carried downstream, thereby improving overall utilization of organic carbon, and supporting the development of trophic complexity (Elwood et al. 1983). The availability of prey, substrate complexity for shelter, and substrate stability for macrophyte production likely support large numbers of crayfish and benthic fish such as darters and madtoms, key prey items for adult Shoal Bass (Wheeler and Allen 2003; Marcinek et al. 2005; Sammons 2012).

If typical of other years, our results suggested that migration may be minimal between the mainstem and Ichawaynochaway Creek, which has implications for management and conservation. Limited migration between the mainstem and tributaries can result in reduced dispersal among populations, which can act as vectors for gene flow and increase overall fitness of a population (Northcote 1997). Fitness can further be reduced because of hybridization among species, which has been documented in Ichawaynochaway Creek, with Shoal Bass hybridizing with nonnative Spotted Bass *Micropterus punctulatus* (Alvarez et al. 2015). Because Shoal Bass in Ichawaynochaway Creek appear to be segregated to some degree from the mainstem Flint River population, opportunities may exist for specific regulations and/or management for each stock if warranted or desired. Overharvest of disjunct and vulnerable fish populations can have severe impacts by minimizing potential for genetic rescue (if migration does occur) and increasing potential for genetic bottlenecks (Diaz et al. 2000). Catch and harvest rates of Shoal Bass in the nearby Flint River have been documented in excess of 20%, and population dynamics models demonstrate harvest at this level can affect size structure of these populations (Ingram and Kilpatrick 2015; Sammons 2016). If harvest rates in smaller streams are similar, there may be need for more restrictive harvest regulations relative to larger mainstem stocks.

Acknowledgments

This research was funded through the Federal Aid in Sportfish Restoration Act. The findings and conclusions in this article are those of the authors and do not necessarily represent the views of the U.S. Fish and Wildlife Service. We gratefully acknowledge the following personnel of the Georgia Department of Natural Resources, Fisheries Management Section for support and assistance: Josh Tannehill, Craig Robbins, Rob Weller, and Ethan Lovelace. Thanks to all of the guys at Georgia River Fishing, the Shoal Bass Alliance, and River Bassin for their continued support of Shoal Bass management. A special thanks to the Joseph W. Jones Ecological Research Center for allowing us access and assisting with boat shuttles.

References

Alvarez, A. C., D. Peterson, A. T. Taylor, M. D. Tringali, and B. L. Barthel. 2015. Distribution and amount of hybridization between Shoal Bass and the invasive Spotted Bass in the lower Flint River, Georgia. Pages 503–521 *in* M. D. Tringali, J. M. Long, T. W. Birdsong, and M. S. Allen, editors. Black bass diversity: multidisciplinary science for conservation. American Fisheries Society, Symposium 82, Bethesda, Maryland.

Bailey, R. M., and C. L. Hubbs. 1949. The black basses (*Micropterus*) of Florida, with description of a new species. Occasional papers of the Museum of Zoology 516.

Beyer, H. L., D. T. Haydon, J. M. Morales, J. L. Frair, M. Hebblewhite, M. Mitchell, and J. Matthiopoulos. 2010. The interpretation of habitat preference metrics under use-availability designs. Philosophical Transactions of the Royal Society B: Biological Sciences 365:2245–2254.

Bitz, R. D., P. A. Strickland, T. J. Alfermann, C. R. Middaugh, and J. A. Bock. 2015. Shoal Bass nesting and associated habitat in the Chipola River, Florida. Pages 237–248 *in* M. D. Tringali, J. M. Long, T. W. Birdsong, and M. S. Allen, editors. Black bass diversity: multidisciplinary science for conservation. American Fisheries Society, Symposium 82, Bethesda, Maryland.

Bogan, A. E., J. M. Pierson, and P. Hartfield. 1995. Decline in the freshwater gastropod fauna in the Mobile Bay Basin. Pages 249–252 *in* E. T. LaRoe, G. S. Farris, C. E. Puckett, P. D. Doran, and J. J. Mac, editors. Our living resources: a report to the nation on the distribution, abundance, and health of U.S. plants, animals, and ecosystems. U.S. Department of the Interior, National Biological Service, Washington D. C.

Brim Box, J., and J. D. Williams. 2000. Unionid mollusks of the Apalachicola Basin in Alabama, Florida, and Georgia. Alabama Museum of Natural History Bulletin 21:1–143.

Burnham, K. P., and D. R. Anderson. 2002. Model Selection and Multimodel Inference: A Practical Information-Theoretical Approach. 2nd edition. Springer, New York.

Conner, L. M., and B. W. Plowman. 2001. Using Euclidean distances to assess nonrandom habitat use. Pages 275–290 *in* J. J. Millspaugh and J. M. Marzluff, editors. Radio Tracking and Animal Populations. Academic Press, San Diego.

Conner, L. M., M. D. Smith, and L. W. Burger. 2003. A comparison of distance-based and classification based analyses of habitat use. Ecology 84:526–531.

Couch, C. A., E. H. Hopkins, and P. S. Hardy. 1996. Influences of environmental settings on aquatic ecosystems in the Apalachicola-Chattahoochee-Flint River Basin. Water-Resources Investigations Report 95–4278. U. S. Geological Survey, Denver.

Diaz, M., D. Wethey, J. Bulak, and B. Ely. 2000. Effects of harvest and effective population size on genetic diversity in a Striped Bass population. Transactions of the American Fisheries Society 129:1367–1372.

Elwood, J. W., J. D. Newbold, R. V. O'Neill, and W. VanWinkle. 1983. Resource spiraling: an operational paradigm for analyzing lotic ecosystems. Pages 3–27 *in* T. D. Fontaine and S. M. Bartell, editors. Dynamics of Lotic Ecosystems. Ann Arbor Science Publishers, Ann Arbor, Michigan.

Ettinger-Dietzel, S. A., H. R. Dodd, J. T. Westhoff, and M. J. Siepker. 2016. Movement and habitat selection patterns of Smallmouth Bass *Micropterus dolomieu* in an Ozark river. Journal of Freshwater Ecology 31:61–75.

Gelman, A., and J. Hill. 2007. Data Analysis Using Regression and Multilevel/Hierarchical Models. 1st Edition. Cambridge University Press, New York.

Gelman, A., and D. B. Rubin. 1992. Inference from iterative simulation using multiple sequences. Statistical Science 7:457–472.

Gillies, C. S., M. Hebblewhite, S. E. Nielsen, M. A. Krawchuk, C. L. Aldridge, J. L. Frair, D. J. Saher, C. E. Stevens, and C. L. Jerde. 2006. Application of random effects to the study of resource selection by animals. Journal of Animal Ecology 75:887–898.

Goclowski, M. R., A. J. Kaeser, and S. M. Sammons. 2013. Movement and habitat differentiation among adult Shoal Bass, Largemouth

Bass, and Spotted Bass in the upper Flint River, Georgia North American Journal of Fisheries Management 33:56–70.

Golladay, S. W., D. W. Hicks, and T. K. Muenz. 2007. Streamflow changes associated with water use and climatic variation in the lower Flint River Basin, southwest Georgia. *In*: Proceedings of the 2007 Georgia Water Resources Conference. University of Georgia, Athens.

Grubaugh, J. W., J. B. Wallace, and E. S. Houston. 1997. Production of benthic macroinvertebrate communities along a southern Appalachian river continuum. Freshwater Biology 34:581–476.

Hosmer, D. W., Jr., and S. Lemeshow. 2000. Applied logistic regression. 2nd edition. Wiley, New York.

Ingram, T. I., and J. M. Kilpatrick. 2015. Assessment of the Shoal Bass population in the Lower Flint River, Georgia. Pages 503–521 *in* M. D. Tringali, J. M. Long, T. W. Birdsong, and M. S. Allen, editors. Black bass diversity: multidisciplinary science for conservation. American Fisheries Society, Symposium 82, Bethesda, Maryland.

Ingram, T. I., J. E. Tannehill, and S. P. Young. 2013. Post-release survival and behavior of adult Shoal Bass in the Flint River, Georgia. North American Journal of Fisheries Management 33:717–722.

Johnson, D. H. 1980. The comparison of usage and availability measurements for evaluating resources preference. Ecology 61:65–71.

Jonsson, B., and N. Jonsson. 1993. Partial migration: niche shift versus sexual maturation in fishes. Reviews in Fish Biology and Fisheries 3:348–365.

Kaeser, A. J., and T. L. Litts. 2008. An assessment of deadhead logs and large woody debris using side scan sonar and field surveys in streams of southwest Georgia. Fisheries 33:589–597.

Kaeser, A. J., and T. L. Litts. 2010. A novel technique for mapping habitat in navigable streams using low-cost side scan sonar. Fisheries 35:163–174.

Katechis C. T. 2015. Habitat associations and distribution of black bass and imperiled fish species in tributary streams of the middle Chattahoochee River, Georgia and Alabama. Master's Thesis, Auburn University, Alabama.

Kincaid T, T. Olsen, D. Stevens, C. Platt, D. White, and R. Remington. 2013. "spsurvey: Spatial survey design and analysis." R package version 2.0.

Langhurst, R. W. and, D. L. Schoenike. 1990. Seasonal migration of Smallmouth Bass in the Embarrass and Wolf rivers, Wisconsin. North American Journal of Fisheries Management 10:224–227.

Maceina, M. J., J. W. Slipke, and J. M. Grizzle. 1999. Effectiveness of three barrier types for confining Grass Carp in embayments of Lake Seminole, Georgia. North American Journal of Fisheries Management 19:968–976.

Marcinek, P. A., P. J. Gagnon, M. C. Freeman, C. Straight, M. D. Merill, and B. J. Freeman. 2005. Ecological importance and conservation status of southeastern river shoal habitat. Final Report, U. S. Fish and Wildlife Service, Atlanta.

Mesing, C. L. and, A. M. Wicker. 1986. Home range, spawning migrations, and homing of radio-tagged Florida Largemouth Bass in two central Florida lakes. Transactions of the American Fisheries Society 115:286–295.

Northcote, T. G. 1997. Potadromy in Salmonidae—living and moving in the fast lane. North American Journal of Fisheries Management 17:1029–1045.

Northrup, J. M., M. B. Hooten, C. R. Anderson, and G. Wittemyer. 2013. Practical guidance on characterizing availability in resource selection functions under a use-availability design. Ecology 94:1456–1463.

O'Connell, T. R., and R. S. Campbell. 1953. The benthos of the Black River and Clearwater Lake, Missouri. University of Missouri Studies 26:25–41.

Ramsey, J. S. 1975. Taxonomic history and systematic relationships among species of *Micropterus*. Pages 67–75 *in* R. H. Stroud and H. Clepper, editors. Black bass biology and management. Sport Fishing Institute, Washington, D.C.

Ridgway, M. S., and B. J. Shuter. 1996. Effects of displacement on the seasonal movements of and home range characteristics of Smallmouth Bass in Lake Opeongo. North American Journal of Fisheries Management 16:371–377.

Sammons, S. M. 2012. Diets of juvenile and subadult size classes of three Micropterus spp. in the Flint River, Georgia: Potential for trophic competition. Southeastern Naturalist 11:387–404.

Sammons, S. M. 2015. First evidence of potadromy and partial migration in black basses: Shoal Bass Micropterus cataractae (Actinopterygii, Centrarchidae) in the Upper Flint River, USA. Hydrobiologia 751:135–146.

Sammons, S. M. 2016. Catch and exploitation of Shoal Bass in the Flint River, Georgia, USA: Implications for harvest restrictions. North American Journal of Fisheries Management 36:606–620.

Sammons, S. M., and L. A. Earley. 2015. Movement and habitat use of Shoal Bass *Micropterus cataractae* in a regulated portion of the Chattahoochee River, Alabama-Georgia, USA. Pages 249–261 *in* M. D. Tringali, J. M. Long, T. W. Birdsong, and M. S. Allen, editors. Black bass diversity: multidisciplinary science for conservation. American Fisheries Society, Symposium 82, Bethesda, Maryland.

Sammons, S. M., and M. R. Goclowski. 2012. Relations between Shoal Bass and sympatric congeneric black bass species in Georgia rivers with emphasis on movement patterns, habitat use, and recruitment. Final Report. Georgia Department of Natural Resources, Social Circle.

Sammons, S. M., M. J. Maceina, and D. G. Partridge. 2003. Changes in behavior, movement, and home ranges of Largemouth Bass following large-scale hydrilla removal in Lake Seminole, Georgia. Journal of Aquatic Plant Management 41:31–38.

Sammons, S. M., K. L. Woodside, and C. J. Paxton. 2015. Shoal Bass Micropterus cataractae. Pages 75–81 *in* M. D. Tringali, J. M. Long, T. W. Birdsong, and M. S. Allen, editors. Black bass diversity: multidisciplinary science for conservation. American Fisheries Society, Symposium 82, Bethesda, Maryland.

SAS Institute. 2000. User's guide, version 8 Cary, North Carolina: SAS Institute.

Sterrett, S. C., L. L. Smith, A. Kaeser, R. A. Katz, J. C. Brock, and J. C. Maerz. 2015. Spatial ecology of female Barbour's map turtle (*Graptemys barbouri*) in Ichawaynochaway Creek. Copeia 2:263–271.

Stormer, D. G., and M. J. Maceina. 2009. Habitat use, home range, and movement of Shoal Bass in Alabama. North American Journal of Fisheries Management 29:604–613.

Westhoff, J. T., C. Paukert, Ettinger-Dietzel, S., H. Dodd, and M. Siepker. 2016. Behavioural thermoregulation and bioenergetics of riverine smallmouth bass associated with ambient cold-period thermal refuge. Ecology of Freshwater Fish 25:72–85.

Wheeler, A. P., and M. S. Allen. 2003. Habitat and diet partitioning between shoal bass and largemouth bass in the Chipola River, Florida. Transactions of the American Fisheries Society 132:438–449.

Williams, J. D., and G. H. Burgess. 1999. A new species of bass, *Micropterus cataractae* (Teleostei: Centrarchidae), from the Apalachicola River basin in Alabama, Florida, and Georgia. Bulletin of the Florida Museum of Natural History 42:80–114.

Winter, J. D. 1970. Summer home range movements and habitat use by four Largemouth Bass in Mary Lake, Minnesota. Transactions of the American Fisheries Society 106:323–330.

American Fisheries Society Symposium 87:213–230, 2019

Repatriation of Guadalupe Bass in the Blanco River, Texas: A Case Study in the Opportunistic Use of Drought as a Fisheries Management Tool

STEPHAN MAGNELIA*, GORDON LINAM, RYAN MCGILLICUDDY, KEN SAUNDERS, AND MELISSA PARKER
Texas Parks and Wildlife Department
Post Office Box 1685, San Marcos, Texas 78667, USA

TIMOTHY BIRDSONG
Texas Parks and Wildlife Department
4200 Smith School Road, Austin, Texas 78744, USA

DIJAR LUTZ-CARRILLO, JÁNAYE WILLIAMSON
Texas Parks and Wildlife Department
505 Staples Road, San Marcos, Texas 78666, USA

RACHAEL LINDSEY RANFT[1]
The Nature Conservancy
23345 Bat Cave Road, San Antonio, Texas, 78266, USA

TIMOTHY BONNER
Texas State University, Department of Biology/Aquatic Station
601 University Drive, San Marcos, Texas 78666, USA

*Corresponding author: stephan.magnelia@tpwd.texas.gov
[1]Current address: Eco-vative Associates
223 Paisano Drive, Victoria, TX 77904, USA

Abstract.—Repatriation of Guadalupe Bass *Micropterus treculii* was undertaken in the Blanco River, Texas, after extirpation due to hybridization with non-native, introduced Smallmouth Bass *Micropterus dolomieu*. In summer 2011, much of the river was reduced to a series of disconnected, enduring pools due to severe drought. Electrofishing and seines were used to remove Smallmouth Bass from fifty-two enduring pools in a 24-km reach of the river upstream of a natural fish passage barrier. After the removal effort was completed, stockings of non-introgressed, hatchery-produced Guadalupe Bass fingerlings (*n* = 316,016) were initiated in spring 2012 when river flows returned. Successful recruitment of wild, naturally-produced Guadalupe Bass was documented in fall 2014 and 2015. In the reach where Smallmouth Bass were removed, only Guadalupe Bass were detected, despite a historic flood event in spring 2015, which provided an opportunity for upstream movement of Smallmouth Bass and Guadalupe Bass x Smallmouth Bass hybrids past the barrier.

Introduction

Guadalupe Bass (GB) *Micropterus treculii*, the official state fish of Texas, is endemic to streams that originate in the Edwards Plateau ecoregion of central Texas, including portions of the Brazos, Colorado, Guadalupe and San Antonio river basins (Figure 1; Curtis et al. 2015). They inhabit spring-fed streams, with larger individuals associated with moderate depths (1.0 m) and slow current velocities (<0.05 m/s) downstream from riffles with instream cover such as log complexes, boulders, and undercut banks (Edwards 1980; Perkin et al. 2010; Curtis et al. 2015). Juveniles initially seek swift habitats, but transition to deeper depths and moderately flowing eddies as they near age-1 (Edwards 1980; Curtis et al. 2015). They are a highly sought after sport fish for river anglers (Thomas et al. 2015) and also a species of conservation concern (Hubbs et al. 2008; Texas Parks and Wildlife Department 2012) due to decreased stream flow (Hurst et al. 1975), reservoir construction (Edwards 1978), habitat degradation, and hybridization with nonnative Smallmouth Bass (SMB) *M. dolomieu* (Edwards 1980). Extirpation of GB has occurred in portions of their historical range as a result of hypolimnetic dam releases (Edwards 1978) and hybridization with SMB (Littrell et al. 2007; Bean et al. 2013). Since 1990, Texas Parks and Wildlife Department (TPWD) restoration efforts have included a genetic refuge restoration strategy (Araguas et al. 2008, 2009; Caudron et al. 2011), where SMB are not stocked in any location that provides it access to GB populations whose genetic purity is to be maintained or restored (Garrett 1991). Despite this effort, the proportion of SMB and GB × SMB hybrids (hybrids) within some streams has been shown to continually increase until Guadalupe Bass are extirpated (Littrell et al. 2007; Bean et al. 2013; Fleming et al. 2015). In response to the threat of additional extirpations of GB populations, TPWD initiated the Guadalupe Bass Restoration Initiative (GBRI) (Birdsong et al. 2010; Garrett et al. 2015) in 2010.

The Blanco River is located along the eastern edge of the Edwards Plateau ecoregion in south-central Texas in an area commonly referred to as the Texas Hill Country. The watershed (Figure 1) is within the historical range of GB and is of high ecological importance given that it hosts 36 species of

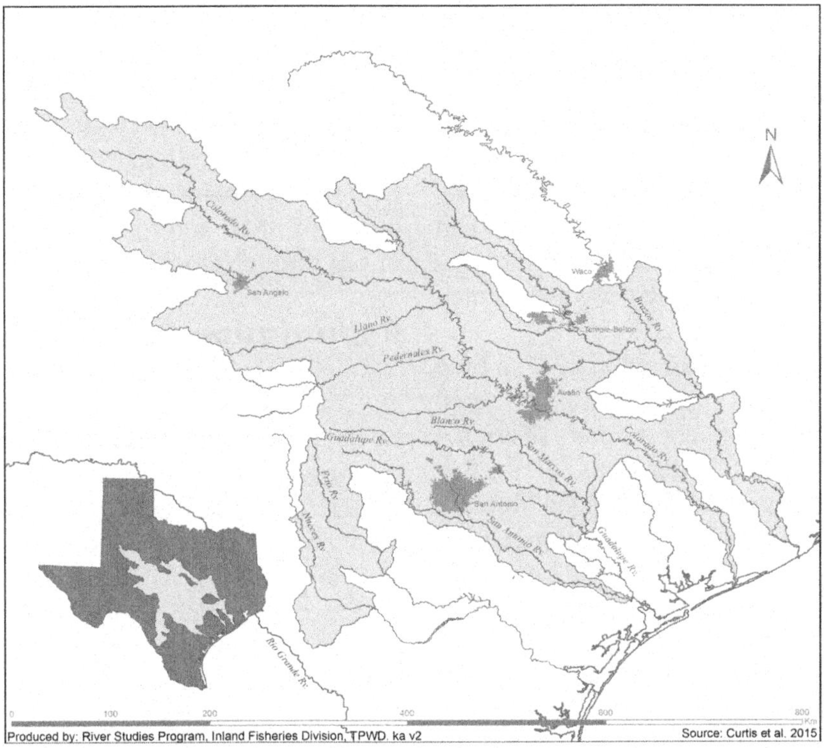

Figure 1. Distribution of Guadalupe Bass including historical and introduced populations. Map adapted from Curtis et al. 2015.

conservation concern including five species of fish (TPWD 2012). The watershed is also a priority for conservation investments by The Nature Conservancy (TNC) who has established relationships with a network of landowners interested in wildlife conservation (TNC 2016). Furthermore, the Blanco River is at high risk for habitat alteration due to its location just west of the Austin-San Antonio urban corridor (Crawford et al. 2016). These factors would have otherwise propelled the Blanco River to be considered as a priority for the GBRI; however, stockings of SMB in the Blanco River in 1977 and 1980 resulted in GB × SMB hybridization and genetic introgression (Garrett 1991). By 1991, 30% of the micropterids collected from the Blanco River were genetically identified as GB × SMB hybrids (Morizot et al. 1991). Supplemental stockings of 80,014 genetically pure GB fingerlings occurred in the Blanco River in 1994 and 1995, a conservation strategy that was initiated in the Guadalupe River headwaters in 1992 intended to reduce hybridization and genetic introgression of wild populations of GB (Fleming et al. 2015); however, because of state fish hatchery production resource constraints, supplemental stocking of GB in the Blanco River was abandoned after 1995 (G. Garrett, TPWD retired, personal communication), and the apparent extirpation of GB from the Blanco River was subsequently documented in 2005 (Littrell et al. 2007).

The continued persistence of SMB and hybrids in the Blanco River following GB extirpation provided evidence of the suitability of stream habitats to support SMB in the long-term. Bean et al. (2013) observed variable levels of GB × SMB hybridization in streams where nonnative SMB had been introduced, with the highest levels of hybridization found in streams where SMB persisted long-term. Fisheries managers believed that supplemental stockings of GB, such as those made in the Guadalupe River (Fleming et al. 2015), would likely be an ineffective repatriation approach in the Blanco River in the absence of SMB removal. Furthermore, a large-scale removal program for SMB and hybrids did not seem feasible, so efforts to repatriate GB to the Blanco River were deemed futile.

This perspective changed in 2011, a time during which most of Texas experienced exceptional drought conditions (National Drought Mitigation Center et al. 2016). Record high temperatures from June through August combined with minimal precipitation contributed to record low stream flows throughout much of the State (Winters 2013). The Blanco River was reduced to a series of disconnected enduring pools. Conditions appeared to offer an opportunity for the targeted removal of SMB and hybrids seeking refuge in those pools. Opportunistic removal of invasive fish species that aggregate as a result of weather conditions has been suggested as an effective population control strategy (Shepard et al. 2014; Koizumi et al. 2017). A limitation of this strategy for the Blanco River was a constant source of SMB and hybrids known to persist within the San Marcos River, a spring-fed river that had stable flows throughout the 2011 drought. The Blanco River is a tributary of the San Marcos River (Figure 1), and in the absence of an effective barrier to upstream fish movement, we believed that SMB and hybrids would eventually move upstream and recolonize upon the return of normal flows. However, data from past fish collections revealed a discontinuity in the Blanco River fish assemblage upstream from a gorge containing a series of waterfalls (Texas Game and Fish Commission 1957; Bean et al. 2007). It was thought this natural feature might provide an effective barrier to upstream movement of SMB and hybrids and the area upstream from this to the headwaters would be a reach where removal efforts might be effective. The objective of this project was to implement an isolation management strategy (Novinger and Rahel 2003) and establish a self-sustaining population of pure GB upstream of the barrier, through removal of SMB and hybrids, and repatriation of GB through stocking.

Repatriation Area Description

The Blanco River is intermittently fed by shallow springs and seeps, and in some reaches the channel gains water, while in others water is lost to the aquifer through the karst landscape (Slade et al. 2002; Wierman et al. 2010). Although streamflow was absent during the summer of 2011, these springs as well as approximately 90 low-head dams and low-water crossings influenced the amount of water (i.e., number of enduring pools and depth of those pools) that remained. While the entirety of the Blanco River and its tributaries were considered for the repatriation effort, for planning purposes the river was divided into four distinct reaches (Figure 2).

Reach 1 was the upper 38 km of the mainstem Blanco River and tributaries from the headwaters to a low-head dam just downstream from the city of Blanco (Figure 2). This reach was highly fragmented with 37 low-head dams and 17 low-water crossings. It also contained 36 main channel small impoundments, many of which still held large areas of water during summer 2011. The lower 6 km of this reach entering and in the city of Blanco was characterized by a series of 10 consecutive impoundments. While some of these were dry others had as much as 2–3 ha of water remaining (Figure 2). The lower portion of this reach is reported to be a gaining reach (Wierman et al. 2010), likely due to the influence of springs such as Koch Springs (Brune 1975), which enters the river at one of the lower reservoirs and was flowing during

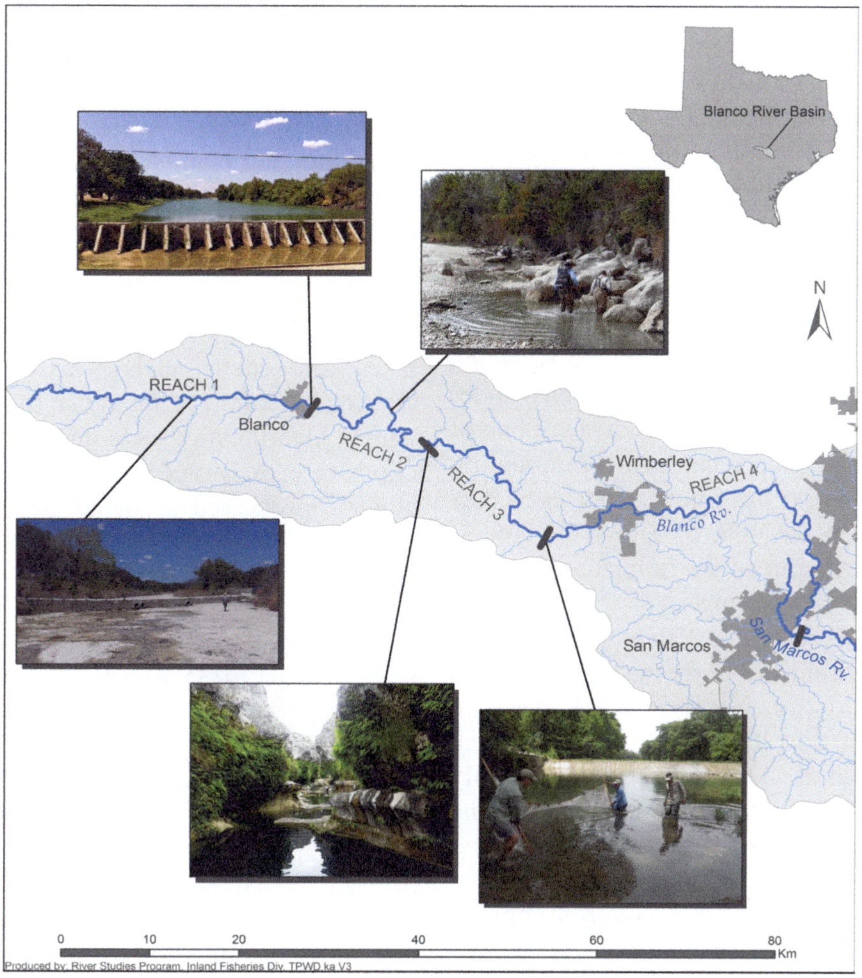

Figure 2. Location of Reaches 1 through 4 for the Blanco River, TX Guadalupe Bass repatriation project. Photographs (starting at top left and proceeding clockwise) are of: 1) A large enduring pool upstream of a low-head dam at the lower portion of Reach 1 in summer 2011, 2) A small enduring pool in Reach 2 in summer 2011, 3) The dam at the lower end of Reach 3 in fall 2014, 4) The bottom of the gorge looking upstream in fall 2014, 5) A dewatered portion of Reach 1 in summer 2011.

the project. Portions of this reach not immediately upstream of a dam or crossing were largely dewatered in summer 2011 (Figure 2).

Reach 2 was the 24 km of the mainstem Blanco River and tributaries from the low-water dam located just upstream of the FM 165 bridge crossing to a 0.4-km long in-channel narrow gorge bordered by 23-m high cliffs (Texas State Historical Association 2016) located at approximately river km 62. In the most upstream portion of the gorge was a series of five consecutive natural waterfalls ranging in height from 0.6 to 3.3 m in height (Figure 2). Unlike the numerous low-head dams and low-water crossings on the Blanco River the gorge was suspected to be an effective fish passage barrier at least among small fishes (e.g., percids; Bean et al. 2007), but its effectiveness as a barrier to larger and more mobile fishes (e.g., black basses) was unknown. Fifty-two enduring pools were identified, and while this reach contained seven dams and seven low-water crossings, all but one of the enduring pools were small (≤0.1 ha), and all but three were shallow (<1 m) (Figure 2). The one large pool was located at the very upper portion of the reach and covered approximately 1.6 ha. Approximately 8 km of this area was reported as a losing reach (Wierman et al. 2010), which likely contributed to the lack of water under the severe drought conditions. At least one of the larger and deeper pools was reported by the landowner to have held water even during the 1950s drought of record for Texas, which is likely due to the influence of an undescribed seep or spring.

Reach 3 was the Blanco River and tributaries from the gorge to an in-channel dam at approximately river km 85 (Figure 2). Most of this stretch is a gaining reach (Wierman et al. 2010) influenced not only by springs, but also a major tributary, the Little Blanco River. Thirty-two enduring pools were identified, with the largest being an impoundment that covered approximately 6.5 ha located at the very downstream boundary. A number of the other pools were between 2 and 3 ha.

Reach 4 consisted of the remaining 56 km of the Blanco River to the confluence with the San Marcos River. The approximate upper half of this portion is a gaining reach, influenced by input from a major spring, Jacobs Well (Brune 1975), which feeds the Blanco River tributary Cypress Creek (Wierman et al. 2010). While we did not enumerate or measure the size of enduring pools in this reach, on-the-ground observations from public access points indicated the upper half contained many large and deep pools while the lower portion (a losing reach) was predominately dry.

Detection of Smallmouth Bass and Hybrids

In summer 2011 seining, backpack electrofishing, and boat electrofishing were used in a targeted collection of micropterids for the purpose of detecting the presence of SMB and hybrids within each of the four reaches. All fish were identified to species in the field and released, except a pelvic fin clip was taken for genetic analysis to confirm species identification of small fish (<75 mm) for which identification was inconclusive. Fin clips were preserved in 95% nondenatured ethanol, processed to isolate genomic DNA (Lutz-Carrillo et al. 2015), and then evaluated for taxonomic status using a 10-locus subset of taxa-informative microsatellite loci (MPX4, MPX5, MPX6, and MPX17; Lutz-Carrillo et al. 2015). The detection of SMB or hybrids in Reach 1 would have likely ended the repatriation effort, as we felt removal efforts would have been futile due to the relatively large amount of water remaining in impoundments, and downstream movement of SMB and hybrids was likely. More than 400 micropterids were collected from 16 locations within Reach 1 in an effort to detect the presence of SMB and hybrids. Locations included all private lands accessible by TNC staff, including small impoundments in the headwaters, and all public bridge and low-water crossings, where water remained. All micropterids from this reach were identified either visually in the field, or through genetic analysis (n = 55) as Largemouth Bass *Micropterus salmoides* (LMB). This finding was consistent with previous collections of micropterids in this reach between 2003 and 2005 (Bean et al. 2007). Smallmouth Bass and hybrids were encountered in surveys of Reaches 2, 3, and 4.

Reach-Specific Repatriation Strategies

Although SMB and hybrids were not detected in Reach 1, affording a seemingly simple opportunity for repatriation of GB, the number of impoundments in this reach made it appear habitat was more suitable for LMB than GB. There was also concern that high levels of predation from LMB, which were prevalent in this reach, could occur on stocked GB fingerlings. We were unsure if GB could persist in this reach long-term and it was not initially considered for the repatriation effort.

Reach 2 afforded what appeared to be the highest likelihood for successful removal of SMB and hybrids. Almost all the enduring pools in this reach were relatively small (≤0.1 ha) and it was thought electrofishing and seining would be effective removal techniques. If removal of SMB and hybrids was successful in Reach 2, we believed the risk of SMB recolonization from downstream of the gorge would be minimized. It was also thought habitat in this reach was conducive for establishing a naturally reproducing population of pure GB. The decision was made to conduct georeferenced aerial mapping of enduring pools in Reach 2 and then send removal crews to each of the pools.

Reaches 3 and 4 lacked a significant barrier to prevent upstream migration of SMB and hybrids from the San Marcos River. Given the large enduring pools in these reaches we also believed that removal of SMB and hybrids would be nearly impossible and as such determined that efforts to repatriate GB to these reaches would be limited to supplemental stockings similar to GB restoration efforts in the upper Guadalupe River (Fleming et al. 2015).

Mapping of Enduring Pools

To increase efficiency of ground crews for removal efforts, we collected georeferenced imagery of the river channel using an unmanned aerial system (UAS). The specific UAS platform, data collection, and image processing techniques are described by Birdsong et al. (2015). The UAS mapping produced a 20-cm georeferenced photomosaic of the river channel of Reaches 2 and 3. Nine enduring pools in Reach 2 were encountered in the field that were not visible in the UAS imagery because of obstruction by tree canopy or shadows. Pools visible in the photomosaic were delineated in Google Earth Pro (Google Inc., Mountain View, CA, USA).

Removal of Smallmouth Bass and Hybrids

Pulsed direct current boat, barge (Smith Root pulsator model GPP-5.0), and backpack electrofishing (Smith Root model 12-B POW), as well as seines, were used to target SMB and hybrids from enduring pools in Reach 2. Boat and barge pulsators were set to the low voltage range (50–500 V), 60 Hz, and voltage increased until fish were stunned (approximately 10–15 amps). The backpack shocker was set to 60 Hz, 6 micro seconds pulse width (setting I5) and an output voltage of 300. Electrofishing has been reported to be a successful technique for eradicating (Shepard et al. 2014), or reducing the number of nonnative salmonids in small streams during genetic restoration projects (Caudron and Champigneulle 2011). Seines were comprised of delta weave mesh and had double lead weights to help maintain contact with the bottom. Seine sizes were: 4.6 m × 1.8 m with 4.8-mm mesh and 9.1 m x 1.8 m with 6.4-mm mesh. The 9.1-m seine was also used as a block net when deemed appropriate.

Each pool was examined on-site, the sampling gear(s) best suited for the situation determined, and repeated sampling conducted until we were satisfied the pool had been adequately sampled, and no SMB or hybrids were caught. Approximately 270 h were spent seining and electrofishing the fifty-two enduring pools in Reach 2. Individuals identified or suspected to be a SMB or hybrid were removed. A pelvic fin-clip was preserved in ethanol, and species identification determined with genetic analysis, using the methodology previously described in the detection sampling portion of the project. Nontarget species were not dipped while electrofishing, and those collected in seines were released back into the pool.

A total of 45 micropterids were removed from enduring pools in Reach 2. Of those, 25 were genetically identified as SMB (mean length = 181.5 mm; range = 135–460 mm) while the remainder were identified as LMB. No hybrids were collected. Thousands of fishes, including many species previously collected in this portion of the Blanco River (Bean et al. 2007) and common to central Texas, were collected and returned to enduring pools, suggesting removal methods were effective. We were surprised by the small number of SMB collected, given the relatively large number of other species in pools, and surmised their abundance had been reduced as a result of the drought and low flow conditions, and/or they were not particularly abundant in Reach 2. A previous study indicated relative abundance of GB and SMB in the overall fish assemblage of the Blanco River was <1% (Bean et al. 2007). At the conclusion of the fish removal portion of the project, we were confident that all pools in Reach 2 had been located and sampled.

Repatriation of Guadalupe Bass

By spring 2012, drought conditions had subsided throughout much of Texas and typical flow conditions returned to the Blanco River. Guadalupe Bass fingerlings cultured at the TPWD A.E. Wood State Fish Hatchery were stocked in spring 2012 (*n* = 102,616; mean length = 40 mm; range = 35–51 mm), 2013 (*n* = 133,817; mean length = 39 mm; range = 35–48 mm), 2015 (*n* = 23,805; mean length = 40 mm; range = 35–51 mm) and 2016 (*n* = 55,778; mean length = 46 mm; range = 36–58 mm). Broodstock used for production of fingerlings were acquired from the wild (South Llano River, TX) and were supplemented annually with additional wild fish from the same source. Each year approximately 100 pairs of broodfish were penned for spawning with nearly half of penned fish participating in spawning based on genetic analysis of production fingerings (Lutz-Carrillo et al. 2015; TPWD, unpublished data).

In 2012 and 2013, 72% of the fingerlings were stocked at 8 sites located in Reach 2 where removal efforts were conducted. The remaining fingerlings were stocked at three sites located in Reach 3, and one site in Reach 4 to see if supplemental stocking might reduce introgression without removal of SMB and hybrids, similar to results reported from the upper Guadalupe River (Fleming et al. 2015). No fish were stocked in spring 2014 as we were interested in documenting if natural reproduction from the 2012 and/or 2013 stockings had occurred. Guadalupe Bass are reported to attain sexual maturity at Age 1 (Edwards 1980). All GB produced in 2015 and 2016 were stocked in Reach 1. While the habitat in Reach 1 appeared more conducive to LMB, lotic habitat was present under higher flow conditions and we were interested to see if this area might also sustain a population of GB.

Post-Stocking Monitoring

In fall 2014 and 2015, micropterids appearing to be GB, SMB, or hybrids were collected using seines and boat and backpack electrofishing for genetic analysis. An effort was made in both years to collect 100 fish for analysis. Seine and electrofishing specifications/settings previously described for removal efforts were also used for monitoring collections. A total of 119 seine hauls covering 1.45 river km, and 1.9 h of boat and backpack electrofishing effort were expended for both monitoring collections. A pelvic fin clip was removed and preserved in 95% nondenatured ethanol for genetic analysis to confirm species identification. In fall 2014, individuals were collected from six sites in Reach 2 (n = 69) and two sites in Reach 4 (n = 31). In fall 2015, individuals were collected from four sites in Reach 2 (n = 59) and a total of five sites in Reaches 3 and 4 (n = 32).

Each post-stocking sample was first genotyped using the taxa-specific marker panel previously described to verify species status. In order to verify the origins of recovered fish, potential parental broodfish and wild collected individuals were genotyped at 22 microsatellite loci (Table 1; Lutz-Carrillo et al. 2008; Seyoum et al. 2013). Reactions were performed as multiplexes in 10-μL volumes with Platinum Multiplex PCR Master Mix (1X), equimolar amounts of labeled and unlabeled primer pairs (0.02–0.15 μM), and 50 ng template DNA (Table 1). Polymerase chain reaction (PCR) conditions were 95°C (120 s) followed by 28 cycles of 95°C (30 s), annealing (55.6–57.9°C) for 90 s, and 72°C (90 s), ending with a 60°C final extension (30 min). Products from each reaction were diluted in formamide and resolved on an ABI 3500 genetic analyzer (ThermoFisher Scientific). Allele sizes were quantified using GeneMapper Software 5 (ThermoFisher Scientific).

Hatchery broodfish were then assigned or excluded as potential parents using the likelihood approach implemented in CERVUS v.3.0.3 (Marshall et al. 1998; Kalinowski et al. 2007). Simulations (parent-pair, sexes known), assuming 85% of candidate parents were sampled, to allow for documentation errors and typing errors of 1%, were used to determine critical logarithm of odds (LOD) thresholds. The LOD threshold value was then used to assign parentage with 95% and then 80% confidence. Mismatches can arise due to a true parent missing from the parent pool, mis-assigned genders, de novo mutations in the fingerlings, scoring errors, or failed PCRs (Araki and Blouin 2005), thus prior to final assignments any trio of offspring and parental pair with mismatches but a positive LOD score were also visually inspected. Individuals excluded as offspring of all hatchery parent pairs were classified as having wild origins. Individuals assigned to a hatchery parent pair were classified as having primary hatchery origins. Using the measured total lengths of recovered samples, annual pen assignments for broodfish, and detected spawning pairs (based on pre-stocking fingerling sampling and genotyping; Lutz-Carrillo et al. 2015) these individuals were further classified to year-class.

During the first post-stocking collection (fall 2014) genetic analyses resolved 86 GB, 13 hybrids, and a single LMB. Ignoring the LMB (not a target species), all fish (100%) recovered from Reach 2 (n = 68) were identified as GB and 32% recovered downstream of the gorge were identified as GB (n = 10) (Figure 3). Among the GB recovered, 28% (n = 13) had primary hatchery origins. All of these were recovered from Reach 2 (Figure 4). Among individuals identified as having primary hatchery origins, 15% (n = 2) were stocked in 2012 and 85% (n = 11) were stocked in 2013. The remaining GB (n = 33), including all GB collected in Reach 3 (n = 10), were not assigned to hatchery parents.

Table 1. Parentage marker panel. Primer sequences, concentrations (µM), labels (5'-mod), multiplex groups (rxn), and annealing temperatures (T_a). 5'-PIG tail additions to promote 3'-A-addition are denoted by lower case bases. TPW prefix indicates primers taken directly or modified from Lutz-Carrillo et al. 2008. Msaf prefix indicates primers taken directly or modified from Seyoum et al. 2013.

Primer	Primer Sequence (5'-3')	µM	5'-mod	rxn	Ta
TPW130f	GTCTCACCTGTGCCAATGTGA	0.05		1	57.9
TPW130r	TAGGAGTTGGGGTCACTTGCTT		6FAM	1	
TPW38f	ACATTCAAACCACTTTCCCACA	0.1	6FAM	1	
TPW38r	gtttGTGACAGACTTTGTTGTCCCAT			1	
TPW51f	gttTCATTGCAACACAACACAGAGA	0.05		1	
TPW51r	ATGGGACTATAGGCATCACACA		6FAM	1	
TPW70f	ATATCCACTTCGGTGATCCAGT0.15		VIC	1	
TPW70r	gtttCAGGAGTATGTTTCCCTCATGC			1	
TPW11f	ACCTTGCAACCCAAAGGTTATT	0.08	PET	2	57.1
TPW11r	gttTCTGCCTCAGAGACACAACATA			2	
TPW12f	GTAGGGTTGTTCCCCATGAATC	0.05		2	
TPW12r	TGCTCGCTGTCTCTATCTCTTC		6FAM	2	
TPW55f	CAGCAAGATGTCAACCAGGTAA	0.05	VIC	2	
TPW55r	gttTCCTGAGATACTGAATTAGGGGT			2	
TPW90f	CGATGTTCAAGAAGATGCCAGA	0.03	VIC	2	
TPW90r	gtttCATTTCCCTGTTCCCACTTACC			2	
Msaf14f	GAAAGAAAGGTCCAGTCATGTAGAA	0.1		2	
Msaf14r	ATGTTATTGAGTAGATCCCAATGGA		6FAM	2	
TPW173f	AACCACACAGTGACACAAACTG	0.05	NED	2	
TPW173r	gttTCTTACAGAGGGCAATAGGGAC			2	
Msaf10f	CACACTTCACTCAATCCCGTC	0.05	VIC	3	57.9
Msaf10r	gtttACAGCTGTAAAGATTTTGGGCA			3	
Msaf32f	gttTGCCTCACATCCTCTTATCCAT	0.1		3	
Msaf32r	CCTGTCTCACAAGTCTCACAAG		6FAM	3	
TPW96f	gTTGTGTTTGATTGAGCGAAGGA	0.05		3	
TPW96r	GCATAAAGACTGGGAACAGAGG		6FAM	3	
TPW62f	gtttATGCTTTCTGCTGAAGTGC	0.05		3	
TPW62r	TCCCTTTTGGAAGTAGAAGT		PET	3	
TPW107f	gtttACCTTGTTGGCCTGATGTTGGAGC	0.05		3	
TPW107r	AAGGGGTAAATCCATCTGTCCACC		VIC	3	
TPW112f	GTCCACCAGAGACAACCTGCGT	0.05		3	
TPW112r	CTAGTGCGACCCAGAGAACG		6FAM	3	
TPW132f	TCAGCAAAGTTTCTCCAGTGGAAC	0.05	NED	3	
TPW132r	GCATCTCTGAGCCAGGACAT			3	
Msaf22f	GCAGCAGGAGAGGCAAG	0.02	NED	4	55.6
Msaf22r	gttTGTCTGAAGAGCAGTGACA			4	
TPW25f	CCTGCACCCACATACGT	0.05	VIC	4	
TPW25r	gtttACCTTTGTGCTGTTCTGTC			4	
TPW50f	gttTGGCAGGTCGTAATGTATG	0.04		4	
TPW50r	GATTTGAGCTTCTACTTTACG		VIC	4	
TPW68f	CATCAACTTTAATGCAAATACAG	0.05	6FAM	4	
TPW68r	gtttAAGAGGAGGATGCAACTAAG			4	
TPW76f	ACACAGTGTCAGTTCTGCA	0.05	PET	4	
TPW76r	GTGAATACCTCAGCAAGCAT			4	

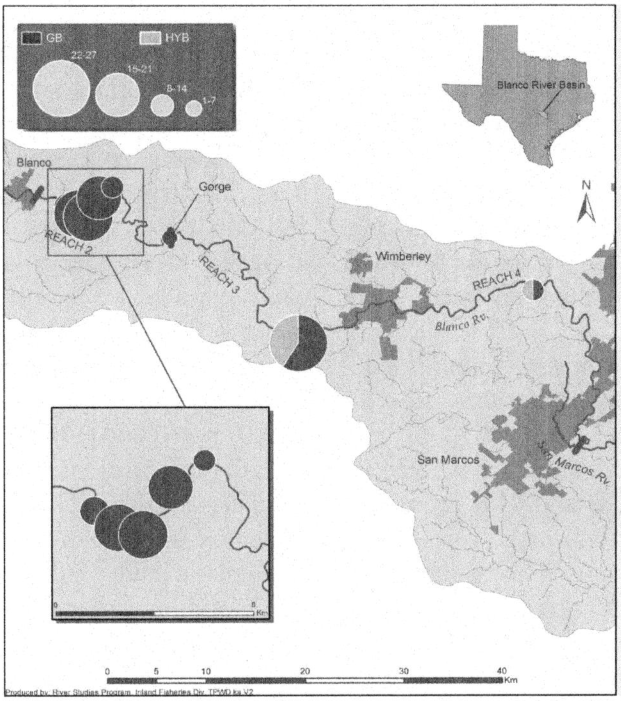

Figure 3. Guadalupe Bass (GB) and Smallmouth Bass or hybrids (HYB) detected in fall 2014 at Blanco River, TX sampling sites in the post-stocking monitoring effort upstream and downstream of the gorge barrier. Circles are scaled by sample size.

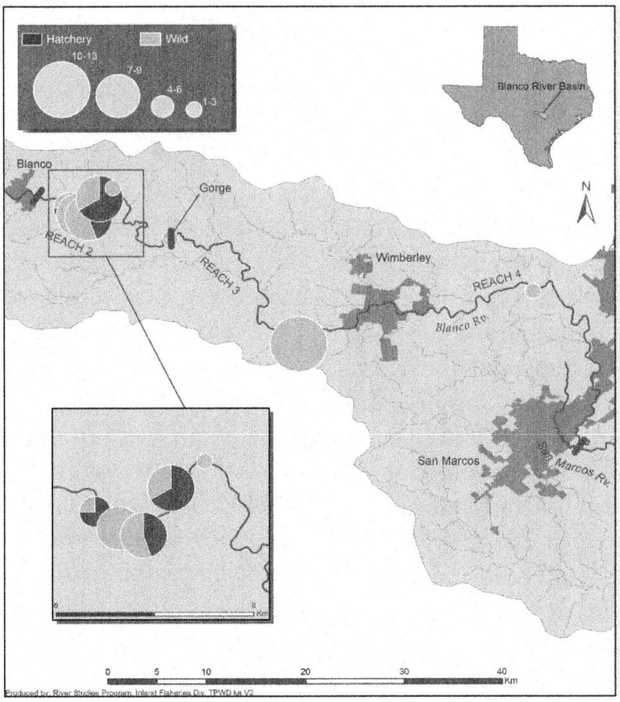

Figure 4. Stocked fish (HATCHERY) and wild fish (WILD) detected among Guadalupe Bass in fall 2014 at Blanco River, TX sampling sites in the post-stocking monitoring effort upstream and downstream of the gorge barrier. Circles are scaled by sample size.

On May 24, 2015, the Blanco River rose approximately 12.2 m above the base flow stage of 1.5 m (U.S. Geological Survey 2016). As a reference, the National Weather Service flood stage is 3.96 m. Floodwaters spread into the riparian area at the gorge (C. Furl, Edwards Aquifer Authority, personal communication) making it possible for SMB or hybrids to have moved upstream from Reach 3 into Reach 2.

During the second post-stocking collection (Fall 2015), 91 micropterids were collected. Genetic analyses resolved 72 GB and 19 hybrids. All fish recovered from Reach 2 (n = 59) were again GB and 41% recovered from downstream of the gorge were GB (n = 13) (Figure 5). Among GB recovered, 4% had primary hatchery origins (n = 3) with one of these fish recovered from Reach 3 (Figure 6). Among individuals identified as having primary hatchery origins, two were from the 2013 stocking and one was from the 2015 stocking. The remaining GB (n = 69) were not assigned to hatchery parents.

Discussion

Following extirpation of GB from the Blanco River (Littrell et al. 2007) and failure to detect GB in our pre-stocking sampling, the detection of only nonintrogressed GB in post-stocking collections from the target reach in 2014 suggested that both the target species removal and GB repatriation efforts were successful. It also appeared the gorge fish passage barrier was effective in preventing upstream movement of SMB and hybrids to the newly established GB population in Reach 2. The detection of only nonintrogressed GB in this reach in 2015, following a historic flood event, reinforced these conclusions. The detection of some nonintrogressed GB downstream of the barrier was also encouraging, but a more tentative result, as these individuals are expected to be more susceptible to hybridization than those upstream of the barrier given the enduring SMB and hybrids in these reaches.

While multiple sites were stocked with hatchery offspring, most GB recovered in post-stocking collections did not have primary hatchery origins. In post-stocking collections individuals resolved as GB, but not assigned to hatchery parents were significantly smaller by length (mean = 90.2 mm; range = 77–130 mm) than the GB assigned to hatchery parents (mean = 182.5 mm; range = 148–223 mm) (Mann–Whitney U-test, U = 429, $p < 0.0001$) indicating they were potentially offspring of stocked fish (secondary hatchery origins). This hypothesis was supported by the lack of GB detected in a previous genetic study in this system (Littrell et al. 2007) and our pre-stocking screening, which suggested no other source was present to produce these GB. Even assuming these unassigned GB were derivatives of an undetected refuge population, we would expect to resolve population substructure between these fish and hatchery broodfish derived from an independent source (i.e., South Llano River); however, Bayesian clustering of multi-locus genotypes (Pritchard et al. 2000; Falush et al. 2003) suggested all samples were most likely derived from a single panmictic population. This suggests that not only were the initial stocking efforts successful in reintroducing GB to a portion of the Blanco River, but they have been successfully reproducing.

Previous efforts to restore GB have required regular supplemental stockings to maintain GB influence in the face of introgression pressure from congeners (Fleming et al. 2015); however, in upper Johnson Creek, isolated from congeners by downstream barriers, GB have persisted for several years after the cessation of stocking efforts (Fleming and Smith 2019, this volume). By comparison, the gorge at the lower end of Reach 2 in the Blanco River is much larger than the barriers separating upper and lower Johnson Creek. Thus, if a sustaining GB population

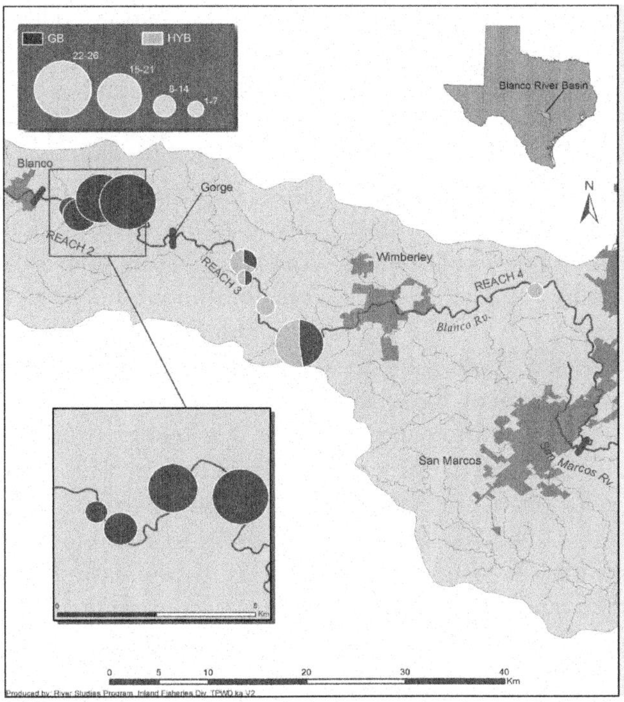

Figure 5. Guadalupe Bass (GB) and Smallmouth Bass or hybrids (HYB) detected in fall 2015 at Blanco River, TX sampling sites in the post-stocking monitoring effort upstream and downstream from the gorge barrier. Circles are scaled by sample size.

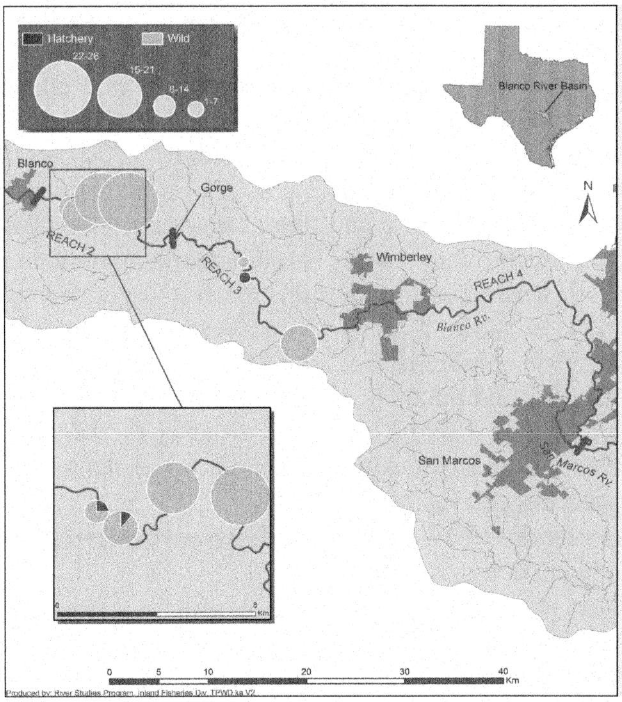

Figure 6. Stocked fish (HATCHERY) and wild fish (WILD) detected among Guadalupe Bass in fall 2015 at Blanco River, TX sampling sites in the post-stocking monitoring effort upstream and downstream of the gorge barrier. Circles are scaled by sample size.

can persist in upper Johnson Creek, we are relatively confident in the persistence of GB upstream of the gorge in the Blanco River. Natural reproduction should also relieve the need for additional hatchery stockings, except if necessary to introduce novel genetic variation (Carlson et al. 2014). In addition, the presence of a sustaining population in Reach 2 may beneficially impact reaches downstream of the barrier.

Barriers have been used, with varying levels of success, as management tools for native fish protection and restoration in the southwest (Rinne and Turner 1991; Carpenter and Terrell 2005) and throughout North America (Meronek et al. 1996; Harig et al. 2000); however, the primary measure of effectiveness has been the ability to prevent interactions with populations downstream of the barrier (Carpenter and Terrell 2005). The literature is lacking with respect to the seeding effect of restored populations, upstream of a barrier, on the prevalence of nonnatives or magnitude of introgression in downstream populations; however, gene flow in lotic systems is commonly asymmetric with greater dispersal and migration in the downstream direction (Müller 1954; Hänfling and Weetman 2006; Caldera and Bolnick 2008; Raeymaekers et al. 2009). The presence of barriers amplifies this asymmetry making upstream movement more difficult, but in many cases having little effect on downstream movement (Junker et al. 2012). In addition, even low levels of gene flow, from highly divergent populations, can have substantial demographic affects across a barrier (Fitzpatrick et al. 2016). Most protective barriers are located on smaller order lotic stretches and reproduction from restored populations may easily exceed a limited carrying capacity providing an excess of progeny that may move downstream (Waters 1972). Thus, it is reasonable to assume that the restored above barrier-population established in the Blanco River would annually integrate progeny into below barrier populations. This should, to some degree, suppress the frequency of nonnatives in downstream stretches by occupying available niches and providing resource competition. This is an idea that requires careful evaluation, however the same-source stocking of downstream sites, in addition to above barrier sites, in this study does not allow for conclusions about this potential seeding effect.

Future efforts for the Blanco River include periodic genetic monitoring throughout the basin, and stockings upstream of barriers in tributary streams in Reaches 3 and 4. If SMB and hybrids are identified upstream of tributary barriers, concerted efforts could be made to remove them and create additional GB populations for seeding the mainstem without additional hatchery stockings. Drought conditions are sure to return to Texas and this may again provide optimal conditions to conduct such efforts. Future monitoring results from the 2015 and 2016 stockings in Reach 1 are also of particular interest. During years with average or above average rainfall, portions of this reach are free running and could provide suitable habitat for GB. While we and others (Bean et al. 2007) did not detect GB in this portion of the river, we wonder if it is because they were never present, or if habitat is truly not conducive to their long-term persistence. If a naturally reproducing population could be established in this reach it might be maintained during even more severe droughts than that experienced in 2011. Individuals taking refuge in the many impoundments in this reach might provide progeny for downstream seeding even if all enduring pools in Reach 2 went dry.

Acknowledgments

Funding for this project was provided through the Federal Sport Fish Restoration Act. The authors acknowledge TPWD Inland Fisheries Division staff Kevin Mayes, Clint Robertson, Karim Aziz, John Botros,

Roy Kleinsasser (retired), Marcos De Jesus, Greg Cummings, Darin Gossett, Archis Grubh, Steve Boles, Kevin Kolodziejcyk, Gary Garrett (retired), Beth Bendik, Tom Heger, Megan Bean; and Texas State University staff and students Stephen Curtis, Virginia Dautreuil, Kristy Kollaus, and Sarah Robertson, for assistance in SMB removal efforts; TPWD A.E. Wood State Fish Hatchery staff Carl Kittel, the late Rob Schmid, Michael Matthews, Chris Thibodeaux, and Hugh Glenewinkel for producing the GB used for stocking; TPWD River Studies Program Cartographer Karim Aziz for producing manuscript figures; and, The Meadows Center for Water and the Environment staff Thom Hardy, Kristy Kollaus, and Tom Heard for aerial imagery used to locate enduring pools. Gratitude goes to the landowners along the Blanco River who gave us access to their property during all phases of the project.

References

Araguas, R. M., Sanz, N., R. Fernandez, F. M. Utter, C. Pla, and J. L. Garcia-Martin. 2008. Genetic refuges for a self-sustained fishery: experience in wild Brown Trout populations in the eastern Pyrenees. Ecology of Freshwater Fish 17:610–616.

Araguas, R. M., N. Sanz, R. Fernandez, F. M. Utter, C. Pla, and J. L. Garcia-Martin. 2009. Role of genetic refuge in the restoration of native gene pool of Brown Trout. Conservation Biology 23:871–878.

Araki, H., and M. S. Blouin. 2005. Unbiased estimation of relative reproductive success of different groups: evaluation and correction of bias caused by parentage assignment errors. Molecular Ecology 14:4097–4109.

Bean, P. T., T. H. Bonner, and B. M. Littrell. 2007. Spatial and temporal patterns in the fish assemblage of the Blanco River, Texas. Texas Journal of Science 59:179–200.

Bean, P. T., D. J. Lutz-Carillo, and T. H. Bonner. 2013. Rangewide survey of the introgressive status of Guadalupe Bass: implications for conservation and management. Transactions of the American Fisheries Society 142:681–689.

Birdsong, T., D. Krause, J. Leitner, J. Long, S. Robinson, S. Sammons, and J. Sedell. 2010. A business plan for conservation of native black bass species in the southeastern United States: a ten year plan. National Fish and Wildlife Foundation, Washington, D.C.

Birdsong, T., M. Bean, T. B. Grabowski, T. B. Hardy, T. Heard, D. Holdstock, K. Kollaus, S. Magnelia, K. Tolman, and L. Clark. 2015. Application and utility of a low-cost unmanned aerial system to manage and conserve aquatic resources in four Texas rivers. Journal of the Southeastern Association of Fish and Wildlife Agencies 2:80–85.

Brune, G. 1975. Major and historical springs of Texas. Report 189. Texas Water Development Board, Austin, Texas. Available: https://www.twdb.texas.gov/publications/reports/numbered_reports/doc/R189/R189.pdf. (September 2018).

Caldera, E. J., and D. I. Bolnick. 2008. Effects of colonization history and landscape structure on genetic variation within and among Threespine Stickleback (*Gasterosteus aculeatus*) populations in a single watershed. Evolutionary Ecology Research 10:575–598.

Carlson, S. M., C. J. Cunningham, and P. A. H. Westley. 2014. Evolutionary rescue in a changing world. Trends in Ecology and Evolution 29:521–530.

Carpenter, J. and J. W. Terrell. 2005. Effectiveness of fish barriers and renovations for maintaining and enhancing populations of native southwestern fishes. Final Report to U.S. Fish and Wildlife Service, Arizona Ecological Services Office, Interagency Agreement No. 201814N756. U.S. Geological Survey, Fort Collins Science Center, Colorado.

Caudron, A., A. Champigneulle, R. Guyomard, and C. R. Largiader. 2011. Assessment of three strategies for restoring native Brown Trout (*Salmo trutta*) populations in northern French alpine streams. Ecology of Freshwater Fish 20:478–491.

Caudron, A., and A. Champigneulle. 2011. Multiple electrofishing as a mitigate tool for removing nonnative Atlantic Brown Trout (*Salmo trutta* L.) threatening a native Mediterranean Brown Trout population. European Journal of Wildlife Research 57:575–583.

Crawford, S., G. Whelan, D. M. Infante, K. Blackhart, W. M. Daniel, P. L. Fuller, T. Birdsong, D. J Wieferich, R. McClees-Funinan, S. M. Stedman, K. Herreman, and P. Ruhl. 2016. Through a fish's eye: the status of fish habitats in the United States 2015. National Fish Habitat Partnership. Available: http://assessment.fishhabitat.org/. (October 2018).

Curtis, S. G., J. S. Perkin, P. T. Bean, M. L. Sullivan and T. H. Bonner. 2015. Guadalupe Bass (*Micropterus treculli*) (Vaillant and Bocourt, 1874). Pages 55–66 *in* M. D. Tringali, J. M. Long, T. M. Birdsong, and M. J. Allen, editors. Black bass diversity: multidisciplinary science for conservation. American Fisheries Society, Symposium 82, Bethesda, Maryland.

Edwards, R. J. 1978. The effect of hypolimnion reservoir releases on fish distribution and species diversity. Transactions of the American Fisheries Society 107:71–77.

Edwards, R. J. 1980. The ecology and geographic variation of the Guadalupe Bass, *Micropterus treculii*. Doctoral dissertation. University of Texas, Austin, Texas.

Falush, D., M. Stephens, and J. K. Pritchard. 2003. Inference of population structure using multilocus genotype data: linked loci and correlated allele frequencies. Genetics 164:1567–1587.

Fitzpatrick, S. W., J. C. Gerberich, L. M. Angeloni, L. L. Bailey, E. D. Broder, J. Torres-Dowdall, C. A. Handelsman, A. López-Sepulcre, D. N. Reznick, C. K. Ghalambor, and W. C. Funk. 2016. Gene flow from an adaptively divergent source causes rescue through genetic and demographic factors in two wild populations of Trinidadian guppies. Evolutionary Applications 9:879–891.

Fleming, B. P., G. P. Garrett, and N. G. Smith. 2015. Reducing hybridization and introgression in wild populations of Guadalupe Bass through supplemental stocking. Pages 537–548 *in* M. D. Tringali, J. M. Long, T. M. Birdsong, and M. J. Allen, editors. Black bass diversity: multidisciplinary science for conservation. American Fisheries Society, Symposium 82, Bethesda, Maryland.

Fleming, B. P., and N. G. Smith. 2019. Spatial ditribution and hybridization in Guadalupe Bass five years after remedial stocking. Pages 231–244 *in* M. Siepker and J. Quinn, editors. Managing centrarchid fisheries in rivers and streams. American Fisheries Society, Symposium 87, Bethesda, Maryland.

Garrett, G. P. 1991. Guidelines for the management of Guadalupe Bass. PWD-RP-N3200–367-11/91. Texas Parks and Wildlife Department, Austin, Texas.

Garrett, G. P., T. W. Birdsong, M. G. Bean, and R. McGillicuddy. 2015. Guadalupe Bass restoration initiative. Pages 379–385 *in* M. D. Tringali, J. M. Long, T. W. Birdsong, and M. J. Allen, editors. Black bass diversity: multidisciplinary science for conservation. American Fisheries Society, Symposium 82, Bethesda, Maryland.

Hänfling, B., and D. Weetman. 2006. Concordant genetic estimators of migration reveal anthropogenically enhanced source-sink population structure in the River Sculpin, *Cottus gobio*. Genetics 173:1487–1501.

Harig, A. L., K. D. Fausch, and M. K. Young. 2000. Factors influencing success of greenback Cutthroat Trout translocations. North American Journal of Fisheries Management 20:994–1004.

Hubbs, C., R. J. Edwards, and G. P. Garrett. 2008. An annotated checklist of the freshwater fishes of Texas, with keys to identification of species. Texas Academy of Science. Available: https://repositories.lib.utexas.edu/bitstream/handle/2152/6290/Hubbs_et_al_2008_checklist.pdf?sequence=2&isAllowed=y. (September 2018).

Hurst, H., G. Bass, and C. Hubbs. 1975. The biology of the Guadalupe, Suwanee, and Redeye basses. Pages 47–53 *in* R. Stroud and H. Clepper, editors. Black bass biology and management. Sport Fishing Institute, Washington, D.C.

Junker, J., A. Peter, C. E. Wagner, S. Mwaiko, B. Germann, O. Seehausen, and I. Keller. 2012. River fragmentation increases localized population genetic structure and enhances asymmetry of dispersal in Bullhead (*Cottus gobio*). Conservation Genetics 13:545–556.

Kalinowski, S. T., M. L. Taper, and T. C. Marshall. 2007. Revising how the computer program CERVUS accommodates genotyping error increases success in paternity assignment. Molecular Ecology 16:1099–1106.

Koizumi, I., Y. Kanazawa, C. Yamazaki, Y. Tanaka, and K. Takaya. 2017. Extreme winter aggregation of invasive Rainbow Trout in small tributaries: implications for effective control. Ichthyological Research 64:197–203.

Littrell, B. M., D. J. Lutz-Carrillo, T.H. Bonner, and L. T. Fries. 2007. Status of an introgressed Guadalupe Bass population in a central Texas stream. North American Journal of Fisheries Management 27:785–791.

Lutz-Carrillo, D. J., C. Hagan, L. A. Dueck, and T. C. Glenn. 2008. Isolation and characterization of microsatellite loci for Florida Largemouth Bass, *Micropterus salmoides floridanus*, and other micropterids Molecular Ecology Resources 8:178–184.

Lutz-Carrillo, D. J., C. Thibodeaux, M. Elliott, N. A. Rathjen, C. Kittel, L. T. Fries, and G. P. Garrett. 2015. Inferred reproductive behavior of captive Guadalupe Bass. Pages 549–583 *in* M. D. Tringali, J. M. Long, T. M. Birdsong, and M. J. Allen, editors. Black bass diversity: multidisciplinary science for conservation. American Fisheries Society, Symposium 82, Bethesda, Maryland.

Marshall, T. C., J. Slate, L. E. B. Kruuk, and J. M. Pemberton. 1998. Statistical confidence for likelihood-based paternity inference in natural populations. Molecular Ecology 7:639–655.

Meronek, T., P. Bouchard, E. Buckner, T. Burri, K. Demmerly, D. Hatleli, R. Klumb, S. Schmidt, and D. Coble. 1996. A review of fish control projects. North American Journal of Fisheries Management 16:63–74.

Morizot, D. C., S. W. Calhoun, L. L. Clepper, M. E. Schmidt, J. H. Williamson, and G. J. Carmichael. 1991. Multi-species hybridization among native and introduced centrarchid basses in central Texas. Transactions of the American Fisheries Society 120:283–289.

Müller, K. 1954. Investigations on the organic drift in North Swedish streams. Reports of the Institute of Freshwater Research Drottningholm 34:133–148.

National Drought Mitigation Center, United States Department of Agriculture, and National Oceanic and Atmospheric Administration. 2016. United States Drought Monitor. National Drought Mitigation Center, Lincoln, Nebraska. Available: http://droughtmonitor.unl.edu/. (September 2018).

Novinger, D. C., and F. J. Rahel. 2003. Isolation management with artificial barriers as a conservation strategy for Cutthroat Trout in headwater streams. Conservation Biology 17:772–781.

Perkin, J. S., Z. R. Shattuck, P. T. Bean, and T. H. Bonner. 2010. Movement and microhabitat associations of Guadalupe Bass in two Texas rivers. North American Journal of Fisheries Management 30:33–46.

Pritchard, J. K., M. Stephens, and P. Donnelly. 2000. Inference of population structure using multilocus genotype data. Genetics 155:945–959.

Raeymaekers, J. A. M., D. Raeymaekers, I. Koizumi, S. Geldof, F.A.M. Volckaert. 2009. Guidelines for restoring connectivity around water mills: a population genetic approach to the management of riverine fish. Journal of Applied Ecology 46:562–571.

Rinne, N. and P. R. Turner. 1991. Reclamation and alteration as management techniques, and a review of methodology in stream renovation. Pages 219–246 *in* W. L. Minckley and J. E. Deacon, editors. Battle against extinction: native fish management in the American west. University of Arizona Press, Tucson, Arizona.

Seyoum, S., B. L. Barthel, M. D. Tringali, M. C. Davis, S. L. Schmitt, P. S. Bellotti, and W. F. Porak. 2013. Isolation and characterization of eighteen microsatellite loci for the Largemouth Bass, *Micropterus salmoides*, and cross amplification in congeneric species. Conservation Genetics Resources 5:697–701.

Shepard, B. B., L. M. Nelson, M. L. Taper, and A. V. Zale. 2014. Factors influencing successful eradication of nonnative Brook Trout from four small Rock Mountain streams using electrofishing. North American Journal of Fisheries Management 34:988–997.

Slade, R. M., Jr., J. T. Bentley, and D. Michaud. 2002. Results of streamflow gain-loss studies in Texas, with emphasis on gains from and losses to major and minor aquifers. United States Geological Survey open-file report 02–068. Available: https://pubs.usgs.gov/of/2002/ofr02-068/. (September 2018).

Texas Game and Fish Commission. 1957. Basic survey and inventory of fish species present, as well their distribution in the Blanco River, its tributaries and watershed lying within Blanco, Kendall and Hays counties, Texas. Project No. F9R3, Job B-10.

Texas Parks and Wildlife Department. 2012. Texas Conservation Action Plan 2012–2016: Overview. Wendy Connally, editor. Texas Parks and Wildlife Department, Austin, Texas. Available: http://tpwd.texas.gov/landwater/land/tcap/handbooks.phtml. (September 2018).

Texas State Historical Association. 2016. Description of the Narrows (Hays County). Available: https://tshaonline.org/handbook/online/articles/rnn18. (September 2018).

The Nature Conservancy. 2016. Blanco River project. Available: http://www.nature.org/ourinitiatives/regions/northamerica/unitedstates/texas/placesweprotect/blanco-river-project.xml. (September 2018).

Thomas, Z. A., T. L. Arsuffi, and S. J. Magnelia. 2015. Fishing warmwater streams with limited public access: angling behavior, economic impact, and the role of Guadalupe Bass in a twenty-four county region of Texas. Pages 123–137 *in* M. D. Tringali, J. M. Long, T. M. Birdsong, and M. J. Allen, editors. Black bass diversity: multidisciplinary science for conservation. American Fisheries Society, Symposium 82, Bethesda, Maryland.

United States Geological Survey. 2016. U.S. Geological Survey response to flooding in Texas, May–June 2015. Available: https://pubs.usgs.gov/fs/2016/3027/fs20163027.pdf. (September 2018).

Waters, T. F. 1972. The drift of stream insects. Annual Reviews of Entomology 17:253–272.

Wierman D. A., Broun A. S., and B. B. Hunt. 2010. Hydrogeologic atlas of the hill country Trinity Aquifer, Blanco, Hays, and Travis counties, central Texas: prepared by the Hays-Trinity, Barton Springs/Edwards Aquifer, and Blanco Pedernales Groundwater Conservation Districts. Available: https://repositories.lib.utexas.edu/handle/2152/8977. (September 2018).

Winters, K. E. 2013. A historical perspective on precipitation, drought severity, and streamflow in Texas during 1951–56 and 2011: U.S. Geological Survey scientific investigations report 2013–5113. Available: http://pubs.usgs.gov/sir/2013/5113/. (September 2018).

American Fisheries Society Symposium 87:231–244, 2019

Spatial Distribution and Hybridization Levels of Guadalupe Bass Five Years after Remedial Stocking

B. PAUL FLEMING* AND NATHAN G. SMITH
Texas Parks and Wildlife Department
Heart of the Hills Fisheries Science Center
5103 Junction Highway
Mountain Home, Texas 78058, USA

Abstract.—We reexamined a local population of Guadalupe Bass *Micropterus treculii* after cessation of a stocking program that significantly reduced the number of individuals that were hybridized with Smallmouth Bass *M. dolomieu*. Analysis of five years of post-stocking fish collection data show that the percentage of hybrids has increased in all four study sections of the upper Guadalupe River since stocking stopped. We found that hybridization had increased from 0.0% to 9.9% in Johnson Creek, 19.8% to 28.8% in North Fork Guadalupe River, 5.6% to 12.0% in South Fork Guadalupe River, and 24.2% to 26.5% in Guadalupe River main stem sections. This represents an average increase of 6.8% for all streams combined (range = 2.2–9.9%) from the last year of stocking. We also examined hybridization levels in individual stream fragments created by instream barriers (n = 76) to assess spatial variability along tributary and main stem corridors. Although hybridization appears to be increasing on a stream-wide basis, finer scale analysis show that hybrid distribution was spatially complex and there was a substantial portion of the study area with little or no hybridization. There were gradients of decreasing hybridization in an upstream direction in each of the tributaries and marked break points upstream of which hybridization was considerably lower (<10%). We also found that hybridization was highly variable among individual stream fragments (range 0%–71.4%) including those immediately adjacent to one another. These findings suggest that instream barriers may restrict genetic mixing throughout the system and contribute to partial isolation of populations within stream fragments. As such, further targeted intervention efforts are likely necessary to further reduce hybridization in fragments that still contain high

*Corresponding author: Paul.Fleming@tpwd.texas.gov

proportions of hybrids. The patchy distribution that we observed may be a critical component to consider in assessment and restoration efforts; not only for Guadalupe Bass but also in other endemic black basses threatened by hybridization.

Introduction

Genetic contamination from hybridization with introduced Smallmouth Bass *Micropterus dolomieu* has had detrimental effects on endemic Guadalupe Bass *M. treculii* populations for nearly 40 years (Edwards 1979). Stocking of Smallmouth Bass began in the 1950s (Lowman 1958), and by the 1980s it was apparent that hybridization between the species posed a serious threat to Guadalupe Bass persistence (Whitmore 1983). In response, the Texas Parks and Wildlife Department (TPWD) conducted a stocking program in three tributaries and a section of main stem river in the headwaters of the Guadalupe River, Texas. That program dramatically reduced the proportion of hybridized individuals in all study segments (Fleming et al. 2015) with no evidence of depressed genetic variation (Bean et al. 2013). Over a 19 year period, the stocking of roughly 600,000 pure Guadalupe Bass fingerlings nearly eliminated hybrids from Johnson Creek. Additionally, hybridization was reduced by about 45% in the North Fork Guadalupe River and Guadalupe River main stem over a shorter time period with fewer stocked fingerlings (109,000 and 87,000 respectively; Fleming et al. 2015).

Despite the apparent short-term success of the stocking program, predicting if hybridization levels would remain suppressed after stocking ended was difficult and uncertain. Temporal trends in hybridization in other contaminated Guadalupe Bass populations where no stocking has occurred have been antithetical (Garrett 1991; Bean et al. 2013). Hybridization in the Concho River increased to the point that pure Guadalupe Bass were extirpated while hybridization in the San Gabriel River has declined from 46% (Garrett 1991) to undetectable levels (Bean et al. 2013). Oddly, other populations have been more consistent, demonstrating little change over roughly the same time period (Garrett 1991; Bean et al. 2013). Failed stocking efforts in other Guadalupe Bass populations further complicated predictability of post-stocking genetic stability in the Guadalupe River. Stocking in the Blanco River failed to stop hybrid progression despite stocking fingerlings in numbers comparable to those stocked in North Fork Guadalupe River and Guadalupe River main stem (Littrell et al. 2007; Bean et al. 2013). Conflicting temporal trends in non-stocked populations and the failed stocking attempt in the Blanco River necessitated further monitoring in the upper Guadalupe River to assess changes following the rehabilitation effort.

Fleming et al. (2015) assessed hybridization levels in the upper Guadalupe River by pooling sample data and calculating hybridization at large spatial scales (e.g. individual streams)—a practice common in Guadalupe Bass genetic surveys (Garrett 1991; Bean 2012; Bean et al. 2013). However, available data suggests that course-scale assessment may not be adequate to fully characterize hybridization in this system. Garrett (1991) demonstrated clear differences between upper and lower portions of Johnson Creek (0% and 32% respectively). These differences were also present in genetic assessments done concurrent with the stocking program (Fleming et al. 2015). Disparities in Guadalupe Bass hybridization levels between sampling locations are not unique to the Guadalupe River as other researchers have reported similar spatial variability between sampling locations in several other streams (e.g., Morizot et al. 1991; Bean 2012; Bean et al. 2013). This indicates that hybridization levels are not spatially homogenous and assessment at finer spatial scales may be necessary to fully understand hybrid dynamics across longitudinal stream gradients.

It's counterintuitive that hybridization varies so widely in presumably panmictic populations. However, there are numerous instream barriers in the upper Guadalupe River that may be hindering fish movement and subsequent genetic homogenization. Barriers are known to significantly restrict fish movement (Porto et al. 1999) and genetic dispersal (Wofford et al. 2005). As a result, instream barriers have been used to help protect against genetic contamination from encroaching non-natives. This practice is known as isolation management (Thompson and Rahel 1998, Novinger and Rahel 2003, Fausch et al. 2009) and has been particularly useful in small headwater streams (Muhlfeld et al. 2012). Although existing barriers in the Guadalupe River were not intended for fish conservation, we suspect they are inhibiting the spread of non-native Smallmouth Bass genes. Obstacles restricting genetic mixing may explain the heterogeneity observed by Garrett (1991) and Fleming et al. (2015) as well as those reported by other researchers. If so, instream barriers may be critical in shaping genetic structure and thus important to consider in conservation and restoration efforts.

This study had two main objectives. First, to reexamine the upper Guadalupe River at spatial scales comparable to Fleming et al. (2015) to allow for direct comparison between current hybridization levels and those reported at the end of the stocking program. Second, to quantify hybridization within the individual stream fragments to evaluate genetic heterogeneity within the study area.

Methods

The study area consisted of the same portion of the upper Guadalupe River as Fleming et al. (2015; Figure 1). To facilitate ease of comparison, we retained the naming convention of that study; i.e., Johnson Creek, North Fork Guadalupe River (North Fork), South Fork Guadalupe River (South Fork), and Guadalupe River main stem (Main Stem). A complete description of the study area can be found in Fleming et al. (2015). Sampling was constrained to areas downstream of Heart of the Hills Fisheries Science Center on Johnson Creek (N 30.153427; W 99.35079), Wagon Wheel road crossing on the North Fork (N 30.06208; W 99.445870), and Lynx Haven road crossing on the South Fork (N 29.97932; W 99.44286). The lower boundary of the sampling area was marked at the Bear Creek road crossing on the Main Stem (N 30.06788; W 99.19325).

The fish stocking program (Fleming et al. 2015) was completed in the spring of 2009 in Johnson Creek and the spring of 2010 in the three remaining streams. Fleming et al. (2015) continued monitoring for two additional years post-stocking (Fall 2009 and 2010) in Johnson Creek and one additional year (2010) in North Fork, South Fork, and Main Stem. Those post stocking data were reported in the initial analysis of Fleming et al. (2015). We included those data here in order to incorporate all available post-stocking data and add robustness to our analyses. In addition to these data, we collected three annual samples over a four year period; 2011, 2012, and 2014. As in the previous study, fish were collected in the fall and winter (August–January) of each year. Sampling was conducted over the entirety of each stream in an attempt to provide complete spatial coverage. We collected micropterids (except Largemouth Bass *M. salmoides*) from all wadeable areas with a barge-style electrofishing unit. The geographic coordinates of collection locations

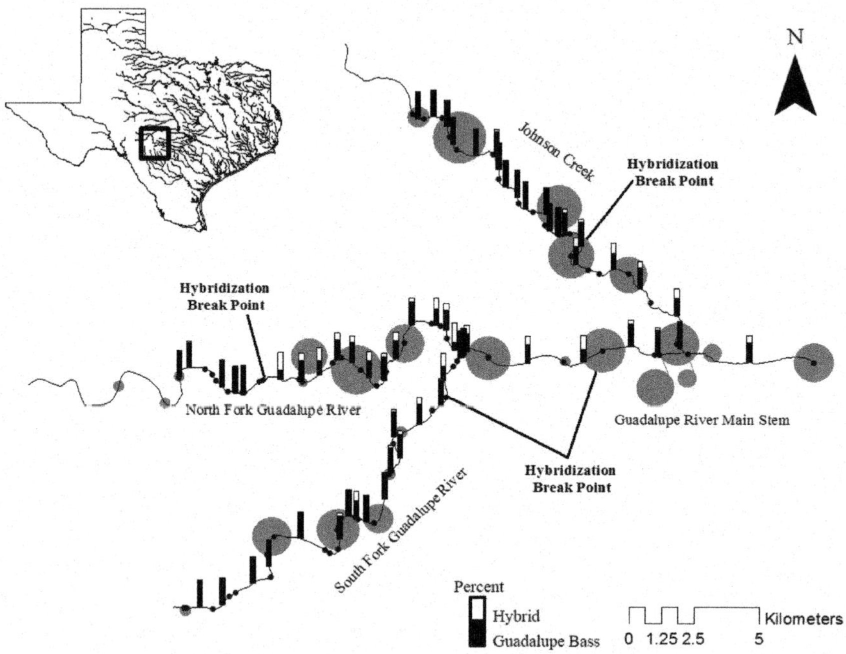

Figure 1. Proportion of pure and hybridized Guadalupe Bass (stacked bar graph), historic stocking locations of Guadalupe Bass fingerlings (gray circles = size proportional to number of individuals stocked), and locations of instream barriers (small black dots) in the upper Guadalupe River, Texas.

were recorded for each individual fish and a fin clip was retained for genetic analysis.

Genetic tests were performed using procedures developed by the Inland Fisheries A. E. Wood Laboratory (TPWD) as described in Littrell et al. (2007). Multilocus microsatellite genotypes were developed based on 14 loci (*lma10*, *lma12*, *lma120*, *lma121*, *mdo1*, *mdo3*, *mdo5*, *mdo6*, *mdo7*, *mdo10*, *msal13*, *msal21*, *msal25*, and *lar7*). The Bayesian inference algorithm implemented in STRUCTURE (version 2.3.4; Pritchard et al. 2000; Falush et al. 2003) was used to probabilistically assign individuals to clusters of reference genotypes representing Guadalupe Bass and Smallmouth Bass. Cluster assignment probabilities represent estimates of admixture proportions and were used to identify hybrids and infer the percentage of alleles derived from each lineage within individuals and groups. Each genotype was also manually inspected to identify any non-native alleles.

We first examined hybridization at the individual stream level (i.e., Johnson Creek, North Fork, South Fork, and Main Stem. In Fleming et al. (2015) Johnson Creek was analyzed in two separate sections (upper and lower) due to previously observed differences in hybridization levels (Garrett 1991). However, at the end of the stocking program, those differences no longer existed (i.e., hybridization was undetectable in both segments) eliminating the need for separate treatment here. We calculated the percentage of hybridized individuals for each stream in two ways. First, samples from all years were combined to calculate an overall mean hybridization for each stream. Second we calculated hybridization for each sampling year.

A geographic information system was used to quantify hybridization within individual stream fragments separated by instream barriers (e.g., road crossings, low-head dams, and waterfalls) throughout the study area. Fish collection data from 2009 (from Fleming et al. 2015) were excluded from this analysis because georeferenced fish collection locations were not available. Potential instream barriers were identified and georeferenced from both in-field observation and satellite imagery. Coordinates of instream barriers were loaded into ArcGIS (version 10.0; Esri, Redlands, California). The 2010 data (from Fleming et al. 2015) were pooled with the new samples collected in 2011, 2012, and 2014 and added into ArcGIS. Individual fish were assigned to the stream fragment from which they were collected and pooled across years to calculate the overall hybridization percentage for each fragment.

Distances between barriers were determined with ArcGIS and subsequently used in linear regression analyses. The flow line from the National Hydrography Data Set for each stream was loaded into ArcGIS and overlaid with geographic coordinates for each barrier. A dynamic segmentation process (linear referencing) was used to calculate linear distance along the flowline from the extreme lower end of each stream to each individual barrier. Distances for barriers in the North Fork, South Fork, and Johnson Creek were calculated in an upstream direction from their confluence with the Main Stem. Distances for the Main Stem segment were calculated from the most downstream barrier in the study area upstream to the confluence of North and South Fork segments.

An iterative linear regression approach was used to look for break points in hybridization in each tributary based on hybridization rates between barriers (weighted for sample size) and the distance each barrier was located from the downstream end (e.g. confluence). We first fit a simple linear regression of hybridization by river kilometer. The second option was fitting a typical piecewise regression (Neter et al. 1996) which allowed for two slopes to be identified along with a change point where the slope changes. Finally, we used Proc GLM (SAS Institute 2010) to fit a model with predefined break points at each barrier, letting the slope and intercept of each line be independent of each other. This process fit 20 separate models for Johnson Creek, 5 for the Main Stem, 15 for the North Fork, and 17 for the South Fork. The best fitting model was chosen among the candidate models based on Akaike Information Criterion (AIC; Akaike 1974) and R^2 values. Once inflection points were identified, we calculated the percentage of hybrids above and below those points.

Results

Our analyses of hybridization at the individual stream level included 1,085 fish: 711 fish from the new samples (2011–2014) plus 374 fish from Fleming et al. (2015). In total, 205 hybrids and one pure Smallmouth Bass were collected (19.0%). The total number of fish from each individual stream was 352 in Johnson Creek, 320 in North Fork, 209 in South Fork, and 204 in Main Stem. Mean hybridization rates over all post-stocking years were 9.9% in Johnson Creek, 28.8% in North Fork, 12.0% in South Fork, and 26.5% in Main Stem. Annual sample sizes ranged from 19 (South Fork, 2011) to 111 (North Fork, 2010) fish with a mean of 63.8. In Johnson Creek, no hybrids were found in the final year of stocking (2009; Fleming et al. 2015), but hybridization percentage has rebounded to as high as 17.7% in 2014 (Figure 2a). Similarly, hybridization rate was 5.6% in the South Fork during the final year of stocking in 2010 but increased to 21.0% in 2012 (Figure 2c). The Main Stem and North Fork both historically had overall higher hybridization rates than the other study segments. In the North Fork, hybridization went from 19.8% in the last year of stocking (2010) to

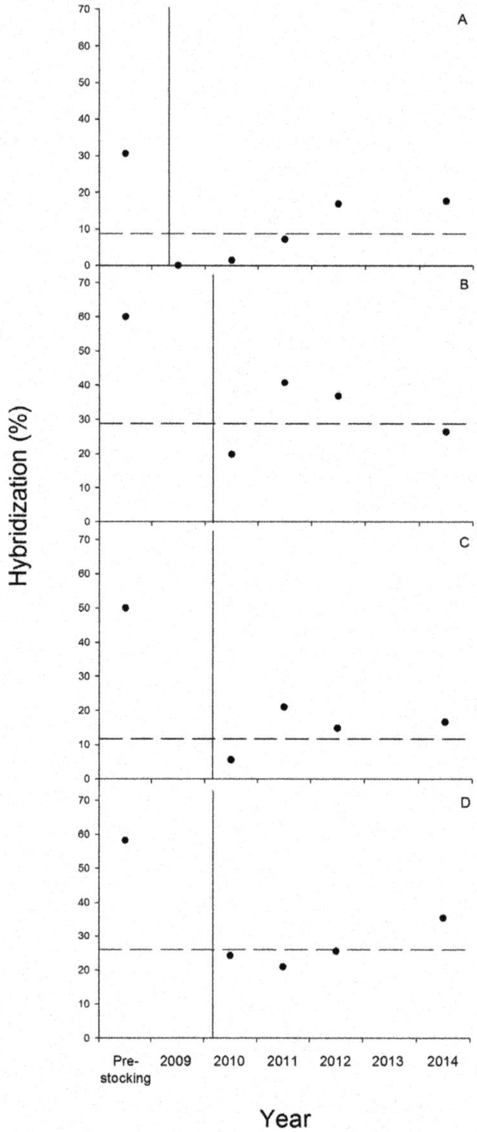

Figure 2. The percentage of hybridized Guadalupe Bass collected in four study streams; a) Johnson Creek, b) North Fork Guadalupe River, c) South Fork Guadalupe River, and d) Guadalupe River main stem. Vertical line represents the termination of remedial stocking of pure Guadalupe Bass fingerlings and horizontal dashed line is mean hybridization among all post-stocking years.

as high as 40.7% in 2011 (Figure 2b). Finally, the Main Stem has increased from 24.2% in the last year of stocking (2010) to 35.4% in 2014 (Figure 2d).

We identified 76 instream barriers within the area where sampling occurred. However, fish were only collected from 61 (Table 1) individual stream fragments (Figure 1) with sample sizes ranging from 1 to 90 fish (mean = 15.2 fish per segment). Among all streams, hybridization rates within individual stream fragments varied considerably (range = 0–71.4%). There were multiple fragments where no hybrids were collected in Johnson Creek (10), North Fork (4), and South Fork (9) but all fragments in the Main Stem contained hybrids (range = 6.3–53.3% hybridization).

Linear regression analyses revealed higher concentration of hybrids in the lower portions of Johnson Creek, North Fork, South Fork, and the entire Main Stem than were present in the headwater portions of the study area (Figure 3). For all four streams, a model that included a fixed break point at a barrier provided the best fit to the data. A breakpoint in Johnson Creek was found at a barrier located 7.5 km upstream from the confluence with the Main Stem. The average hybridization rate above this point was 3.2% versus 29.7% below. Similarly, a break point in hybridization was located at a barrier 2.4 km upstream in the South Fork. Hybridization was 9.4% in the upper area and 35.3% below the barrier. A break point on the North Fork was located at 11.5 km upstream. Only 1 hybrid was found in the upper area (3.44%) while hybridization in the lower stretch was 31.8%. Finally, the best fitting model for the Main Stem included a break point at a barrier near the middle of the sampling reach with mean hybridization of 21.6% below, and 35.5% above.

The barriers located at the breakpoints are all various forms of road crossings and are of typical size and design as others in the sample area (Figure 4). The barrier at Johnson Creek is a public road crossing that was rebuilt during the course of this evaluation (2011). Prior to 2011, the crossing consisted of a solid concrete dam with multiple perched culverts. The current construction consists of bridge that spans the entirety of the stream with natural streambed through the opening. The remaining three barriers are private road crossings. North Fork and South Fork are marked by concrete flow-over road crossings (perched approximately 1.25 and 0.5 m above the water) while the Main Stem has a partially submerged private road crossing with an elevation at the center of the stream of 0.25 m above the water surface.

Discussion

We found that hybrid dynamics in this system are considerably more spatially complicated than previously thought. The distribution of hybrids is not uniform but rather, marked by longitudinal gradients of decreasing hybridization moving upstream in each tributary with patchy inclusions of fragments with both low and high hybridization. Therefore, stream-wide comparison between immediate post-stocking and current hybridization rates are inadequate and only relay a portion of the existing scenario. Drawing inference about current and future hybridization trends for this system necessitates consideration at finer spatial scales. Nevertheless, when calculated on a stream-wide basis, hybridization rates in all four streams of the Guadalupe River are still considerably lower than they were before the stocking program began. Johnson Creek, North Fork, South Fork, and Main Stem streams contain about 32%, 29%, 24%, and 45% respectively fewer hybrids now than they did in 1991 (Garrett 1991).

Our regression analysis demonstrated weak longitudinal gradients along all four stream corridors with marked break points of

Table 1. Marked drops in the percentage of hybridized Guadalupe Bass (Hybridization Break Point) in four stream sections of the upper Guadalupe River identified through regression analysis. River kilometer indicates location of instream barriers where drops occurred. Also included are the number of stream fragments created by instream barriers as well as sample sizes and percentage of hybridized Guadalupe Bass for each fragment.

Stream	Hybridization Break Point		Fish (N)	Percent Hybrids	Stream Fragments		
					N	Sample Size Mean (Range)	Percent Hybrids (Range)
Johnson Creek							
	7.5 Rkm	Downstream	64	29.7	5	12.8 (7–20)	0–38.5
		Upstream	222	3.2	15	14.8 (1–35)	0–14.3
North Fork							
	11.5 Rkm	Downstream	286	31.8	12	23.8 (5–60)	13.5–71.4
		Upstream	29	3.4	5	5.8 (2–17)	0–5.9
South Fork							
	2.4 Rkm	Downstream	17	35.3	4	4.3 (1–8)	0–50.0
		Upstream	191	9.4	15	12.7 (1–42)	0–33.3
Main Stem							
	8.75 Rkm	Downstream	125	21.6	3	41.7 (16–90)	6.3–25.6
		Upstream	76	35.5	2	38.0 (15–61)	31.1–53.3

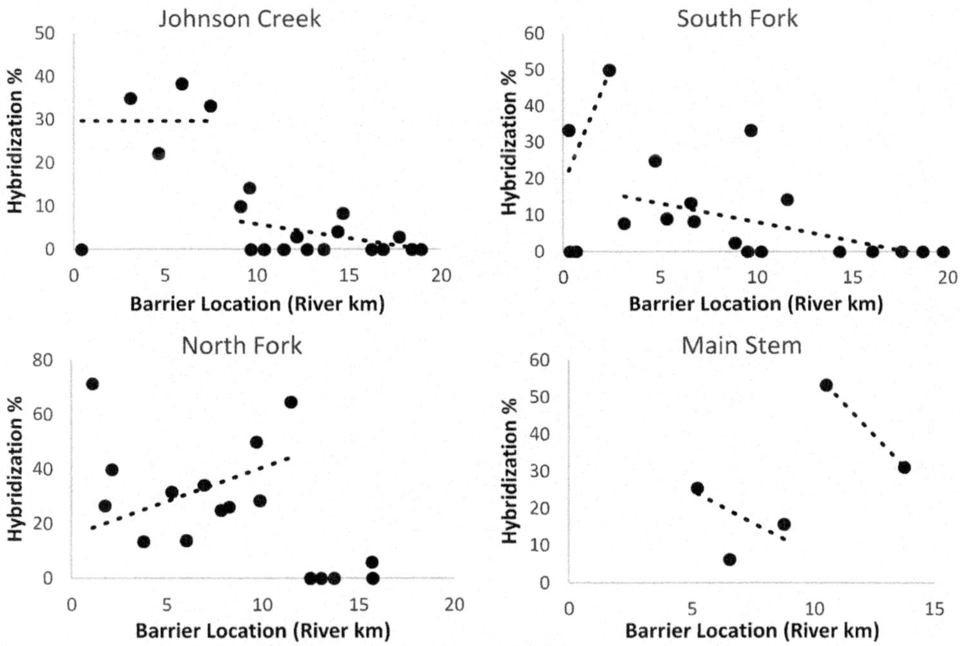

Figure 3. Proportion of hybrid Guadalupe Bass in stream fragments in each of four streams (Johnson Creek; South Fork; North Fork; Main Stem) in the upper Guadalupe River. The *x*-axis depicts barrier locations as the distance upstream from the confluence in the three tributary streams and the distance upstream from the lower extent of the study area in the Main Stem. Fitted lines depict the final regression model with fixed break points representing the best fitting model for each stream segment.

reduced hybridization. Generally, our spatial data indicated that the bulk of hybrids were concentrated in the central part of the system (i.e., in and immediately adjacent to the Main Stem) and gradually decrease upstream in all three tributaries. Johnson Creek had a substantial proportion of hybrids (29.7%) downstream of the barrier located at 7.5 km from its confluence with the Main Stem. Upstream of that break point, hybridization declines rapidly and almost completely disappears above the barrier located at 9.6 km. A similar pattern is apparent in the South Fork where we failed to collect a single hybrid above 12 km. In the North Fork, hybrids were prevalent throughout the lower portion (0–11.5 river km) then, the proportion of hybrids abruptly dropped. We only collected 1 hybrid in 29 fish in the upper reach. In the areas upstream from break point barriers, all three tributaries had low numbers of hybrids (<10%) in relation to areas downstream (29%–35%).

Although the bulk of the hybrids are concentrated in the Main Stem and lower sections of the tributaries, there was also high variability in hybridization among individual stream fragments (0–71.4%). In many cases, adjacent fragments demonstrated wide differences. For example, one fragment in the Main Stem had just over 50% while adjacent fragments up and downstream had only 16% and 31% respectively. Similarly, a fragment in the North Fork with 27% hybrids was flanked by segments with 40% and 71%. Several such examples exist in the remaining two streams as well. As previously noted, this is not unique to the Guadalupe River system. Although more spatially separated, Bean

Figure 4. Instream barriers where marked drops in the percentage of hybridized Guadalupe Bass occurred in (a) Johnson Creek, (b) North Fork Guadalupe River, (c) South Fork Guadalupe River, and (d) main stem Guadalupe River, Texas.

(2012) showed similar inconsistencies in the Llano River. There, hybridization among sampling sites ranged from 0% to over 33% among five individual sample sites. The inclusion of fragments with markedly different hybridization rates suggests interbreeding between fragments may be restricted. Otherwise, we would expect hybridization rates to be much more similar, especially between adjacent fragments.

We believe instream barriers are a primary driving factor influencing the patchy distribution of hybrids in the upper Guadalupe River. On average, a barrier exists every 1.1 km in the study area. They are primarily comprised of road crossings (several construction types including culverts and perched boxes), and low-head dams, but natural waterfalls and two major concrete dams are also present. There is extensive literature demonstrating that these types of barriers restrict fish movement but the extent is largely dependent on construction type (Belford and Gould 1989; Warren and Pardew 1998) and individual fishes swimming ability (Castro-Santos 2006). There is very little published information on how barriers affect movement of Smallmouth Bass and none on Guadalupe Bass or their hybrids. Movement, in general, in Smallmouth Bass is well studied and shown to be extremely variable depending on the type of system where they reside (e.g. small stream, large river, or reservoirs; Ridgway et al. 2002). Although they are often described as sedentary (Funk 1957; Hafs et al. 2010) they do travel long distances through river corridors (up to 61 km; Montgomery et al. 1980). By comparison, at least in small streams, Guadalupe Bass move much shorter distances (i.e., approximately 1.5 km; Perkin et al. 2010). Barrier traversal in Guadalupe Bass and propensity of movement in their hybrids are critical knowledge gaps that are required to better understand dispersal dynamics and will greatly aid in predicting the eventual long-term outcome of restoration efforts, not only for the Guadalupe River, but also in other fragmented systems.

It is apparent that hybridization was decreased by Guadalupe Bass stockings, but establishing a firm relationship between current hybrid distribution and stocking location is not possible with existing data. While Fleming et al. (2015) made attempts to distribute Guadalupe Bass fingerlings throughout the study area, the number of stocked fish was not spatially (Figure 2) or temporally uniform. Further, stocking typically occurred above and below easily accessible road crossings that also mark the separation of individual stream fragments. Because of these two factors, we were unable to formally analyze current hybridization rates relative to location-specific stocking abundances. However, influence is discernable in a general sense. Some stream fragments that received high numbers of stocked fish, have far fewer hybrids and lower genetic admixture than proximal areas that were never stocked. This contrast is most apparent in the lower portion of the North Fork where higher numbers of hybrids (roughly 65%) occur in segments that never received stocking compared to fewer hybrids (25%) in segments that were stocked.

One concern with this study was whether sample sizes were adequate to reliably estimate true hybridization levels. Hybrid detection is highly dependent on sample size (Della Croce et al. 2017). Genetic surveys of Guadalupe Bass are frequently marked with low samples sizes from individual locations (i.e., usually <30; Whitmore and Butler 1982; Garrett 1991; Morizot et al. 1991; Fleming et al. 2015). The major difficulty is that typical Guadalupe Bass streams are small, especially head-water reaches, supporting proportionally few fish; rendering obtaining large sample sizes difficult if not impossible. There should be little doubt though, that the low proportion of hybrids that we observed in the upper reaches of the tributaries above the break points was real. Cumulatively, the fewest fish were collected in the North Fork (29) but sample sizes from Johnson Creek and South Fork were quite large (222 and 191 respectively) and still yielded very few hybrids. Less certainty does exist with hybrid proportions in individual stream fragments. In many cases stream fragment sample sizes consisted of less than 30 fish. Nonetheless, many fragments in close proximity to one another had moderate to large sample sizes (e.g. 37 and 60) and clearly showed inconsistencies. Hybrid detection requires larger sample sizes when populations are newly invaded (nonnatives and F-1 present), but as admixture increases fewer fish are required (Della Croce et al. 2017). In our study area, Smallmouth Bass and F-1 hybrids are exceedingly rare indicating that our sample sizes were adequate (Della Croce et al. 2017) and the patchiness we observed was real and unlikely to be a sampling artifact.

Although there are patches within the study area that contain a high proportion of hybrids, the overall outlook for this system remains good. Wofford et al. (2005) demonstrated that in salmonid populations, barriers create partially independent populations, restrict gene flow, and alter genetic structure. Clearly, these fragments are only partially isolated but instream barriers are serving to dramatically reduce genetic dispersal. The fact that many stream fragments within this system exhibit little or no hybridization five years post-stocking illustrates that dispersal of fish and concurrent foreign genes are at least partially contained. But, because fish are not completely contained, further remedial stocking may be necessary to ensure the advances from the initial stocking program are not lost over time. With our new understanding of spatial dynamics we can now prioritize individual stream fragments to simultaneously reduce local hybrid numbers and lessen risk to adjacent fragments from leaking genetics. A major advantage for future efforts is that most of the highly contaminated fragments are small. This will facilitate more focused effort while reducing manpower needs and requiring the production of far fewer fingerlings. Another positive sign was the discovery of low hybridization rates across a larger portion of the study area than was realized in Fleming et al. (2015). Low levels of hybridization in the upper portions of North and South Forks significantly increases the amount of area where high proportions of pure Guadalupe Bass persist.

Hybridization with introduced non-natives is a problem that frustrates conservation of native black basses across the southern USA and has caused widespread threats to the natural diversity of *Micropterus* (Koppelman 2015). As many as six endemic species with small geographic ranges have been impacted by hybridization with non-native congeners (Koppelman 2015) and, to date, remedies are extremely limited. Like salmonid species, the restriction on homogenous genetic mixing caused by instream barriers may prove to be a useful tool in conservation and management of black bass populations threatened with hybridization. Given the ubiquity of instream barriers, it is likely that isolated populations exist in other river systems as well, and should be prioritized for preservation and restoration. Further, our work demonstrates the importance of sampling on an extensive spatial scale because variability in hybridization can fluctuate widely over relatively short distances. Pooling fish on larger spatial scales provides a reasonable overall assessment and has been used historically in several projects, but should be considered a coarse tool. When concerted restoration efforts are underway we recommend identifying potential barriers to gene flow and sampling accordingly to avoid misleading results.

Acknowledgments

We thank the technicians and staff at Heart of the Hills Fisheries Science Center for assistance with data collection over the course of the study, especially S. Pavlicek, D. Wilson, and B. Wienecke. Genetic analyses, testing, and guidance was provided by D. Lutz-Carrillo. G. Garrett was instrumental in pioneering efforts at genetic restoration in Johnson Creek and subsequent expansion into the North and South Forks and upper Guadalupe River main stem. Previous versions of this manuscript benefited by review from P. Bean, D. Buckmeier, K. Bodine, D. Daugherty, D. Lutz-Carrillo, and three anonymous reviewers. Funding was provided by Texas Parks & Wildlife Department and Federal Aid in Sport Fish Restoration grants F-231-R1 and F-231-R2.

References

Akaike, H. 1974. A new look at the statistical model identification. Institute of Electrical and Electronics Engineers Transactions on Automatic Control 19:716–723.

Bean, P. T. 2012. Introgressive status, population genetic structure, phylogeographic history, and individual-level resource specialization of the Guadalupe Bass *Micropterus treculii*. Doctoral Dissertation, Texas State University, San Marcos.

Bean, P. T., D. J. Lutz-Carrillo, and T. H. Bonner. 2013. Rangewide survey of the introgressive status of Guadalupe Bass: implications for conservation and management. Transactions of the American Fisheries Society 142:681–689.

Belford, D. A., and W. R. Gould. 1989. An evaluation of trout passage through six highway culverts in Montana. North American Journal of Fisheries Management 9:437–445.

Castro-Santos, T. 2006. Modeling the effect of varying swim speeds on fish passage through velocity barriers. Transactions of the American Fisheries Society 135:1230–1237.

Della Croce, P., G. C. Poole, R. A. Payn, and R. E. Gresswell. 2017. Early detection of nonnative alleles in fish populations: when sample size actually matters. Fisheries 42:44–56.

Edwards, R. J. 1979. A report of Guadalupe Bass (*Micropterus treculi*) X Smallmouth Bass (*M. Dolomieui*) hybrids from two localities in the Guadalupe River, Texas. Texas Journal of Science 31:231–238.

Falush, D., M. Stephens, and J. K. Pritchard. 2003. Inference of population structure using multilocus genotype data: linked loci and correlated allele frequencies. Genetics 164:1567–1587.

Fausch, K. D., B. E. Rieman, J. B. Dunham, M. K. Young, and D. P. Peterson. 2009. Invasion versus isolation: trade-offs in managing native salmonids with barriers to upstream movement. Conservation Biology 23:859–870.

Fleming, B. P., G. P. Garrett, and N. G. Smith. 2015. Reducing hybridization and introgression in wild populations of Guadalupe bass through supplemental stocking. Pages 527–547 *in* M. D. Tringali, J. M. Long, T. W. Birdsong, and M. S. Allen, editors. Black bass diversity: multidisciplinary science for conservation. American Fisheries Society, Symposium 82, Bethesda, Maryland.

Funk, J. L. 1957. Movement of stream fishes in Missouri. Transactions of the American Fisheries Society 85:39–57.

Garrett, G. P. 1991. Guidelines for management of Guadalupe Bass. Texas Parks and Wildlife Department, Special Publication N3200-367, Austin.

Hafs, A. W., C. J. Gagen, and J. K. Whalen. 2010. Smallmouth Bass summer habitat use, movement, and survival in response to low flow in the Illinois Bayou, Arkansas. North American Journal of Fisheries Management 30:604–612.

Koppelman, J. B. 2015. Black bass hybrids: a natural phenomenon in an unnatural world. Pages 467–479 *in* M. D. Tringali, J. M. Long, T. W. Birdsong, and M. S. Allen, editors. Black bass diversity: multidisciplinary science for conservation. American Fisheries Society, Symposium 82, Bethesda, Maryland.

Littrell, B. M., D. J. Lutz-Carrillo, T. H. Bonner, and L. T. Fries. 2007. Status of an introgressed Guadalupe Bass population in a central Texas stream. North American Journal of Fisheries Management 27:785–791.

Lowman, F. G. 1958. Introduction of Smallmouth black bass, *Micropterus dolomieu* in the upper regions of the Llano River lying within Sutton, Edwards, and Kimble counties, Texas. Texas Game and Fish Commission, Dingell-Johnson Project F-9-R-5, Job F-1, Austin.

Montgomery, J. C., D. H. Fickeisen, and C. D. Becker. 1980. Factors influencing Smallmouth Bass production in the Hanford area, Columbia River. Northwest Science 54:296–305.

Morizot, D. C., S. W. Calhoun, L. L. Clepper, and M. E. Schmidt. 1991. Multispecies hybridization among native and introduced centrarchid basses in central Texas. Transactions of the American Fisheries Society 120:283–289.

Muhlfeld, C. C., V. D'Angelo, S. T. Kalinowski, E. L. Landguth, C. C. downs, J. Tohtz, and J. L. Kershner. 2012. A fine-scale assessment of using barriers to conserve native stream salmonids: a case study in Akokala Creek, Glacier National Park, USA. The Open Fish Science Journal 5:9–20.

Neter, J. A., M. H. Kutner, C. J. Nachtsheim, and W. Wasserman. 1996. Applied linear statistical models, 4th edition. McGraw-Hill, New York.

Novinger, D. C., and F. J. Rahel. 2003. Isolation management with artificial barriers as a conservation strategy for Cutthroat Trout in headwater streams. Conservation Biology 17:772–781.

Perkin, J. S., Z. R. Shattuck, P. T. Bean, T. H. Bonner, E. Saraeva, and T. B. Hardy. 2010. Movement and microhabitat associations of Guadalupe Bass in two Texas Rivers. North American Journal of Fisheries Management 30:33–46.

Porto, L. M., R. L. McLaughlin, and D. L. G. Noakes. 1999. Low-head barrier dams restrict the movements of fishes in two Lake Ontario streams. North American Journal of Fisheries Management 19:1028–1036.

Pritchard, J. K., M. Stephens, and P. Donnelly. 2000. Inference of population structure using multilocus genotype data. Genetics 155:945–959.

Ridgway, M. S., B. Shuter, T. A. Middel, and M. L. Gross. 2002. Spatial ecology and density-dependent process in Smallmouth Bass: the juvenile transition hypothesis. Pages 47–60 *in* D. P. Phillip and M. S. Ridgway, editors. Black bass: ecology, conservation, and management. American Fisheries Society, Symposium 31, Bethesda, Maryland.

SAS Institute. 2010. SAS/STAT 9.3 User's Guide. SAS Institute, Cary, North Carolina.

Thompson, P. D. and F. J. Rahel. 1998. Evaluation of artificial barriers in small Rocky Mountain streams for preventing the upstream movement of Brook Trout. North American Journal of Fisheries Management 18:206–210.

Warren M. L. and M. G. Pardew. 1998. Road crossings as barriers to small-stream fish movement. Transactions of the American Fisheries Society 127:637–644.

Whitmore, D. H. 1983. Introgressive hybridization of Smallmouth Bass (*Micropterus dolomieui*) and Guadalupe Bass (*M. treculi*). Copeia 1983:672–679.

Whitmore, D. H. and W. Butler. 1982. Interspecific hybridization of Smallmouth and Guadalupe Bass (*Micropterus*): evidence based on biochemical genetic and morphological analyses. The Southwestern Naturalist 27:99–106.

Wofford, J. E. B., R. E. Gresswell, and M. A. Banks. 2005. Influence of barriers to movement on within-watershed genetic variation of coastal Cutthroat Trout. Ecological Applications 15:628–637.

American Fisheries Society Symposium 87:245–270, 2019

Management of Choctaw Bass in the Choctawhatchee River Drainage Based on Novel Life History Information

Katie L. Woodside*
Florida Fish and Wildlife Conservation Commission
Division of Freshwater Fisheries Management
3911 Highway 2321, Panama City, Florida 32409, USA

Matthew G. Wegener and Neil J. Branson
Florida Fish and Wildlife Conservation Commission
Blackwater Fisheries Research and Development Center
8384 Fish Hatchery Road, Holt, Florida 32564, USA

Abstract.—The Choctaw Bass *Micropterus cf. punctulatus* is an undescribed black bass found in coastal river systems of northwest Florida and southern Alabama. The Choctaw Bass is morphologically similar to the Spotted Bass *M. punctulatus*. Prior to this study, there was no published life history information specific to this black bass. The need to address these information gaps was further influenced by the popularity of black bass fishing and potential statewide black bass regulation changes in Florida. Our research examined relative abundance, age, mortality, growth, and diet of Choctaw Bass collected from the Choctawhatchee River, Florida and Holmes Creek, a major tributary of the Choctawhatchee River in fall 2012 and 2013. Choctaw Bass were compared to sympatric Largemouth Bass *M. salmoides* × Florida Bass *M. floridanus* intergrades. Relative abundance was lower for Choctaw Bass than Largemouth Bass in both the Choctawhatchee River and Holmes Creek. The oldest Choctaw Bass from the Choctawhatchee River was age-5 and the oldest Choctaw Bass from Holmes Creek was age-11. Total instantaneous mortality of Choctaw Bass was statistically similar between the study areas. Total instantaneous mortality for Choctaw Bass from Holmes Creek was also similar to sympatric Largemouth Bass. Growth differed for Choctawhatchee River and Holmes Creek Choctaw Bass. Diets of Choctaw Bass collected from the Choctawhatchee River

*Corresponding author: katie.woodside@myfwc.com

consisted primarily of fish while diets of Choctaw Bass and Largemouth Bass collected from Holmes Creek consisted primarily of crayfish. The observed restricted distribution, limited abundance relative to Largemouth Bass, limited number of Choctaw Bass collected greater than 406 mm TL, and the lack of Choctaw Bass older than age-5 in the Choctawhatchee River led to the recommendation that Choctaw Bass be managed separately from Largemouth Bass.

Introduction

Choctaw Bass *Micropterus cf. punctulatus* is an undescribed black bass found in coastal river systems of northwest Florida and southern Alabama (Tringali et al. 2015b). The Choctaw Bass was once thought to be Northern Spotted Bass *M. punctulatus punctulatus* due to its similarity in appearance and presumed geographic range (Page and Burr 2011). Bailey and Hubbs (1949) examined 10 specimens from the East Gulf Coastal Plain in Florida and concluded that these populations represented Spotted Bass and Alabama Bass intergrades (*M. punctulatus* × *M. henshalli*). However, Tringali et al. (2015b) determined that the fish in drainages east of the Mobile River up to, and including, the Choctawhatchee River had autapomorphies at two nuclear gene introns and a mitochondrial DNA gene that distinguished them from all the recognized black bass species. Nuclear gene based phylogenies indicated that these fish formed a reciprocally monophyletic group that shared a common ancestor with a clade that included Spotted Bass and Guadalupe Bass (Tringali et al. 2015b). Tringali et al. (2015b) determined that Choctaw Bass can also be discriminated from the other black bass species based on differences in the allele frequency distributions present at microsatellite DNA loci. In summary, a suite of morphologic and genetic traits demonstrated that Choctaw Bass should be elevated to full species rank.

In Florida, Choctaw Bass share their range with Largemouth Bass intergrades (*M. salmoides* × *M. floridanus*). Choctaw Bass can be found in rivers and streams from the Choctawhatchee River drainage and west to the Perdido River (Tringali et al. 2015b; Figure 1). Current data indicates that the only Florida river possessing both Choctaw Bass and Spotted Bass is the Chipola River (Tringali et al. 2015a, 2015b). However, Choctaw Bass can be distinguished from Spotted Bass through mean/modal counts of soft anal-fin and dorsal-fin rays or through genetic samples. Choctaw Bass and Spotted Bass in Florida differ from Largemouth Bass by having a tooth patch on the tongue versus no tooth patch, parellel rows of spots below the lateral line versus no spots, a membranous connection between the spinous and soft dorsal fin versus separation between the fins, and small scales near the base of the soft dorsal fin versus no scales (Boschung and Mayden 2004).

Prior to this study, there was a paucity of life history information specific to Choctaw Bass, but anecdotal field observations suggest Choctaw Bass avoid headwater streams and occur in flowing stretches of medium depth and velocity, often in association with woody debris (Florida Fish and Wildlife Conservation Commission [FWCC], unpublished data). Relative to the sympatric Largemouth Bass, Choctaw Bass are generally smaller in size with the largest Choctaw Bass collected in Florida to date measuring 428 mm total length (TL). Due to the lack of life history information, Choctaw Bass were previously regulated as all other black bass in Florida with a five-fish aggregate bag limit and a 305

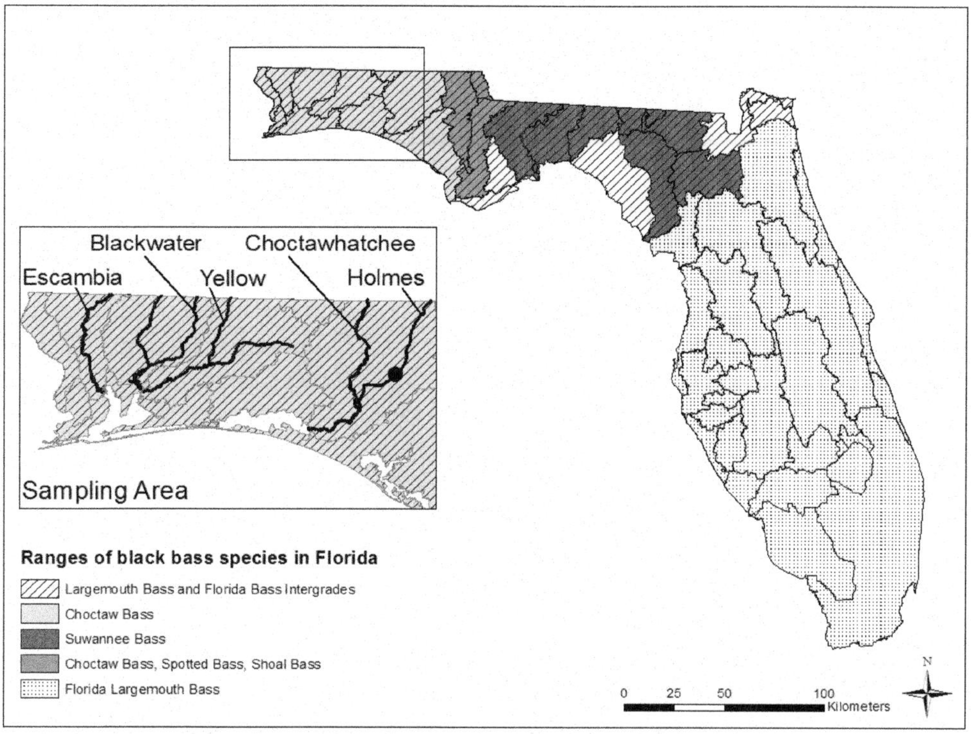

Figure 1. Ranges of black bass species in Florida including Choctaw Bass. Map shows location of the Choctawhatchee River and Holmes Creek. Sampling on the Choctawhatchee River occurred throughout the river from the Alabama border to the Choctawhatchee Bay. Sampling on Holmes Creek occurred downstream (south) of the large black dot.

mm (12 in) minimum length limit. Regulations for Florida's black bass changed July 1, 2016 to permit the take of up to five black bass per person per day (including Largemouth Bass (*M. salmoides* × *M. floridanus* intergrades and *M. floridanus*), Suwannee Bass *M. notius*, Spotted Bass, Choctaw Bass, and Shoal Bass *M. cataractae* individually or in total) and only one black bass may be 406 mm (16 in) or longer in total length. There is no minimum length limit for Largemouth Bass, but information collected in this study and others lead to a retention of a 305 mm minimum length limit for riverine black bass such as Suwannee Bass, Spotted Bass, Choctaw Bass and Shoal Bass.

Black bass are the most popular freshwater sportfish in the United States, and Florida is one of the states where fishing participation directed towards black bass is the highest (U.S. Fish and Wildlife Service 1999). Unique riverine black bass species have been gaining local and national attention. Interest in fishing for these species has grown with an increase in kayak fishing and promotion of the Bass Anglers Sportsman Society (B.A.S.S.) Slam, a B.A.S.S. program where anglers are encouraged to catch each of nine recognized species of black bass (Long et al. 2015). The FWCC offers its own Bass Slam challenge where anglers are encouraged to document catch of different Florida black bass species in one year. The expected result of bass slam programs and increased kayak fishing is increased fishing effort directed at riverine black bass species, including those with restricted distributions such as the Choctaw Bass. FWCC's Black Bass Management Plan (2011) promotes optimal sustainable use of Florida's bass fisheries while still providing diverse angling opportunities that result in high angler satisfaction. Therefore, understanding the population dynamics of Choctaw Bass and filling life history information gaps is necessary to develop appropriate management strategies. Our main and first study objective was to describe relative abundance, age, mortality, growth, and diet of Choctaw Bass from the Choctawhatchee River and Holmes Creek, a tributary of the Choctawhatchee River. Since the two study areas differ, an additional and second study objective was designed to determine whether or not differences exist between Choctaw Bass collected from the Choctawhatchee River compared to Holmes Creek. The final and third objective was to compare Choctaw Bass and sympatric Largemouth Bass; this comparison was primarily conducted in Holmes Creek.

Methods

Study Area

The Choctawhatchee River orginates in Alabama and is considered an unimpounded, alluvial stream. It has a total length of 280 km, 201 km of which lie within Florida (Figure 1). In terms of discharge, it is the third largest stream in Florida, and its watershed drains 12,033 km^2 of swamps, forests, and agricultural lands (Bass and Cox 1985). Considered one of the most turbid streams in Florida (Bass 1983), the Choctawhatchee River continually carries a heavy load of sediment to the Choctawhatchee Bay. There are at least 13 springs connected to the Floridan Aquifer in the Choctawhatchee River basin. During low to average flow conditions, groundwater (primarily discharged from springs) contributes a significant portion of the Choctawhatchee River's discharge (Barrios 2005). Upstream habitat is characterized by sand and gravel substrate and shallow maximum water depth (<3 m). As the river flows downstream through the lower segments, maximum water depth increases to 10 m, sediment type transitions to sand and silt, and off-channel habitats become numerous. Silt is the dominant sediment type and the water is brackish (0.5–9.3 ppt, FWCC unpublished data) as the lower Choctawhatchee River flows downstream into the Choctawhatchee Bay.

Holmes Creek is the largest tributary of the Choctawhatchee River in Florida and is considered a spring fed stream (Figure 1). It originates in Alabama with approximately 90 km of river located in Florida. In Florida, Holmes Creek has a surface water drainage basin of approximately 1,181 km^2 and there are at least 51 springs connected to the Floridan Aquifer in the Holmes Creek basin (Barrios and Chelette 2008). There is a greater estimated abundance of benthos (i.e., invertebrates, including crayfish) from shoreline substrates and from mid-river substrates in Holmes Creek compared to the Choctawhatchee River (Bass and Cox 1985). Holmes Creek also differs from the Choctawhatchee River in that turbidity (NTU) and water color is typically lower (the Choctawhatchee River is more tannin-stained) and water temperatures in winter are generally warmer (Livingston 1991). Holmes Creek is narrower than the Choctawhatchee River and is characterized by sandy substrate and logjams which often span the entire width of the creek. As it flows downstream and reaches its confluence with the Choctawhatchee River, Holmes Creek widens and water clarity is reduced. Given the hydrological connection between the Choctawhatchee River and Holmes Creek, it should be noted that *Micropterus* spp. populations are not geographically isolated between study areas. However, it is unknown whether *Micropterus* spp.

move between the Choctawhatchee River and Holmes Creek.

Sampling

Choctaw Bass and Largemouth Bass were sampled from the Choctawhatchee River and Holmes Creek during standardized sampling or supplemental, targeted sampling events. Standardized sampling was used to asses and compare relative abundance between the study areas and species. Standardized sampling and supplemental sampling were used to collect black bass for age, mortality, and growth. Sampling events and available data determined what analyses were conducted and what years those analyses included (Table 1). All sampling occurred between October and December 2012 and 2013 using Smith Root pulsed DC electrofishing boats (7.5 generator powered pulsator [GPP]). Electrofisher settings were adjusted based on temperature and conductivity to obtain a peak applied power of 3,000 W. All Choctaw Bass and Largemouth Bass were weighed (g) and measured (mm, TL). Choctaw Bass were identified in the field using characteristics as described in the introduction of this manuscript and in Boschung and Mayden (2004). Fin clips from Choctaw Bass were collected, analyzed to confirm field identifications, and determined inclusion of Choctaw Bass in analyses. Shallow depths combined with consecutive logjams spanning the width of the river restricted sampling in Holmes Creek to the lower portion of the river, an approximately 43-km stretch below Burnt Sock Landing (the northern most access point for hand launching of small craft) to the confluence with the Choctawhatchee River (Figure 1). Sampling on the Choctawhatchee River extended from the Alabama/Florida border to the Choctawhatchee Bay.

Black bass collected during standardized sampling were sampled according to standard FWCC long-term fish community monitoring protocols for lotic systems from the Choctawhatchee River in 2012 and from Holmes Creek in 2012 and 2013 (Bonvechio 2017). This standardized protocol was used to assess and compare relative abundance estimates between the study areas and species. Thirty transects of 0.1 km each were considered adequate to characterize fish communities and relative abundance of most sport fish; therefore, each randomly selected site was sampled for 0.1 km (Bonvechio 2017). A geographical grid was overlaid onto aerial photography using ArcGIS 9.3 (Environmental Systems Research Institute, Redlands California). Each microgrid that contacted the stream shoreline was assigned a mesohabitat attribute (e.g., stream run, inside bend, outside bend, or marsh if applicable). Each microgrid for the Choctawhatchee River was assigned to a segment of the river: upper, middle, or lower. Microgrids assigned to the upper section of the river were located below the Alabama border and above Highway 90 while microgrids assigned to the middle section were located between Highway 90 and Highway 20, and lower river microgrids were located below Highway 20 to the river mouth. Given the shorter distance of Holmes Creek that could be sampled via electrofishing, no river segments were assigned to Holmes Creek. The relative availability of each segment length and mesohabitat type was determined and used to proportionally weigh the number of sample locations. Thirty to thirty-two random locations were selected from the available microgrids to monitor the sport fish populations and community compositions. Thirty-two electrofishing transects were sampled using standardized protocols for 0.1 km each in the Choctawhatchee River in 2012 (Table 2). Thirty electrofishing transects were sampled using standardized protocols for 0.1 km each in Holmes Creek in 2012 and 2013.

Limited numbers of Choctaw Bass were collected using standardized fish community monitoring protocols therefore supplemen-

Table 1. Information collected and analyzed by river, species, and year with X indicating information used in this study. Relative abundance was estimated from standardized sampling only. Mortality, age, growth, and diet were estimated from fish collected during standardized sampling and supplemental sampling.

River	Species	Year	Number of sacrificed fish	Relative Abundance	Mortality	Growth	Diet
Choctawhatchee River	LMB	2012	NA	X			
	CHBA	2012	41	X	X	X	
	CHBA	2013	66		X	X	X
Holmes Creek	LMB	2012	140*	X	X	X	X
	LMB	2013	NA	X			
	CHBA	2012	49	X	X	X	X
	CHBA	2013	60	X	X	X	X

*426 LMB were measured and weighed from Holmes Creek in 2012; 140 of the 426 LMB were sacrificed.

Table 2. Median and mean ± SE electrofishing catch per river km (fish/km) of Choctaw Bass (CHBA CPUE) and Largemouth Bass (LMB CPUE) in the Choctawhatchee River (2012) and Holmes Creek (2012 and 2013) collected during standardized sampling. Standardized sampling was not conducted on the Choctawhatchee River in 2013. N Choctaw Bass indicates the total number of Choctaw Bass collected while N LMB indicates the total number of Largemouth Bass collected during standardized sampling. Ratio LMB:CHBA is the number of Largemouth Bass to Choctaw Bass obtained.

River	Choctawhatchee River	Holmes Creek	
Year	2012	2012	2013
Number transects	32	30	30
Mean LMB/km	33.13 ± 5.93	65.00 ± 9.33	38.67 ± 5.71
Mean CHBA/km	6.25 ± 1.66	8.00 ± 2.06	3.00 ± 1.19
Median LMB/km	30	40	30
Median CHBA/km	0	5	0
N Choctaw Bass	20	24	9
N LMB	106	195	116
Ratio LMB:CHBA	5.30:1	8.13:1	12.89:1

tal, habitat-targeted sampling was used to increase the sample size of individuals used for mortality, age, growth, and diet. Supplemental sampling was conducted on the Choctawhatchee River and Holmes Creek in 2012 and 2013. Supplemental sampling targeted woody debris in areas that appeared to have medium velocity with respect to the rest of the river. These habitats often occurred in river bends and were located within sections of the rivers where Choctaw Bass had previously been located. In 2013, limited resources prevented fish community sampling using standardized monitoring protocols in the Choctawhatchee River, therefore targeted sampling was used to collect all Choctaw Bass from the Choctawhatchee River in that year.

Relative Abundance, Age, Mortality, and Growth

Mean electrofishing catch per river km (catch per unit effort [CPUE]) for Choctaw Bass and Largemouth Bass collected during standardized electrofishing sampling was used as an index of relative abundance in the Choctawhatchee River and Holmes Creek. There were a large number of zero-catch transects for Choctaw Bass, causing the data to be highly skewed. Therefore, a nonparametric Wilcoxon rank sum test was used to test for differences in median relative abundance of Choctaw Bass and Largemouth Bass within each study area and year (Choctawhatchee River in 2012 and Holmes Creek in 2012 and 2013). The test determined whether the median catch per river km was similar between species and results were considered significant when $P < 0.05$.

All Choctaw Bass (except one age-0 fish) and a subsample of Largemouth Bass were sacrificed for sagittal otoliths. Otoliths were used to estimate age and then determine mortality and growth. Choctawhatchee River Choctaw Bass were sacrificed in 2012 ($N = 41$; 93–342 mm TL) and 2013 ($N = 66$;

68–384 mm TL; Figure 2; Table 3). Holmes Creek Choctaw Bass were sacrificed in 2012 (N = 49; 61–408 mm TL) and 2013 (N = 60; 75–418 mm TL). Largemouth Bass were sampled in Holmes Creek in 2012 (N = 426; 42–483 mm TL) and a subsample of these Largemouth Bass (n = 140; 59–483 mm TL; Figure 3) were sacrificed. Black bass possessing a definitive tooth patch were considered Choctaw Bass in the field; however, any black bass that could possibly be a hybrid based on physical features was removed from age, mortality, growth, and diet analysis. Genetic information from fin clips was then utilized to determine final inclusion in analyses as Choctaw Bass. Sex was determined by dissecting sacrificed fish and visually inspecting reproductive organs; immature reproductive structures prevented gender assignment in small individuals.

Two independent readers estimated age by counting annuli. When fish were older than age-2 or when there was disagreement or uncertainty on the number of rings, otoliths were sectioned using a low speed diamond wheel saw (South Bay Tech Inc., model 650), mounted on glass slides, and then annuli were counted under a dissecting microscope (Buckmeier and Howells 2003). *Micropterus* spp. not sacrificed were assigned ages using an age-length key through the S-FAT Add-in tool in Microsoft Excel (Devries and Frie 1996; Slipke 2005). Lengths of assigned fish were compared to lengths of aged fish to validate age assignments.

Total instantaneous mortality (Z) was estimated using a weighted catch-curve regression (Ricker 1975). Examination of length frequencies (Figure 2) suggested that catch-curve assumptions such as constant recruitment may have been violated and a small number of older fish were collected per year. Two consecutive years of catch-at-age data were therefore pooled and used in

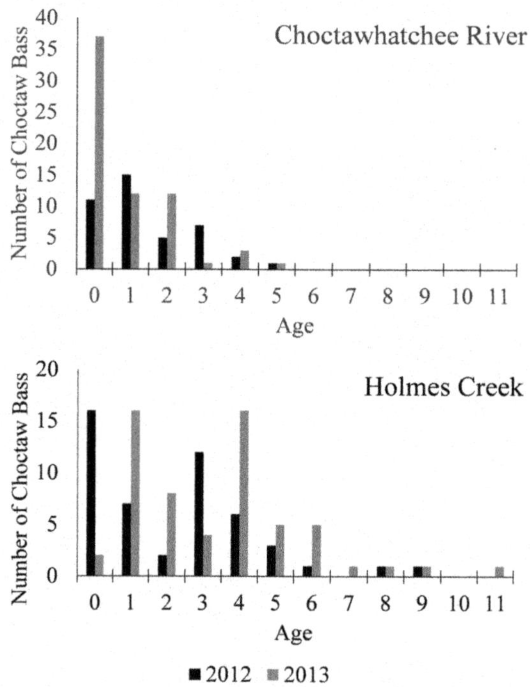

Figure 2. Number of Choctaw Bass collected from the Choctawhatchee River (top) and Holmes Creek (bottom) in 2012 and 2013 by age.

Table 3. Predicted total length (L_t) based on Choctaw Bass von Bertalanffy growth equations (Predicted Choctawhatchee River $L_t = 351[1 - e^{(-0.57(t+0.65)})]$, Predicted Holmes Creek $L_t = 420[1 - e^{(-0.28(t+1.01)})]$, number of fish ($n$), mean TL, and ranges of TL for Choctaw Bass by age and river.

Age	Choctawhatchee River				Holmes Creek			
	n	Mean TL (mm)	L_t (mm)	Range TL (mm)	n	Mean TL (mm)	Lt (mm)	Range TL (mm)
0	48	109	109	68–156	18	98	102	61–156
1	27	210	214	70–284	23	181	179	70–284
2	17	282	273	232–332	10	257	237	161–320
3	8	294	307	245–384	16	262	281	225–326
4	5	333	326	281–374	22	321	314	260–381
5	2	348	337	342–354	8	333	339	310–359
6	-	-	-	-	6	353	359	305–394
7	-	-	-	-	1	345	373	-
8	-	-	-	-	2	390	385	380–399
9	-	-	-	-	2	413	393	408–418
10	-	-	-	-	-	-	-	-
11	-	-	-	-	1	406	404	406

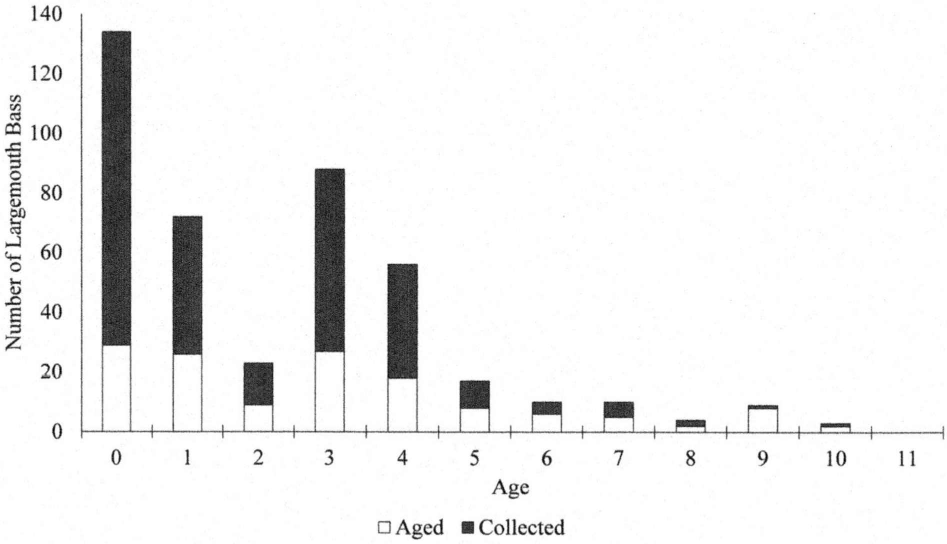

Figure 3. Number of Largemouth Bass from Holmes Creek (Holmes) in 2012 by age. Aged indicates aged fish and fish used in diet analysis. Collected indicates fish that were assigned ages.

the construction of Choctawhatchee River and Holmes Creek catch-curves in an effort to further increase sample size, dampen any possible effects of variable recruitment, and improve precision and accuracy of estimates (Miranda and Bettoli 2007). Despite supplemental sampling and pooling two years of data, a lower total sample size of Choctaw Bass was collected each year than the minimum sample size of 500 fish recommended by Coggins et al. (2013). All catch-curves included age-2 or older fish. Age-2 fish were considered fully recruited to the gear since all total lengths were greater than 150 mm TL (Jackson and Noble 1995; Bayley and Austen 2002); therefore, the catch-curves did not include fish less than age-2. Smith et al. (2012) recommended the use of weighted catch-curve regression with the Peak criterion and no right truncation or the Chapman Robson Peak Plus estimator over the Heincke estimator, particularly when data suggested variable recruitment. Therefore, there was no right truncation even when the number of fish per aged group was less than five fish.

Instantaneous mortality rates (i.e., slopes of the catch-curve regressions, Z) were compared using an analysis of covariance (ANCOVA) via the general linear model (GLM) procedure (Miranda and Bettoli 2007; SAS 2010). Instantaneous mortality rates of Choctaw Bass were compared between the Choctawhatchee River and Holmes Creek, testing the hypothesis that the slopes of the catch-curves were similar. Instantaneous mortality rates of Choctaw Bass from Holmes Creek collected in 2012 and 2013 were also compared to Largemouth Bass collected from Holmes Creek in 2012. Total annual mortality ($A = 1 - e^{-Z}$) for Choctaw Bass and Largemouth Bass was estimated from total instantaneous mortality determined from weighted catch-curves (Ricker 1975). Total annual mortality was estimated for each species and study area. Weighted catch-curves for Choctaw Bass from the Choctawhatchee River included age-2 through age-5 fish ($R^2 = 0.99$). Holmes Creek Choctaw Bass catch-curves included age-2 through age-11 fish ($R^2 = 0.59$) and Holmes Creek Largemouth Bass catch-curves included age-2 fish through age-10 fish ($R^2 = 0.65$).

The nonlinear von Bertalanffy growth equation is

$$L_t = L_\infty[1 - e^{-k(t + t_0)}]$$

where L_t indicates the length at time t (in years), L_∞ indicates length infinite or asymptotic length, k is a growth rate, and t_0 indicates the theoretical age in years at a length zero. We built von Bertalanffy growth equations from 2012 and 2013 Choctaw Bass and used Akaike's Information Criterion (AICc) to select the most probable equation describing growth of Choctaw Bass. We built all models using PROC NLIN in SAS v9.4 (Cary, NY). Residuals were examined and models met all assumptions. We compared six different models: 1) the same k, L_∞, and t_0 parameters by study area (Choctawhatchee River versus Holmes Creek, 3 parameters, null model), 2) different L_∞, but the same k and t_0 parameters by study area (4 parameters), 3) different k but the same L_∞ and t_0 parameters by study area (4 parameters), 4) different t_0 but the same k and L_∞ by study area (4 parameters), 5) different L_∞ and k but the same t_0 parameters by study area (5 parameters), and 6) different L_∞, k, and t_0 parameters by study area (6 parameters, full model). If the null model was selected as the most probable model, that could indicate that asymptotic length, growth rate, and theoretical age in years at length zero did not differ between study areas and one equation could be used to describe growth of all Choctaw Bass from the Choctawhatchee River drainage. By comparison, if the most probable model was the full model and contained different L_∞, k, and t_0 parameters by study area, separate study area equations should be used to describe growth of Choctaw Bass. In other words, it would be most appropriate to use an equation with specific Choctawhatchee River L_∞, k, and t_0 parameters to describe growth of Choctaw Bass from the Choctawhatchee River and a different equation with Holmes Creek specific L_∞, k, and t_0 parameters to describe growth of Choctaw Bass from Holmes Creek.

The AICc scores from the models were compared and the differences (Delta AICc) between the best model (lowest AICc score) and the other models were calculated (Anderson et al. 2001; Burnham and Anderson 2001). A difference greater than two between the AICc scores provided statistical evidence of a difference between models (Anderson 2008). We also considered the issue of the "pretending variable" as described by Anderson (2008). Two AICc models may be two units apart through the addition of one parameter (since one parameter incurs a penalty of two). Both models are good models, however the additional parameter is considered a "pretending variable" or noninformative variable and the added variable is therefore not considered important. The strength of evidence for a model (model probability) was examined by comparing model weights.

We built a von Bertalanffy model with six terms: L_∞, k, and t_0, as well as three difference parameters ($L_{\infty Choctawhatchee} - L_{\infty Holmes}$, $k_{Choctawhatchee} - k_{Holmes}$, and $t_{0Choctawhatchee} - t_{0Holmes}$) in order to compare growth of Choctaw Bass between the Choctawhatchee River and Holmes Creek. If these difference parameters were greater than zero then bass grew at different rates between the two study areas. We used the full model to predict the size of age-2 bass for each study area (the age at which all examined bass had reached maturity). We then estimated the age at which bass reached harvestable size (305 mm TL) for each study area. We also built a von Bertalanffy growth model for Largemouth Bass from Holmes Creek (with the traditional three parameters). The same size and age comparisons were made between Largemouth Bass and Choctaw Bass in Holmes Creek. We built all models using PROC NLIN in SAS v9.4 (Cary, NY). Residuals were examined and models met all assumptions. Results were considered significant when $P < 0.05$.

Diet

Stomach contents from fish sacrificed for age, growth, and mortality estimates were examined (66 Choctaw Bass from the Choctawhatchee River, 108 Choctaw Bass from Holmes Creek, and 138 Largemouth Bass from Holmes Creek). Diets were not examined from Choctaw Bass or Largemouth Bass collected in the Choctawhatchee River during 2012 nor were they examined for Largemouth Bass collected from Holmes Creek in 2013 (Table 1). Stomach contents were identified, counted, and wet weighed to the nearest 0.001 g. Detritus (i.e., leaf litter, sand) was considered accidental ingestion and was not included in the diet analysis. Diet items were grouped into prey categories and then quantified in terms of frequency of occurrence and percent weight (Bowen 1996). The percentage similarity (percentage overlap) measure developed by Renkonen (1938) was calculated between rivers and species. Sixteen prey categories (Oligocheata, Hirundinea, Amphipoda, Ictaluridae, Centrarchidae, Percidae, Cyprinidae, unidentified fish, Decapoda, Trichoptera, Emphemeroptera, Diptera, Odonata, unidentified aquatic invertebrate, terrestrial invertebrate, and Mollusca [Gastropoda]) were used when comparing percentage similarity. Seven prey categories (Annelida, Amphipoda, fish, Decapoda, Mollusca, other invertebrates, and unidentified organic material) were used when examining percent by weight of prey items consumed by different length categories (less than 200 mm TL, 200–304 mm TL, and equal to or greater than 305 mm TL). Unidentified organic material was material which was not considered detritus since it was digested organic matter, but lack of bones or exoskeletons prevented further identification of the material as fish, Decapoda, or other invertebrate.

Results

Relative Abundance, Age, Mortality, and Growth

The median relative abundance of Largemouth Bass exceeded the median relative abundance of Choctaw Bass in both Holmes Creek and the Choctawhatchee River (Table 2, P-value <0.001 for all tests). Catch ratios of Largemouth Bass: Choctaw Bass varied by river and year and Largemouth Bass outnumbered Choctaw Bass in all rivers and years (Table 2). Choctaw Bass relative abundance did not differ between study areas in 2012, however Largemouth Bass relative abundance was greater in Holmes Creek than the Choctawhatchee River. Choctaw Bass and Largemouth Bass relative abundance was greater in Holmes Creek 2012 than in Holmes Creek 2013.

Choctaw Bass in the Choctawhatchee River ranged from age-0 to age-5 in both 2012 and 2013 with the dominant age being age-1 in 2012 and age-0 in 2013 (Table 3; Figure 2). The dominant age for all Choctaw Bass collected from the Choctawhatchee River in 2012 and 2013 was age-0 (45% of all Choctawhatchee River Choctaw Bass). No fish older than age-5 were collected from the Choctawhatchee River in 2012 or 2013. Sex ratios for all Choctaw Bass greater than 305 mm TL from the Choctawhatchee River (N = 14) were 71% female to 29% male. Choctaw Bass in Holmes Creek ranged from age-0 to age-9 in 2012 and age-0 to age-11 in 2013 with the dominant age being age-0 and age-3 in 2012 and age-1 and age-4 in 2013 (Table 3; Figure 2). The dominant ages for all Choctaw Bass collected from Holmes Creek in 2012 and 2013 was age-1 and age-4 (21% and 20% of all Holmes Creek Choctaw Bass respectively). The oldest Choctaw Bass collected from Holmes Creek was age-11; no age-10 fish were collected. Sex ratios for all

Choctaw Bass greater than 305 mm TL (all legal sized fish) from Holmes Creek ($N = 35$) were 54% female to 46% male. Ages based on the subsample of sacrificed Largemouth Bass were assigned to all Largemouth Bass collected from Holmes Creek which resulted in estimated ages up to age-10. The dominant ages of all Largemouth Bass were age-0 and age-3 (31% and 21% of all Largemouth Bass respectively; Figure 3). Length frequencies suggested that a strong year-class of Choctaw Bass was produced in the Choctawhatchee River in 2013 and a strong year-class of Largemouth Bass and Choctaw Bass was produced in Holmes Creek in 2012 (Figure 2, Figure 3).

Total annual mortality ranged from 35 to 60% for Choctaw Bass in the Choctawhatchee River (A = 49%). Total annual mortality ranged from 6 to 47% for Choctaw Bass in Holmes Creek (A = 29%) and total annual mortality ranged from 11 to 43% for Largemouth Bass in Holmes Creek (A = 29%). Total instantaneous mortality was similar between Choctaw Bass from the Choctawhatchee River and Choctaw Bass from Holmes Creek ($F_{1,9} = 1.76$, $P = 0.22$). Total instantaneous mortality was also similar between Choctaw Bass and Largemouth Bass from Holmes Creek ($F_{1,14} = 0.00$, $P = 0.99$).

We compared von Bertalanffy growth models for Choctaw Bass using AICc and the top three growth models were plausible given the Delta AICc values were less than 2 (Table 4). Therefore, the model with the lowest Delta AICc was selected as the best model to describe growth of Choctaw Bass. The selected model utilized different parameters for L_∞, k, and t_0 by study area (Choctawhatchee River and Holmes Creek). The model weight indicated a 40% probability that the selected model was the most plausible.

The von Bertalanffy growth equation for Choctaw Bass from the Choctawhatchee River from 2012 and 2013 was:

$$L_t = 351\,[1 - e^{-0.57(t + 0.65)}]$$

whereas the equation for Choctaw Bass from Holmes Creek from 2012 and 2013 was:

$$L_t = 420\,[1 - e^{-0.28(t + 1.01)}]$$

The von Bertalanffy growth equation for Largemouth Bass from Holmes Creek from 2012 was:

$$L_t = 443\,[1 - e^{-0.26(t + 0.90)}]$$

There was a significant difference in L_∞ ($t_{216} = 2.02$, $P = 0.04$) and k ($t_{216} = -2.76$, $P = 0.01$), but not in t_0 ($t_{216} = -1.91$, $P = 0.06$). While Choctaw Bass start off the same size (similar t_0) they grow at different rates to different asymptotic sizes between the study areas.

Choctaw Bass would reach the minimum size limit (305 mm TL) at 3.0 years of age in the Choctawhatchee River and 3.8 years of age in Holmes Creek (Figure 4). Largemouth Bass in Holmes Creek would reach 305 mm TL by age 3.7 (Figure 4). The Choctawhatchee River (2012 and 2013) contained a greater percentage of Choctaw Bass less than 305 mm TL than Holmes Creek (2012 and 2013 combined; 87% compared to 66%). Lower numbers of older fish were collected from Holmes Creek in 2012 and the percentage of Choctaw Bass less than 305 mm TL from Holmes Creek in 2012 was 86% compared to 50% from Holmes Creek in 2013. An age-2 bass (the age at which all bass were found to be sexually mature) from the Choctawhatchee River had a predicted size of 273 mm TL (95% CI: 264–283) whereas one from the Holmes Creek is 237 mm TL (95% CI: 229–246, Table 3). At age-2, Largemouth Bass from Holmes Creek were predicted to be an average of 233 mm TL (95% CI: 229–238).

Table 4. Comparison of von Bertalanffy models by study area (Choctawhatchee River (c) and Holmes Creek (h)) developed to describe growth of Choctaw Bass. L_∞ indicates asymptotic length, *k* indicates growth parameter, and t_0 indicates the theoretical age in years at a length of zero. **Bold indicates selected model**.

Model	Model description	Parameters	AICc	DeltaAICc	Weight
Full (Different L_∞, k, t_0)	**$L_\infty h \neq L_\infty c$, Kh≠Kc, $t_0 h \neq t_0 c$**	**6**	**2121.8**	**0.0**	**0.402**
Different *k* (same L_∞ and t_0)	$L_\infty h = L_\infty c$, Kh≠Kc, $t_0 h = t_0 c$	4	2121.9	0.1	0.382
Different L_∞ and *k*	$L_\infty h \neq L_\infty c$, Kh≠Kc, $t_0 h = t_0 c$	5	2123.4	1.6	0.181
Different L_∞	$L_\infty h \neq L_\infty c$, Kh=Kc, $t_0 h = t_0 c$	4	2126.7	4.9	0.035
Different t_0	$L_\infty h = L_\infty c$, Kh=Kc, $t_0 h \neq t_0 c$	4	2134.4	12.6	0.001
Null (Same L_∞, *k*, t_0)	$L_\infty h = L_\infty c$, Kh=Kc, $t_0 h = t_0 c$	3	2147.2	25.4	0.000

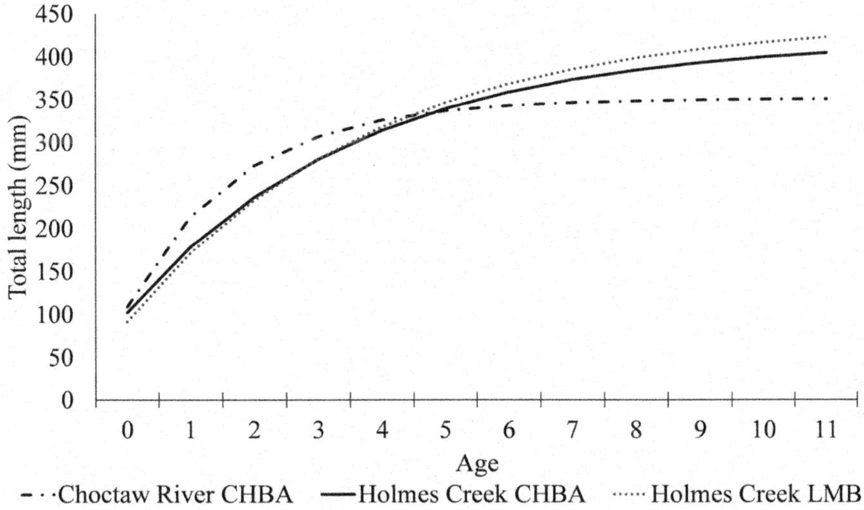

Figure 4. Predicted lengths for Choctaw Bass (CHBA) from the Choctawhatchee River (Choctaw River), Choctaw Bass from Holmes Creek, and Largemouth Bass (LMB) from Holmes Creek.

Diet

Stomach contents were examined for 66 Choctaw Bass from the Choctawhatchee River (19% empty), 48 Choctaw Bass from Holmes Creek in 2012 (35% empty), 60 Choctaw Bass from Holmes Creek in 2013 (20% empty), and 138 Largemouth Bass from Holmes Creek (28% empty).

Fish were the primary prey of Choctaw Bass from the Choctawhatchee River, as indicated by frequency of occurrence and percent by weight (Table 5). Crayfish (Decapoda, primarily Cambaridae) were the primary prey from Holmes Creek Choctaw Bass and Largemouth Bass, as indicated by frequency of occurrence and percent by weight. Fish consumed by Choctaw Bass belonged to the families Ictaluridae, Centrarchidae (*Lepomis* sp.,), Percidae, and Cyprinidae. Invertebrates consumed by Choctaw Bass were of the Orders Trichoptera, Ephemeroptera, Diptera, and Odonata. Amphipoda and Gastropoda were identified in *Micropterus* spp. diets from Holmes Creek, but were not found in Choctawhatchee River Choctaw Bass diets. Palaemonidae and terrestrial insects were identified in Largemouth Bass diets but not Choctaw Bass diets. Twenty-seven Largemouth Bass stomachs contained unidentified organic material which was too highly digested to be assigned to a specific prey category.

Fish represented a higher percent composition by weight across all length-groups of Choctaw Bass diets from the Choctawhatchee River compared to Holmes Creek Choctaw Bass (Figure 5). Decapoda represented the greatest percent composition of the diet by weight in Holmes Creek Choctaw Bass greater than 200 mm TL. It should be noted, however, that a greater percentage (59%) of Choctawhatchee River Choctaw Bass individuals represented in the diet analysis were less than 200 mm TL compared to Holmes Creek (25% for Choctaw Bass and 31% for Largemouth Bass).

There was a greater percent similarity of diets between sympatric Choctaw Bass and Largemouth Bass in Holmes Creek than between Choctaw Bass collected from Holmes Creek and Choctaw Bass collected from the Choctawhatchee River. Holmes Creek

Table 5. Frequency of occurrence (expressed as percent) and percent by weight of diet prey categories by species and river. Choctaw indicates the Choctawhatchee River while Holmes indicated Holmes Creek. CHBA indicates Choctaw Bass while LMB indicates Largemouth Bass. Shaded rows indicate prey categories which may include multiple orders or families from lower taxonomic levels.

	Metric	Frequency of occurrence			Percent by weight		
	River	Choctaw	Holmes	Holmes	Choctaw	Holmes	Holmes
	Species	CHBA	CHBA	LMB	CHBA	CHBA	LMB
	Year	2013	2012–2013	2013	2013	2012–2013	2013
Unidentified organic (fish or invertebrate)		0.00	0.00	27.55	0.00	0.00	16.76
Total Annelida		5.56	2.56	3.06	3.71	0.32	0.10
	Oligocheata	5.56	0.00	3.06	3.71	0.00	0.10
	Hirundinea	0.00	2.56	0.00	0.00	0.22	0.00
Amphipoda		0.00	3.85	2.04	0.00	2.20	0.19
Total Fish		74.07	17.95	36.73	73.37	10.80	33.74
	Ictaluridae	18.52	2.56	0.00	17.42	0.45	0.00
	Centrarchidae (*Lepomis* sp.)	1.85	0.00	3.06	1.85	0.00	1.21
	Percidae	3.70	0.00	2.04	3.41	0.00	1.58
	Cyprinidae	9.26	0.00	3.06	5.77	0.00	1.80
	Unidentified Fish	50.00	15.38	30.61	44.91	10.36	29.15
Total Decapoda		9.26	87.18	54.08	9.26	83.76	47.99
	Cambaridae	0.00	73.08	47.96	0.00	67.23	41.87
	Palaemonidae	0.00	0.00	4.08	0.00	0.00	4.08

Table 5. Continued.

	Metric	Frequency of occurrence			Percent by weight		
	River Species Year	Choctaw CHBA 2013	Holmes CHBA 2012–2013	Holmes LMB 2013	Choctaw CHBA 2013	Holmes CHBA 2012–2013	Holmes LMB 2013
Total invertebrate (not including *Mollusca*)		22.22	5.13	1.02	13.67	2.76	1.02
	Trichoptera	1.85	0.00	0.00	0.04	0.00	0.00
	Ephemeroptera	1.85	2.56	0.00	1.85	1.47	0.00
	Diptera	5.56	0.00	0.00	5.56	1.29	0.00
	Misc. invert (unidentified)	11.11	0.00	0.00	5.75	0.00	0.00
	Misc. invert (terrestrial)	0.00	2.56	1.02	0.00	0.00	1.02
	Odonata	1.85	0.00	0.00	0.46	0.00	0.00
Total Mollusca		0.00	3.85	2.04	0.00	0.17	0.20
	Gastropoda	0.00	3.85	2.04	0.00	0.17	0.20

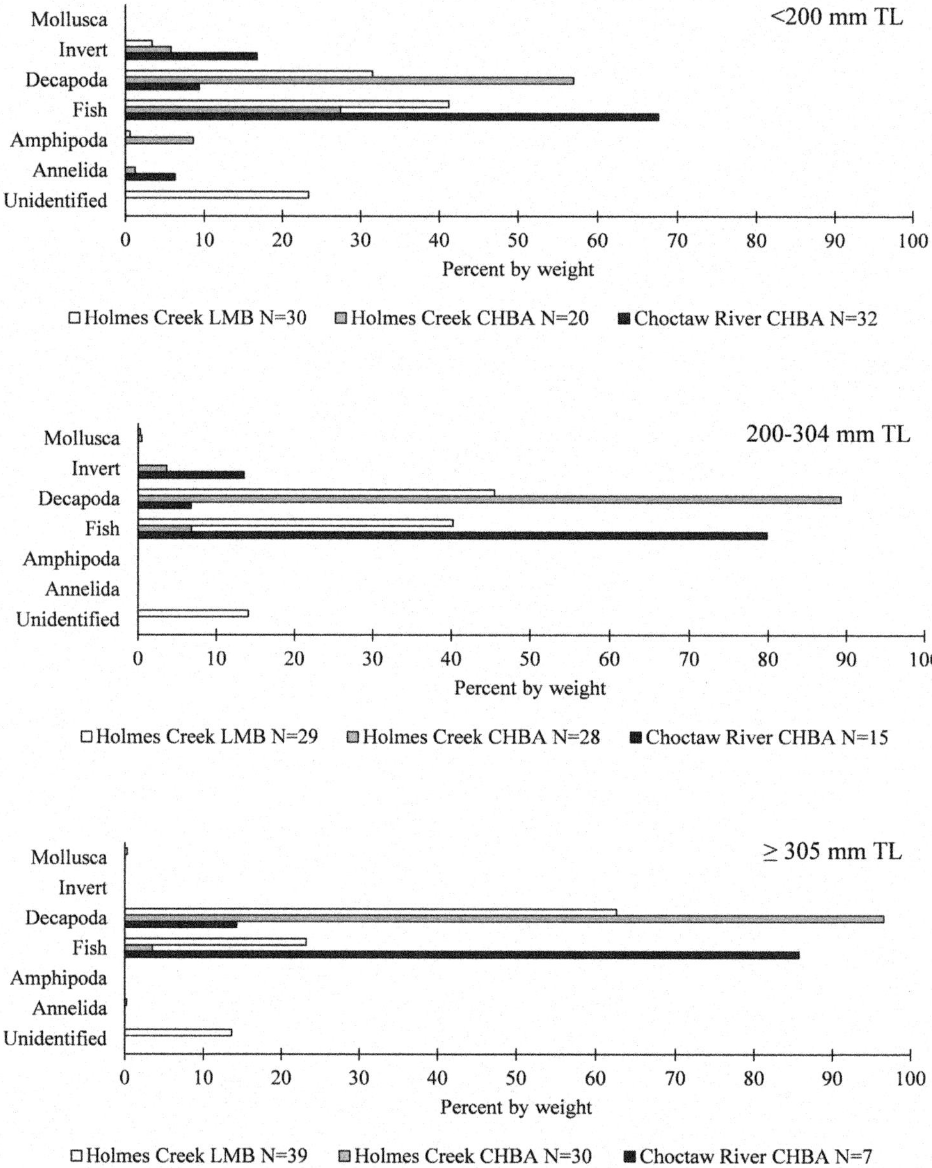

Figure 5. Percent by weight of prey items by study area (Choctaw River indicates Choctawhatchee River) and species (CHBA indicates Choctaw Bass, LMB indicates Largemouth Bass) over length-groups (top: <200 mm TL, middle: 200–304 mm TL, bottom: ≥305 mm TL). N indicates number of fish per length category. Diets were collected for LMB in Holmes Creek in 2012, CHBA in the Choctawhatchee River in 2013, and CHBA in Holmes Creek in 2012 and 2013. Invert indicates invertebrates and unidentified indicates unidentified organic material.

Choctaw Bass diets were 19% similar to Choctawhatchee River Choctaw Bass diets and 71% similar to Holmes Creek Largemouth Bass diets.

Discussion

This study begins to describe the Choctaw Bass by examining the relative abundance, age, mortality, and growth of Choctaw Bass from the Choctawhatchee River drainage. The biology of Choctaw Bass and comparison to sympatric Largemouth Bass assisted in management decisions pertaining to Choctaw Bass when regulation changes were proposed for Florida's black bass. Comparison of relative abundance indicates that Largemouth Bass are more abundant than Choctaw Bass in the Choctawhatchee River and Holmes Creek. However, the median catch rate for Choctaw Bass from the Choctawhatchee River and Holmes Creek in 2013 during standardized sampling was zero. This suggests that Choctaw Bass could inhabit different areas than Largemouth Bass or that they may be less susceptible to capture. For example, Choctaw Bass may be more abundant in complex woody debris habitat located in bends with swift current; these areas are difficult to sample effectively using boat electrofishing. Targeted habitat sampling in the Choctawhatchee River and Holmes Creek also resulted in higher catch rates of Largemouth Bass than Choctaw Bass which provides additional evidence that there are likely fewer Choctaw Bass than Largemouth Bass. A study comparing the capture probabilities or detection probabilities of Choctaw Bass versus Largemouth Bass could provide further clarity on whether Choctaw Bass are better at avoiding capture, less abundant, or if electrofishing catchability varies by another factor such as season or habitat (Hangsleben et al. 2013). Without clear information on detection probabilities, managers presumed that Choctaw Bass abundance was likely lower than Largemouth Bass in many Florida rivers. The low abundance of Choctaw Bass and limited geographic distribution relative to Largemouth Bass in Florida led managers to establish more conservative regulations for Choctaw Bass than Largemouth Bass. There is no longer a minimum length limit for Largemouth Bass, but a 305 mm minimum length limit for Choctaw bass was retained.

Choctaw Bass also appear to be less abundant than Largemouth Bass throughout most of their Florida Range. Ratios of the two species (Largemouth Bass: Choctaw Bass) were examined from other Florida Rivers sampled with standardized FWCC long-term monitoring protocols in 2010 and 2011. These comparisons suggested that Choctaw Bass are less abundant than Largemouth Bass in the Perdido, Escambia, and Yellow Rivers. However, Choctaw Bass may be similar in abundance or more abundant than Largemouth Bass in the Shoal River and Big Escambia Creek (FWCC unpublished data), which are tributaries of the Yellow and Escambia Rivers, respectively. In Florida, Largemouth Bass numerical dominance over other sympatric black bass species has been shown to depend on the river as well as river reach and these variations are likely influenced by habitat (e.g., main-stem river versus tributary, pools versus shoals) and age of fish (Bonvechio et al. 2005; Nagid et al. 2010; Wheeler and Allen 2003). Describing Choctaw Bass habitat associations and comparing them to Largemouth Bass will increase our understanding of Choctaw Bass in order to better manage and protect habitat. Understanding habitat associations may improve targeted sampling efforts and allow for collection of a greater number of individuals or change the gear we use to sample Choctaw Bass. Furthermore, managers could prioritize protection of habitat, restoration projects, or land acquisition based on proximity to Choctaw Bass habitat. If Choctaw Bass show a strong association with a specific habitat feature or suite of habitat features, managers could combine abundance estimates with side-scan sonar maps to generate stream wide abundances (Toepfer et al. 2000). Finally, knowledge of Choctaw Bass habitat and abundance could provide managers with additional information to develop or change regulations. In the case of Shoal Bass, managers used spawning habitat information, population estimates, and angler opinions to establish a conservation, catch and release only zone for Shoal Bass on the Chipola River, Florida (Woodside et al. 2015).

Although total instantaneous mortality was not significantly different for Choctaw Bass between study areas, the difference in mean total annual mortality between Holmes Creek (29%) and the Choctawhatchee River (49%) was nearly two-fold. Older and larger Choctaw Bass were observed in Holmes Creek compared to the Choctawhatchee River. It is not known if Choctaw Bass move between the Choctawhatchee River and Holmes Creek, therefore research into fish movement should also be considered. Genetic research and otolith microchemistry indicates Smallmouth Bass *M. dolomieu* readily move between tributary and main-stem river habitat, both in larval (Humston et al. 2010) and adult (Gross et al. 1994) life stages.

Several factors differ between the two study areas that may influence mortality and result in a lack of older fish collected from the Choctawhatchee River. These factors include environmental and physical habitat characteristics (i.e., temperature, flow, turbidity, surrounding landuse, etc.), diet, movement of fish, older fish not being as susceptible to electrofishing, or exploitation by anglers. Population characteristics of black bass in two Eastern Oklahoma streams were described by Balkenbush and Fisher (2001) and Dauwalter and Fisher (2008). Smallmouth Bass were more abundant and exhibited higher survival of age-1 and older fish in Baron Fork Creek, described as a clear stream maintained by numerous springs with relatively stable flows, whereas Smallmouth Bass exhibited high annual mortality and low recruitment to older ages in the Glover River, described as a moderately turbid river fed primarily by runoff with relatively flashy flows. Baron Fork was also described as having less pools and more wood then the Glover River. Older Smallmouth Bass moved more often and Smallmouth Bass abundances were related to deep pools with moderate amounts of wood on Baron Fork. Densities of Smallmouth Bass were related to wood in Baron Fork, but not in Glover River. Population characteristics of Choctaw Bass may be similarly related to these factors as Holmes Creek is less turbid and may contain more wood than the Choctawhatchee River, therefore spatial and temporal patterns of Choctaw Bass could be examined and habitats adequately described.

Choctaw Bass in the Choctawhatchee River may also be more vulnerable to exploitation by anglers, resulting in populations dominated by individuals under the 305 mm minimum length limit. The presence of log jams in Holmes Creek could limit angler access to areas of high Choctaw Bass abundance where access may not be as restricted in the Choctawhatchee River. No information specific to exploitation or harvest of Choctaw Bass exists; however, exploitation and harvest of black bass in Northwest Florida rivers is typically low and rates have been similar between Largemouth Bass and other riverine black bass species such as Suwannee Bass or Shoal Bass (Middaugh et al. 2016; Woodside et al. 2015). Recent creel information was lacking, therefore fishery information on the two rivers should be considered as a future research priority.

Total instantaneous mortality was similar between Largemouth Bass and Choctaw Bass in Holmes Creek. Comparisons of mortality between other sympatric black bass species in Florida were examined to determine if this was common. Suwannee Bass and Largemouth Bass mortality was similar in the Ochlockonee and Santa Fe Rivers (Bonvechio et al. 2005), however Shoal Bass instantaneous mortality was greater than sympatric Largemouth Bass in the Chipola River (Woodside et al. 2015). Even though instantaneous mortality did not differ between Choctaw Bass and Largemouth Bass in Holmes Creek and exploitation is likely low for black bass across most panhandle rivers, it is unknown if instantaneous mortality differs in the Choctawhatchee River or other Florida rivers. Managers chose to establish more conservative regulations for Choctaw Bass than Largemouth Bass given data gaps, as well as the limited distribution and perceived lower abundance. Anecdotal evidence suggests that some Florida anglers are interested in targeting riverine black bass species other than Largemouth Bass, and that anglers can distinguish between Choctaw Bass or Spotted Bass and Largemouth Bass through the identification of physical features such as the tooth patch. Anglers should be educated regarding the identification of the different black bass present in Florida, as well as what the black bass regulations are and why regulations differ between species.

Managers use growth rates, length frequencies, and location information when providing guidance to anglers seeking to catch Choctaw Bass. Managers also used this information when developing statewide black bass regulations. It was difficult to distinguish between true differences in growth rates for Choctaw Bass in the Choctawhatchee River and Choctaw Bass in Holmes Creek due to the differences in age structure between the two study areas (no Choctaw Bass were collected that were older than age-5 in the Choctawhatchee River). However, there was statistical evidence to support the conclusion that the length infinite and growth parameters of study area specific von Bertalanffy growth equations differed. Choctawhatchee River Choctaw Bass grew faster, but Holmes Creek Choctaw Bass grew to a longer length. Choctaw bass reached the minimum size limit (305 mm TL) by age-3 in the Choctawhatchee River and age-3.8 in Holmes Creek. A high proportion of the Choctaw Bass collected from the Choctawhatchee drainage were less than 305 mm TL (up to 95% of the Choctaw Bass from the Choctawhatchee River in 2012). Relatively few Choctaw Bass lived longer than age-5. Furthermore, only three Choctaw Bass collected from this study were greater than or equal to 406 mm TL. Examination of length frequencies of Choctaw Bass collected during standardized FWCC sampling in 2010 and 2011 from Florida panhandle rivers (Big Escambia Creek, Choctawhatchee River, Holmes Creek, Escambia River, Perdido River, Shoal River, and Yellow River) also suggested a scarcity of Choctaw Bass greater than 406 mm TL (99% of the 107 Choctaw Bass collected in 2010 were less than 406 mm TL and 98% of the 179 Choctaw Bass collected in 2011 were less than 406 mm TL). The 305 mm TL minimum length limit was removed for Largemouth Bass when the regulation was changed to permit take of only one black bass greater than 406 mm TL in 2016. Retaining the 305 mm TL minimum length limit for Choctaw Bass restricts harvest of Choctaw Bass.

Growth rates of *Micropterus* spp. can vary by gender and this information can further influence management decisions. Gender specific growth or mortality rates of Choctaw Bass were not compared due to the limited numbers of collected fish, however the percentage of males and females in Holmes Creek larger than the minimum size limit was roughly equal. Therefore, it is not likely that current regulations would direct fishing mortality towards female bass in Holmes Creek. This contrasts with the Choctawhatchee River where the majority of legal sized Choctaw Bass were females (although a low number of legal sized fish were collected). This also contrasts with Suwannee Bass populations in many Florida Rivers where there was a greater percentage of females at sizes exceeding the minimum length limit (Bonvechio et al. 2005). Exploitation rates of Suwannee Bass did not indicate a risk of over-harvest (Middaugh et al. 2016), but the limited distribution, specific habitat requirements, and limited numbers of Suwannee Bass collected greater than 406 mm TL led to the statewide retention of the 305 mm minimum length limit for that species. Middaugh et al. (2016) suggested that the potential for harvest to be directed at female Suwannee Bass under a 305 mm minimum length limit warranted the additional monitoring of Suwannee Bass populations and angling effort, or consideration of additional protective measures specific to Suwannee Bass. The lack of species and gender-specific information for Choctaw Bass led managers to recommend the retention of the 305 mm minimum length limit given that Choctaw Bass also appeared to be less abundant and have restricted distributions relative to Largemouth Bass. Gender-specific growth and mortality data from multiple Florida rivers combined with exploitation rates would enable managers to determine if further protections or greater restrictions are warranted.

Examination of the number of fish collected by study area, year, and age (particularly for age-0 fish) suggest that factors influencing recruitment may vary between the Choctawhatchee River and Holmes Creek (Figure 2). Flows during or post spawning may influence recruitment and this can vary by river size. For example, Shoal Bass recruitment is negatively influenced by high flows in small rivers such as the Chipola River but does not appear to be impacted in larger rivers such as the Flint River (Sammons et al. 2015). In the Choctawhatchee River, a strong Choctaw Bass year-class was produced in 2013 relative to 2012. In Holmes Creek, a stronger Choctaw Bass year-class was produced in 2012 relative to 2013. Additional data are needed to assess recruitment. Future research should examine factors influencing Choctaw Bass recruitment and these factors could be compared to those influencing sympatric Largemouth Bass. Shoal Bass exhibited a greater susceptibility to variable recruitment than Largemouth Bass in the Chipola River and this was one factor that lead to the establishment of more conservative regulations for Shoal Bass than Largemouth Bass (Woodside et al. 2015).

Fish were the primary prey of Choctaw Bass in the Choctawhatchee River while crayfish were the primary prey of Choctaw Bass from Holmes Creek. Crayfish and fish were both previously described as important diet components for both Spotted Bass and Largemouth Bass (Sammons 2012; Scalet 1977). There are several reasons that Choctaw Bass diets could differ between the Choctawhatchee River and Holmes Creek; two possible explanations are differences in the abundance of crayfish and differences in turbidity. Differences in abundance of crayfish between the two rivers is a likely reason that more diets contained this food item in Holmes Creek compared to the Choctawhatchee River. Although the two systems are connected, Holmes Creek contains greater productivity and a higher biomass of invertebrates (including crayfish) relative to the Choctawhatchee River (Bass and Cox 1985). Black bass will often forage in the most energetically efficient manner, consuming what is most abundant or available (Stein 1977), which could explain why the diets of Choctaw Bass and Largemouth Bass collected in Holmes Creek contained greater numbers of crayfish and diets of Choctaw Bass in the Choctawhatchee River had diets dominated by fish. The availability or abundance of crayfish in these rivers is not known. Turbidity can also influence foraging behavior and prey selection with Largemouth Bass becoming less active, selecting larger prey, or selecting fish over crayfish at higher turbidities (Shoup and Wahl 2009).

Turbidity was not recorded during this study and while turbidity is typically higher in the Choctawhatchee River than Holmes Creek, differences in turbidity between the two rivers may be less during low flow months such as late fall (Livingston 1991).

Continued research on diets, particularly on seasonal variation and variation by fish size is needed. Diets were only collected during the fall; therefore, it is not known if Choctaw Bass diets vary seasonally. Fish size can influence the probability of a diet dominated by fish or crayfish with probabilities increasing as fish size increases (Wheeler and Allen 2003; Sammons 2012). Spotted Bass *M. punctulatus* diets in the Flint River, Georgia varied by season and age (Sammons 2012). Diets of juvenile (less than 200 mm TL) Spotted Bass were piscivorous in fall and winter, but insectivorous in fall and summer while crayfish were important components of subadult (200–300 mm TL) Spotted Bass diets for all seasons but spring. These ontogenetic changes vary with different black bass species found in the same river. Largemouth Bass transitioned to a crayfish dominated diet at a smaller size than Shoal Bass in the Chipola River (Wheeler and Allen 2003). The large amount of unidentified organic material present in Largemouth Bass diets of this study made comparisons to Choctaw Bass difficult since the unidentified material could be either fish or crayfish. It is necessary to examine a greater number of diets across multiple seasons to determine if Choctaw Bass diets also vary seasonally and by fish size.

Choctaw Bass exhibited similar diets to Largemouth Bass within Holmes Creek. The similar diets and morphology of Choctaw Bass and Largemouth Bass make these sympatric species potential competitors for space and prey. It is not known whether Choctaw Bass exhibit similar diets to Largemouth Bass in other Florida rivers but it is common for different sympatric black bass populations to exhibit high diet overlap. For example, Shoal Bass and Largemouth Bass in the Chipola River, Florida exhibited high similarity in food habits but exhibited different habitat associations. Substantial differences in habitat associations between Shoal Bass and Largemouth Bass may facilitate coexistence and reduce interspecific interactions (Wheeler and Allen 2003). Microhabitats of Choctaw Bass and Largemouth Bass were not compared in this study but should be considered for future research. Maintaining habitat diversity could be important to prevent competition between Largemouth Bass and Choctaw Bass.

In conclusion, it is recommended that Choctaw Bass continue to be managed separately from Largemouth Bass. Although mortality did not differ between sympatric Choctaw Bass and Largemouth Bass in Holmes Creek, Choctaw Bass should be managed more conservatively based on their observed restricted distribution, limited abundance relative to Largemouth Bass, limited number of fish collected greater than 406 mm TL, and the lack of fish older than age-5 in the Choctawhatchee River. Future management strategies for Choctaw Bass should focus on public education and habitat conservation. While this study began to address life history information such as relative abundance, age, mortality, growth, and diet, many gaps remain. Habitat associations, distribution, fishery data (exploitation, etc.), spawning, and recruitment information all need to be addressed for proper management and conservation of this species.

Acknowledgments

Thanks to various Florida Fish and Wildlife personnel for field assistance. Nicole Neumann, A. Mattair, B. Arnold, and M. Szatkowski and should receive specific recognition for their substantial contributions in the field and laboratory. Thanks also to J. Knight and C. Paxton for their continued support. Statistical guidance and some analyses were provided by K. Bonvechio and E. Leone. Manuscript reviews were conducted by K. Bonvechio, K. Harriger, and C. Paxton. Genetics information was provided by B. Barthel and M. Tringali. Kate Harriger created the map. This project was partially supported by the U.S. Fish and Wildlife Service Sport Fish Restoration Program.

References

Anderson, D. R. 2008. Model based inference in the life sciences. A primer on evidence. Springer, New York.

Anderson, D. R., W. A. Link, D. H. Johnson, and K. P. Burnham. 2001. Suggestions for presenting the results of data analyses. Journal of Wildlife Management 65:373–378.

Bailey, R. M., and C. L. Hubbs. 1949. The black basses (*Micropterus*) of Florida with description of a new species. University of Michigan Museum of Zoology Occasional Papers 516:1–40.

Balkenbush, P. E., and W. L. Fisher. 2001. Population characteristics and management of black bass in eastern Oklahoma streams. Proceedings of the Annual Conference, Southeastern Association of Fish and Wildlife Agencies 53(1999):130–143.

Bass, D. G. 1983. Rivers of Florida and their fishes. Florida Game and Freshwater Fish Commission. North Florida Streams Research Dingell-Johnson Project F-36, Study 3, Completion Report, Tallahassee, Florida.

Bass, D. G. and D. T. Cox. 1985. River habitat and fishery resources of Florida. Pages 121–187 *in* W. Seaman, Jr. editor. Florida Aquatic Habitat and Fishery Resources. Florida Chapter of American Fisheries Society, Kissimmee.

Barrios, K. 2005. Choctawhatchee River springs inventory. Water resources special report 2005–02. Northwest Florida Water Management District. Havana, Florida.

Barrios, K. and A. Chelette. 2008. Holmes Creek springs inventory. Water resources special report 2008–01. Northwest Florida Water Management District. Havana, Florida.

Bayley, P. B., and D. J. Austen. 2002. Capture efficiency of a boat electrofisher. Transactions of the American Fisheries Society 131:435–451.

Boschung, H. T., and R. L. Mayden. 2004. Fishes of Alabama. Smithsonian Institution Press, Washington, D.C.

Bowen, S. H. 1996. Quantitative description of the diet. Pages 513–532 *in* B. R. Murphy and D. W. Willis, editors. Fisheries Techniques, 2nd edition. American Fisheries Society, Bethesda, Maryland.

Bonvechio, T. F., M. S. Allen, and R. L. Cailteux. 2005. Relative abundance, growth, and mortality of Suwannee Bass in four Florida Rivers. North American Journal of Fisheries Management 25:275–283.

Bonvechio, K. 2017. Standardized Sampling Manual for Freshwater Systems. Florida Fish and Wildlife Conservation Commission, Fish and Wildlife Research Institute, Tallahassee, Florida.

Buckmeier, D. L., and R. G. Howells. 2003. Validation of otoliths for estimating ages of largemouth bass to 16 years. North American Journal of Fisheries Management 23:590–593.

Burnham, K. P., and D. R. Anderson. 2001. Kullback-Leibler information as a basis for strong inference in ecological studies. Wildlife Research 28:111–119.

Coggins, L. G., Jr., D. C. Gwinn, and M. S. Allen. 2013. Evaluation of age-length key samples sizes required to estimate fish total mortality and growth. Transactions of the American Fisheries Society 142:832–840.

Dauwalter, D. C., and W. L. Fisher. 2008. Spatial and temporal patterns in stream habitat and Smallmouth Bass populations in Eastern Oklahoma. Transactions of the American Fisheries Society 137:1072–1088.

Devries, D. R. and R. V. Frie. 1996. Determination of age and growth. Pages 483–512 *in* B. R. Murphy and D. W. Willis, editors. Fisheries Techniques, 2nd edition. American Fisheries Society, Bethesda, Maryland.

Florida Fish and Wildlife Conservation Commission (FWCC). 2011. Florida Black Bass Management Plan. Available: http://myfwc.

com/media/1384317/blackbassmanagement-plan2010-2030.pdf. (September 2018).

Gross, M. L., A. R. Kapuscinski, and A. J. Faras. 1994. Nestspecific DNA fingerprints of smallmouth bass in Lake Opeongo, Ontario. Transactions of the American Fisheries Society 123:449–459.

Hangsleben, M. A., M. S. Allen, and D. C. Gwinn. 2013. Evaluation of electrofishing catch per unit effort for indexing fish abundance in Florida lakes. Transactions of the American Fisheries Society 142:247–256.

Humston, R., B. M. Priest, W. C. Hamilton, and P. E. Bugas. 2010. Dispersal between tributary and main-stem rivers by juvenile Smallmouth Bass evaluated using otolith microchemistry. Transactions of the American Fisheries Society 139:171–184.

Jackson, J. R., and R. L. Noble. 1995. Selectivity of sampling methods for juvenile Largemouth Bass in assessments of recruitment processes. North American Journal of Fisheries Management 15:408–418.

Livingston, R. J., editor. 1991. The rivers of Florida. Ecological Studies, volume 83. Springer, New York.

Long, J. M., M. S. Allen, W. F. Porak, and C. D. Suski. 2015. A historical perspective of black bass management in the United States. Pages 99–122 *in* M. D. Tringali, J. M. Long, T. W. Birdsong, and M. S. Allen, editors. Black bass diversity: multidisciplinary science for conservation. American Fisheries Society, Symposium 82, Bethesda, Maryland.

Middaugh, C. R., T. Alfermann, P. A. Strickland, and P. Nguyen. 2016. A regional evaluation of Suwannee Bass and Largemouth Bass exploitation in north Florida rivers. North American Journal of Fisheries Management 36:958–963.

Miranda, L. E. and P. W. Bettoli. 2007. Mortality. Pages 232–250 *in* C. S. Guy and M. L. Brown, editors. Analysis and interpretation of freshwater fisheries data. American Fisheries Society, Bethesda, Maryland.

Nagid, E. J., W. A. Strong, T. Tuten, R. A. Kiltie. 2010. Habitat use of Suwannee Bass and Largemouth Bass in the Withlacoochee River, Florida. Proceedings of the Annual Conference of Southeastern Associations of Fish and Wildlife Agencies 64:146–153.

Page L. M. and B. M. Burr. 2011. A field guide to freshwater fishes: North America north of Mexico, 2nd edition. Houghton Mifflin Company, Boston.

Renkonen, O. 1938. Statistisch-okologische Untersuchungen uber die terrestrische kaferwelt der finnischen bruchmoore. Annales Botanici Societatis Zoologicae-Botanicae Fennicae Vanamo 6:1–226.

Ricker, W. E. 1975. Computation and interpretation of biological statistics of fish populations. Fisheries Research Board of Canada Bulletin 191.

Sammons, S. 2012. Diets of juvenile and subadult size classes of three *Micropterus* spp. in the Flint River, Georgia: potential for trophic competition. Southeastern Naturalist 11(3):387–404.

Sammons, S. M., K. L. Woodside, and C.J. Paxton. 2015. Shoal Bass *Micropterus cataractae* Williams & Burgess, 1999. Pages 75–81 *in* M. D. Tringali, J. M. Long, T. W. Birdsong, and M. S. Allen, editors. Black bass diversity: multidisciplinary science for conservation. American Fisheries Society, Symposium 82, Bethesda, Maryland.

SAS. 2010. SAS, version 9.4. SAS Institute, Cary, North Carolina.

Scalet, C. G. 1977. Summer food habits of sympatric stream populations of Spotted Bass, *Micropterus punctulatus*, and Largemouth Bass, *M. salmoides*, (Osteichthyes: Centrarchidae). The Southwestern Naturalist 21:493–501.

Shoup, D. E., and D. H. Wahl. 2009. The effects of turbidity on prey selection by piscivorous Largemouth Bass. Transactions of the American Fisheries Society 138:1018–1027.

Slipke, J. W. 2005. Splike's fish aging tools: S-FAT, version 1.0.

Smith, M. T., A. Y. Then, C. Wor, G. Ralph, K. H. Pollock, and J. M. Hoenig. 2012. Recommendations for Catch-Curve Analysis. North American Journal of Fisheries Management 32:956–967.

Stein, R. A. 1977. Selective predation, oprtimal foraging, and the predator-prey interaction between fish and crayfish. Ecology 58:1237–1253.

Toepfer, C. S., W. L. Fisher, and W. D. Warde. 2000. A multistage approach to estimate fish abundance in streams using geographic information systems. North American Journal of Fisheries Management 20:634–645.

Tringali, M. D., P. A. Strickland, R. A. Krause, S. Seyoum, B. L. Barthel, A. Alvarez, and C. Puchulutegui. 2015a. Conservation status of Shoal Bass in the Chipola River, Florida: the threat of hybridization with native and nonnative congeners. Pages 523–536 *in* M. D. Tringali, J. M. Long, T. W. Birdsong, and M. S. Allen, editors. Black bass diversity: multidisciplinary science for conservation. American Fisheries Society, Symposium 82, Bethesda, Maryland.

Tringali, M. D., B. L. Barthel, S. Seyoum, and J. R. Knight. 2015b. The Choctaw Bass: an undescribed species of *Micropterus* in the Gulf Coastal Plain rivers of Florida. Pages 421–448 *in* M. D. Tringali, J. M. Long, T. W. Birdsong, and M. S. Allen, editors. Black bass diversity: multidisciplinary science for conservation. American Fisheries Society, Symposium 82, Bethesda, Maryland.

U.S. Fish and Wildlife Service. 1999. Black bass fishing in the U.S. Addendum to the 1996 national survey of fishing, hunting, and wildlife-associated recreation. U.S. Fish and Wildlife Service, Washington, D.C.

Wheeler, A. P., and M. S. Allen. 2003. Habitat and diet partitioning between Shoal Bass and Largemouth Bass in the Chipola River, Florida. Transactions of the American Fisheries Society 173:438–449.

Woodside, K. L., C. J. Paxton, and N. C. Kierl. 2015. Stock assessment of Shoal Bass in the Chipola River, Florida. Pages 139–155 *in* M. D. Tringali, J. M. Long, T. W. Birdsong, and M. S. Allen, editors. Black bass diversity: multidisciplinary science for conservation. American Fisheries Society, Symposium 82, Bethesda, Maryland.